Exploring The X-ray Universe

Philip A. Charles
Frederick D. Seward

CAMBRIDGE
UNIVERSITY PRESS

Published by the Press Syndicate of the University of Cambridge
The Pitt Building, Trumpington Street, Cambridge CB2 1RP
40 West 20th Street, New York, NY 10011-4211, USA
10 Stamford Road, Oakleigh, Melbourne 3166, Australia

First published 1995

Printed in Great Britain at the University Press, Cambridge

A catalogue record for this book is available from the British Library

Library of Congress cataloguing in publication data
Charles, Philip A.
Exploring the X-ray universe / Philip A. Charles, Frederick D. Seward.
    p.   cm.
Includes index.
1. X-ray astronomy. I. Seward, Frederick D.   II. Title.
QB472.S49   1995
522'.6863–dc20   92-34641 CIP

ISBN 0 521 26182 1 hardback
ISBN 0 521 43712 1 paperback

For Anne, Susan, Robert, Thomas and William

Contents

Preface

Some 30 years ago X-rays from stars other than our Sun were completely unknown and unexpected by all but a few pioneering scientists. Since the discovery of cosmic X-rays in 1962 the field has grown at an astonishing rate. The sensitivity of X-ray observations has increased rapidly, particularly with instrumentation on the X-ray satellites of the 1970s. Many of the major discoveries have been serendipitous in nature, leading to an exciting, but at times chaotic, growth in the science. Information from the spacecraft of the 1990s promises to produce another explosion of knowledge.

The launch of the Einstein Observatory in late 1978 gave the first true X-ray pictures of these objects other than the Sun; and at a resolution and sensitivity comparable to that of optical and radio telescopes. The 1980s saw the *light curves* of binary X-ray sources and active galactic nuclei measured by EXOSAT, which was the closest to an *observatory* style of operation that X-ray astronomy had yet seen. In the 1990s ROSAT has completed the first X-ray imaging all-sky survey, and high resolution spectral observations have been and are being collected by BBXRT and ASCA.

We, the authors, have been observers of celestial X-rays since 1965 (FDS) and 1972 (PAC). We miss the early days when a person could easily name from memory all the known X-ray sources and when the nature of many were unknown. We note with satisfaction that the mechanism of most cosmic sources is now understood, but there are now so many instruments and sources that we are forced to specialise within the field. We have each therefore written chapters concerning our respective special interests.

This book is our attempt to summarise the history and current state of X-ray astronomy and to make it accessible to a wide audience. We follow X-ray astronomy from the 1960s to the 1990s. The material is reasonably complete through the observations of the Einstein and EXOSAT Observatories (1979–1981, 1983–1986), and many Einstein pictures, some never before published, are included. Some results from recent missions (Ginga, ROSAT) are also shown, but it has been impossible to include the latest results from observatories which are up and operating in 1994. We were thus forced to take a somewhat historical perspective. All important areas touched by X-ray astronomy are included. The material is written for someone with an astronomical background, although each chapter starts simply, so someone with no background can understand the basics. For those who do not wish to read all the details, we have separated the more technical discussions into *boxes* that accompany the text. The main text can be followed without the boxes, but an acquaintance with physics and mathematics is helpful in understanding some of the boxed material.

Some of the illustrations have already appeared in *Sky & Telescope* and in the scientific literature. Many illustrations are from our own work and we thank our colleagues who have generously given material generated by their research. In order to make the text more readable (and not outrageously long) we have avoided giving extensive accreditation in the text itself. There is a bibliography for each chapter which lists popular articles, scientific reviews and, in a few cases, detailed technical references.

This book is primarily about science, with excursions into instrumentation. Some scientists who have made key observations or developed key theories are mentioned in the text. We all know, however, that astronomers only contribute part of the effort necessary for a successful astronomy mission. There is planning, funding, building, calibration, launch, control during observations, data transmission and collection, and the data must be put in understandable form and made available to observers. The modern scientific process also includes editors, referees, and those who evaluate proposals. X-ray astronomy now is very much a group effort, a national or international effort for the large missions.

We particularly note our colleagues at NASA and ESA Headquarters, all career scientists, who plan the future of X-ray astronomy and administer the funding for much of the program. This is a difficult and, at times, a thankless job. Successful proposers think winning is their just due. Unsuccessful proposers think the process is flawed. It is easier to make enemies than friends. The NASA Astrophysics Program has been guided by Nancy Roman, Al Opp, and Charles Pellerin. Alan Bunner, Lou Kaluzienski and Guenter Riegler are coping with present and future NASA developments, Brian Taylor and Tony Peacock with those at ESA. We wish them adequate funding and success with future projects.

We would like to acknowledge advice, helpful discussions, and assistance from: the late Yoram Avni, Dan Fabricant, Sharlene Ford, Bill Forman, Paul Gorenstein, Christine Jones, Bob Kirshner, Jonathan McDowell, Frank Primini, John Pye, Rashid Sunyaev and Mike Watson. Finally we would like to thank Susan Bowring for her very careful editing of the manuscript, and Simon Mitton for his support during the book's very long gestation period.

Phil Charles   Fred Seward
Oxford University   Smithsonian Astrophysical Observatory

1 Introduction

1.1 The beginning of X-ray astronomy

The major difficulties getting started

X-rays cannot penetrate the Earth's atmosphere. It is thus impossible to observe X-rays from astronomical sources with ground–based instruments. Even from mountain tops, airplanes, and simple balloons, observations are hopeless. A vehicle capable of placing an instrument above the atmosphere is the first requirement for any observation. To see any X-rays at all, it is necessary to be above 99% of the atmosphere, and to detect X-rays in the band where sources are most prominent, all but one millionth of the atmosphere must be below the instrument.

Cosmic X-ray sources are most clearly detected in the range of 0.5–5 keV in photon energy (or 25–2.5 Å wavelength). By Earthly standards these X-rays are 'soft' and easily stopped by a small amount of material. For example, three sheets of paper or 10 cm of air at one atmosphere pressure will stop 90% of 3 keV X-rays. The higher the energy, however, the more penetrating, or harder, the X-rays. A rocket is needed to observe 3 keV X-rays, which cannot be seen at altitudes below 80 km, whereas 30 keV photons will penetrate to 35 km altitude, which can be reached by the highest-flying balloons. The instrument should be above 200 km to observe X-rays with energies below 1 keV in a direction parallel to the Earth's surface, as would be desirable in a survey.

It was not a trivial matter to build the first instruments that were large enough to be sensitive yet small enough to fit within the rocket or balloon payload. The instruments not only had to withstand the rigours of launch but also had to operate in a vacuum or near vacuum. Time and trial and error were needed to develop the first survey instruments. To detect, for the first time, a phenomenon that many people believe is impossible, takes confidence that the instruments are operating properly.

An X-ray is a quantum of electromagnetic radiation with an energy some 1000 times greater than that of optical photons. So, if it is generated in a thermal process, the temperature must be of the order of 1000 times greater than that in places where light is produced. Thus a search for cosmic X-ray sources is a search for material at temperatures of millions of degrees – in contrast to the familiar stars with surface temperatures of thousands of degrees. Until 1962, very few astronomers believed that the universe contained objects capable of generating detectable amounts of high energy radiation and little was expected from the first observations.

These ideas changed dramatically in the early 1960s with the discovery that there were indeed many discrete, powerful, sources of astronomical X-rays.

The early years (1946–62)

The first technology that enabled significant research to be conducted from above the atmosphere was that of the captured V2 rockets after the Second World War. With these the Naval Research Laboratory (NRL), in Washington, DC, under the direction of Herbert Friedman, was able to reveal the Sun as a powerful source of ultraviolet (UV) and X-radiation. It is curious that this discovery actually caused many scientists to lose interest in the search for sources of X-rays other than the Sun. This was because the Sun appears as a bright source solely due to its proximity to the Earth. A calculation of the intensity of radiation expected at the Earth from the nearest stars (assuming that they are comparable emitters of X-rays to the Sun) shows that the instrumentation available in 1960 would have had to be about a factor of 100 000 more sensitive to detect such objects. Worse still, if the stars were more distant (say at a typical distance of one kiloparsec (kpc) or about 3000 light years) then a 1960 experiment would only have been capable of discovering a process which was producing 100 billion times the X-ray luminosity of our Sun.

It is therefore not surprising that most of the rocket observations of the 1950s were devoted to more detailed studies of the Sun, although NRL did try (without success) to search for other cosmic sources. When a new wavelength band is opened up for study, there are always surprises that nobody anticipates. It was this searching for the unknown that drove several groups to continue experimental development and to improve sensitivity. In the end it was a group at American Science and Engineering (AS&E), led by Riccardo Giacconi (and inspired by Bruno Rossi), that was successful in the first detection of a powerful cosmic source of X-radiation.

The official purpose of the AS&E experiment was to search for X-rays from the Moon produced by the interaction of energetic solar wind particles with the lunar surface. It was also thought that the solar X-ray flux might cause the lunar surface material to fluoresce in X-rays. Such a result would provide valuable information about the nature of the lunar surface; this was an area receiving much publicity and support at the time with America's commitment to a manned lunar landing within the decade. In addition, it was planned to scan a large region of sky in a search for non-solar sources of X-radiation. The first launch of this new experiment took place in October 1961 and illustrates the frustration of the early days when pioneering work was being undertaken with often crude and unreliable equipment. The rocket launch was perfect but the doors, designed to protect the X-ray detectors during launch and passage through the atmosphere, failed to open!

The second launch of the AS&E experiment, on a new Aerobee rocket, took place from White Sands, New Mexico, on June 18, 1962 and this time the doors functioned perfectly. Two of the three X-ray Geiger counters worked well and, although they failed to detect any X-rays from the Moon's surface, they made the now historic first detection of a powerful cosmic X-ray source. This source subsequently became known as Sco X-1, the first-discovered source in the constellation Scorpius. As Richard Hirsch comments in his detailed history of these early days, 'Observing Sco X-1 was the reward nature offered to scientists willing to gamble on a long shot.'

By realising that the observed signal of 100 photons cm^{-2} s^{-1} was caused by an extrasolar source, Giacconi and colleagues captured the interest of the astronomical community and started an exploration which has produced exciting discoveries for over 30 years.

The unusual nature of Sco X-1 was clear as soon as it had been roughly located in the sky. Figures 1.1 and 1.2 contrast the X-ray and optical appearance of

Sco X-1. The source dominates an early rocket X-ray survey. An optical picture, however, containing Sco X-1 shows nothing unusual whatsoever. Until an accurate location of the X-ray source was obtained in 1966, astronomers had not a clue as to the nature of this source. Sco X-1 is an object which stands out like a beacon to a small X-ray detector but is visually four hundred times fainter than the faintest star which can be seen with the naked eye. In every square degree of the sky there are about one hundred stars visually brighter than Sco X-1.

Fig. 1.1 Three minutes of data from a rocket-borne X-ray detector flown in October 1967. This shows the counting rate of the detector as it scanned a great circle containing the source Sco X-1 and a cluster of sources in the direction of the galactic centre. The detector field of view was 5° by 30°. The Sun was below the horizon. The signal from Sco X-1 is very strong.

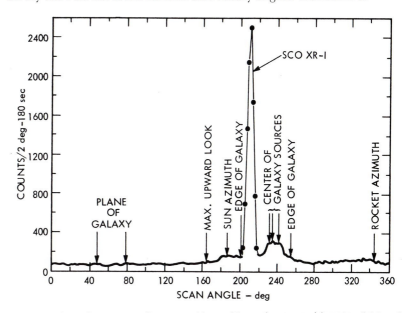

Fig. 1.2 One square degree of the sky from the Palomar Sky Survey. The star indicated with an arrow is the optical counterpart of Sco X-1.

The nature of cosmic X-ray sources and our observations of them is the subject matter of the rest of this book. Many of the sources are truly remarkable.

1.2 An early history of the X-ray sky

After the discovery of Sco X-1, X-ray astronomy progressed rapidly. Evidence for two weaker sources was found on October 12, 1962 by Herbert Gursky and colleagues (AS&E). A team lead by Stuart Bowyer (NRL) confirmed and located one of these sources on April 29, 1963 using a rocket-borne detector. It was identified right away as the Crab Nebula, a well-known young supernova remnant in our galaxy, and high energy X-rays from this source were detected on July 21, 1964 by George Clark of the Massachusetts Institute of Technology (MIT). (This was the first detection of high energy radiation from an extrasolar source with a balloon-borne detector.)

Astronomers were thus forced to recognise that there were many objects at stellar distances which were strong, almost unbelievably strong, sources of high energy photons. The sky was then sporadically explored, with great enthusiasm, using only rockets and balloons until the first all-sky survey with the Uhuru satellite, launched on December 12, 1970.

What were these X-ray sources? How was the energy generated? The answers were not obvious. It was first necessary to obtain precise locations of known X-ray sources, leading to identification with optical or radio objects. The next steps were to measure the X-ray spectra and light curves to determine the emission mechanism, and to survey the sky to find more sources. At the present time thousands of X-ray sources have been discovered. Almost all the bright ones and many of the weaker sources have been identified. The types of objects which emit X-rays in our galaxy, the Milky Way, can be reasonably discussed. Data concerning bright sources in nearby galaxies are also available and useful in these considerations.

Some of the brighter sources in our galaxy radiate ten thousand times as much energy as does the Sun. Almost all (99.9%) of this energy appears as X-rays. Sco X-1 is such a source. The optical counterpart is a 13th magnitude star, invisible to the naked eye and even to small telescopes. The only visual clues to its unusual nature are a blue-violet colour when compared with most stars and an irregular variability marked by occasional rapid flickering. No optical surveys previous to the X-ray detection had indicated anything unusual. Even after the optical counterpart had been identified, the Sco X-1 system was not understood.

A convincing explanation of the nature of Sco X-1 was not found until 1971 when the first satellite devoted to X-ray observations, Uhuru, discovered and measured the peculiar X-ray variation of another source. This object lies in the southern sky in the constellation Centaurus.

The source Cen X-3 (or 4U1118–60 as explained in box 1.1 which covers naming conventions) is an X-ray bright object at a declination of −60°. Although bright enough to be detected by a rocket-borne instrument, it is below the horizon for sounding rockets launched from the main US facility at White Sands. It is clearly accessible, however, to those using launchers in Hawaii and Australia.

In 1967–68 two groups surveyed the southern sky. Charles Swift, Gerry Chodil, and colleagues (LLL, Lawrence Livermore Laboratory) detected Cen X-3 twice and derived a rough location. Figure 1.3 shows data from one of these flights. However, Ken Pounds and co-workers (Leicester) observed twice and did not see it.

In the late 1960s it was no easy task to build detectors, calibrate them, ensure that they survived the quick but hazardous trip into space, and know

Box 1.1 X-ray source names

In the early days, X-ray sources were named as they were discovered and according to the constellation in which they were found. This is how we were given Sco X-1, Cyg X-1, Cyg X-3, etc. This was fine as long as there were only a handful of sources, but, when the first Uhuru results were available, it was clear that this would become cumbersome. Current convention is therefore to give a new X-ray source a name according to its position on the sky, preceded by a letter designating the satellite or instrument with which it was found. An initial number sometimes designates a particular catalogue. For example, H0324+28 was found by HEAO-1 at a location of 3 hours 24 minutes and +28 degrees. 0324 is called the 'right ascension' of the source and is analogous to longitude on Earth's surface. The declination of +28 degrees is exactly equivalent to latitude on Earth. As another example, Sco X-1 is listed in the fourth Uhuru catalogue as 4U1617−15.

where they were pointed. People took pride in their ability to distinguish real sources from the background and expected the source population to be more or less steady, like the stars. In the case of Cen X-3, both groups secretly suspected that the other had not interpreted the data properly! In truth, all these observations were carefully done and correctly interpreted. The source is highly variable. To a small detector, sometimes it appears above background and sometimes not. Furthermore, such variability is a common characteristic of most bright X-ray sources.

The Uhuru observations of Cen X-3, made in 1971 by Ethan Schreier and colleagues (AS&E), were spectacular. The results were unexpected and the X-ray observations alone determined the nature of the source.

The first surprise was the observation of a regular periodicity of 4.84 seconds in X-ray flux from the source. The modulation was high and the pulsations were easily seen during a single scan across the source. Only a rotating neutron star (chapter 3) could produce such rapid pulsations. The period was measured accurately and it was soon discovered that the period varied slightly with time. After several days of data were collected, these variations were recognised as a Doppler shift. The neutron star was moving in a circular

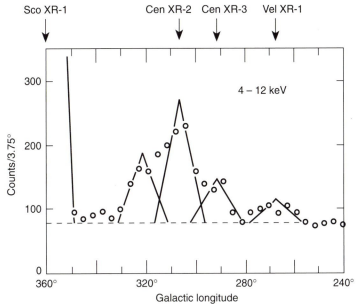

Fig. 1.3 Counting rate from a rocket-borne proportional counter flown in May 1967 by G. Chodil and colleagues (LLL). A slat collimator restricted the field of view to be 10° × 30°. The spinning payload caused the detector to scan a band of the sky repeatedly. By comparing the observed count rate with the expected 'triangular' response, sources were located. Sco X-1 was very bright and is off scale at 360°. The next strongest source at this time was Cen X-2, a 'transient'. The source Cen X-3 was first seen in the data shown here. Note the difficulty of determining source positions in crowded regions.

orbit with a period of only 2.09 days! As frosting on the cake, the X-rays were observed to disappear completely for 11 hours at regular 2.09 day intervals. The source was in an eclipsing binary system!

Here then is a rapidly spinning neutron star, probably emitting X-rays from the near-vicinity of one of its magnetic poles. It orbits a bright B0Ib star (box 4.3 explains this nomenclature), Krzeminski's star (named after the person who identified the optical counterpart). Energy to power the X-ray source comes through accretion of material supplied by the supergiant companion. This matter is captured by the strong gravitational field of the neutron star. It acquires enough energy in the fall to the surface to both heat material to the high temperature required for X-ray emission and to supply the observed luminosity. (See chapter 7 for detail about sources of this type.)

The other bright X-ray sources in the plane of our galaxy were first detected in early rocket surveys (figure 1.4). Most were found by Phil Fischer and co-workers (Lockheed), Hale Bradt and co-workers (MIT), and Friedman and co-workers (NRL). The sources are mostly accretion powered binaries – a normal star and a 'compact star' locked in a close orbit. Some, like Cen X-3, consist of a neutron star and a bright O star (box 4.3). The optical identifications of these have been quickly made. Because the O stars are large, eclipses of the X-ray source associated with the neutron-star companion are common. Other sources consist of dim late-type stars orbiting close to a neutron star (see chapter 8). These optical counterparts are faint and difficult to identify. The accretion-powered sources are the strongest in our galaxy. Some have X-ray luminosities as high as 10^{38} erg s^{-1}.

Some bright sources were found by Uhuru to be within globular clusters. Clark and colleagues (MIT) found more with the third Small Astronomy Satellite (SAS-3) and pointed out that this was an unusual situation. The sources occur with much higher frequency than predicted by calculations based on the ratio of stars to X-ray sources in our galaxy. The high stellar density in globular clusters is apparently favourable for the formation of these binary systems. (Chapter 9 has more detail.)

On March 30, 1973 Saul Rappaport and colleagues (MIT), using rocket-borne detectors, discovered soft X-rays from SS Cygni. This star is a cataclysmic variable or CV. It is one of the brightest and nearest of this class. It has irregular outbursts during which the star brightens from its normal 12th magnitude to 8th magnitude. SS Cyg has been monitored by the American Association of Variable Star Observers since 1896. It has an outburst about every two months, and has been called a dwarf nova. Many CVs are now known to be X-ray sources. They are accreting binary systems consisting of a low-mass normal star and a white dwarf. (See chapter 10.)

Another type of X-ray source, recognised in early surveys, are the supernova remnants (described in chapter 3). The first detected was the Crab Nebula which is quite bright. The X-ray luminosity of most remnants is 10–100 times less than that of the Crab Nebula and the spectrum is soft, so absorption in interstellar gas is more severe. Nevertheless the closer remnants were easily detected and positively identified by their spatial extent (Cygnus Loop and Vela XYZ) or by the spatial coincidence with a non-thermal radio source (Cas A).

Another class of sources are stars, binary perhaps, but without compact companions. The first indication of strong coronal emission from stars was obtained on April 5, 1974 when Richard Catura and colleagues (Lockheed) detected X-rays from the bright star Capella. The X-ray luminosity was ten thousand times the X-ray luminosity of the Sun. The detection occurred by accident! The rocket-borne instruments were pointed at Capella to calibrate star sensors included in the payload for accurate measure of pointing direction. On October 19, 1974, X-ray emission from a second star, the flare star YZ Canis Minoris, was

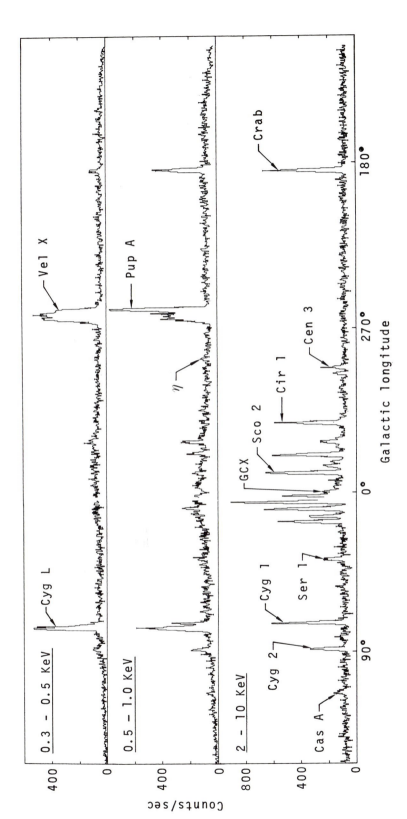

Fig. 1.4 *The entire Milky Way as surveyed with rocket-borne proportional counters in May 1970, May 1971, and October 1972 by R. Hill, G. Burginyon, and colleagues (LLL). Collimation was 1.3° × 20°. Data from three flights have been combined to show counting rate as a function of galactic longitude in three energy bands. There are no soft X-rays observed from the cluster of bright sources around the galactic centre. Intervening gas absorbs the soft X-rays. The nearby supernova remnant Vela XYZ is clearly soft and extended. These data were taken using the payload shown in figure 2.6, which was recovered and refurbished after each flight.*

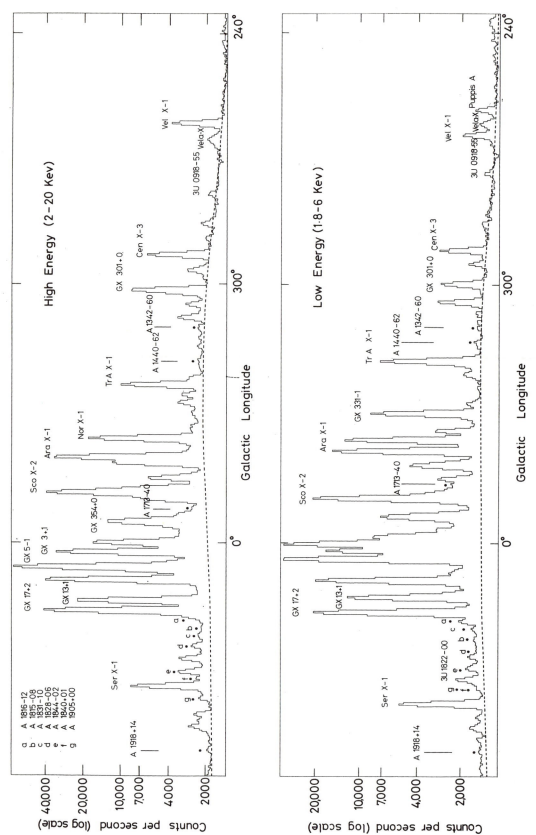

Fig. 1.5 An Ariel V scan of the central half of the galactic plane. Two detectors scanned the sky, each with $0.7° \times 10°$ field of view. The two collimators were inclined at different angles to aid in source location so given sources do not appear at identical longitudes in this figure. Note the improvement in the ability to detect weak sources. (Courtesy of K. Pounds, University of Leicester.)

*Fig. 1.6 (a) The EXOSAT
MED (Medium Energy Detector),
with 0.8° × 0.8° collimation,
scanned the central part of the
galactic plane giving this map of
the bright 'bulge' sources. The
map extends from longitude 60°
through the centre to longitude
320°. In addition to the sources
there is diffuse emission from the
'galactic ridge'. The origin of
these X-rays is at present
unknown. (Courtesy M. Watson,
University of Leicester.) (b) Key
to fig. 1.6 source identifications.
The galactic centre is marked with
a cross and there is no bright
source at this location.*

observed with the ANS satellite by John Heise and collaborators (Utrecht).
(The stars as X-ray sources are discussed at length in chapter 4.)

Since a bright star in an error box was a very tempting identification, false
claims of X-ray detection of bright stars were not uncommon. In spite of this
it was soon evident that the coronal X-ray emission of many active stars was
considerably more intense than that of the Sun.

Figures 1.4, 1.5, and 1.6 show how the galactic plane was surveyed with
increasing sensitivity. As the sensitivity of observations increased, other
sources were discovered which were not in our galaxy. The first extragalactic
source discovered was the active galaxy M87. The observation was made by
Ted Byram and colleagues (NRL) with a rocket launched on April 25, 1965.
In 1971, Uhuru added many quasars, active galaxies (chapter 13), and clusters of galaxies (chapter 14) to the catalogue of extragalactic X-ray sources.

Thus the individual X-ray source populations were recognised as sources
and identified. Twenty years after the first explorations, Kent Wood and collaborators (NRL) completed an all-sky survey using data obtained with a
large array of proportional counters on the HEAO-1 satellite. The result,

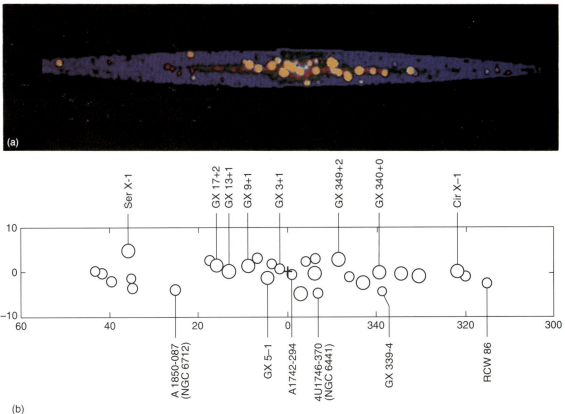

shown in figures 1.7 and 1.8, was a catalogue with limiting sensitivity of
0.003 photons cm^{-2} s^{-1} containing 842 sources. At about the same time, in a
more selective mode of operation, the Einstein X-ray telescope observed 5%
of the sky with a capability of finding sources weaker than 3×10^{-4}
photons cm^{-2} s^{-1}, and the deepest pointings were a factor of 10 more sensitive.
Ten years after Einstein, ROSAT mapped the sky using an imaging telescope and low-background detector. The threshold of this, the most sensitive
X-ray all-sky survey, was 1.5×10^{-4} photons cm^{-2} s^{-1} and the ROSAT deep
pointings go to 1×10^{-5}. Thus cosmic X-ray sources ten million times fainter
than Sco X-1 are known.

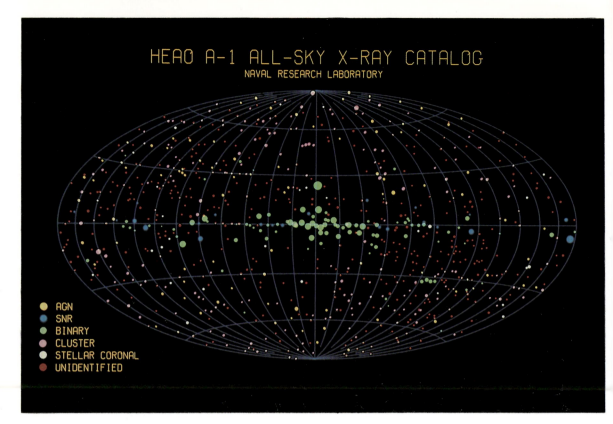

Fig. 1.7 Map showing X-ray
sources from the HEAO-1 all-sky
survey. Size of the dot shows the
brightness of the source. Colours
indicate type of source. (Courtesy
of K. Wood, NRL.)

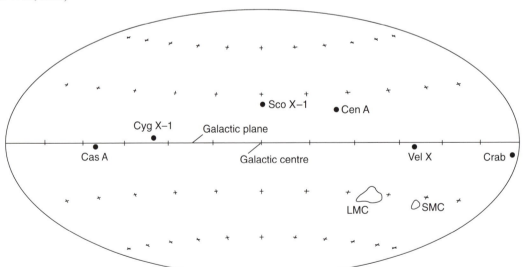

Fig. 1.8 Key to figures 1.7, 1.11,
1.15, and 1.16. These are all-
sky maps in galactic coordinates.
The galactic plane is a horizontal
line with the centre of the galaxy
at the centre of the map. Although
the map is distorted at the edges
the density of sources is not.

Over sixty thousand sources (80% of these thanks to ROSAT), both in and far outside of our galaxy, have been detected and located. The brightest source in the sky is usually Sco X-1, the first discovered, although transient sources up to two times brighter have been observed. The faintest source known before ROSAT was a distant quasar detected in the Einstein deep survey with a flux of 3×10^{-5} photons cm^{-2} s^{-1}. (The Sun, which is very weak in X-ray luminosity compared with other sources but extremely bright in soft X-rays because of its proximity, is excluded. Below a photon energy of about 1 keV the Sun is really the brightest source in the sky.)

Since 1962, X-ray observations have revealed a continuous series of new objects and phenomena. It is interesting to note that the general progress has been similar to that in the radio waveband which came to the attention of astronomers thirty years earlier.

1.3 Radio astronomy, an historical parallel

The exploration of the sky in the radio band started in 1931 with Karl Jansky's discovery of 15 m radio emission from the direction of the galactic centre. This discovery was made at Bell Laboratories with a directional antenna set up to identify sources of interference to radio communication. In those days, practical matters took precedence over astronomical affairs and observations of the sky proceeded slowly. The first radio map of the sky (at 2 m wavelength) was published in 1944 by Grote Reber, who had collected the data with an antenna, designed and built by himself, in a vacant lot next to his house in Wheaton, Illinois.

During the Second World War radio technology was rapidly improved and applied to radar. During wartime operations in Britain, the Sun was discovered to be an intermittent source of radio energy, occasionally strong enough to 'jam' the radar; so, as was to be the case in the X-ray band, the first astronomical source identified was the nearest and most obvious candidate. Radar echos, feared to be from flights of bombers which never materialised, were discovered to be from meteor trails in the upper atmosphere, leading quickly to measurements of meteor direction and velocity. Thus, with the astronomical worth of radio observations well established, there was great enthusiasm in the immediate postwar years for application of the new techniques, particularly in England, Australia, and the Netherlands.

The technology developed for radar was applied to astronomical instruments, and new observations not only gave new insight into the nature of known sources but found new sources, bright in the radio band and unimpressive or invisible on optical photographs. The parallel with X-ray astronomy is very close. Exploration of the sky in a new waveband leads to new, unexpected, discoveries.

The first radio observations were at wavelengths of a few metres and sky maps produced from these observations showed both emission concentrated in the plane of the galaxy and some strong high galactic latitude sources. The new radio sources were given alphabetical names which are now commonly used in the astronomical literature. The brightest radio source in Cassiopeia was Cas A (Figure 1.9), the brightest in Cygnus was Cyg A, etc., demonstrating a lack of foresight, common in astronomical nomenclature, concerning the number of sources which might be discovered in future observations. It is rather curious that the second-brightest radio sources in each constellation are not remembered. Who has heard, for example, of Cygnus B?

The first system for naming X-ray sources: Cyg XR-1, Cyg XR-2, etc. (or sometimes Cyg X-1, Cyg X-2) was a little better because numbers allowed expansion to more than 26 sources per constellation. The ranking in order of brightness, however, was doomed to failure by the early discovery of strong variability in the brightest sources.

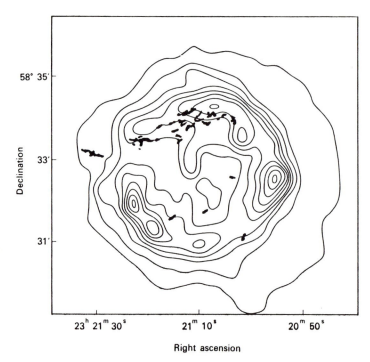

58° 35′

Declination

33′

31′

23^h 21^m 30^s　21^m 10^s　20^m 50^s

Right ascension

Fig. 1.9 A 21 cm map of radio emission from the supernova remnant Cas A made in 1965 with the Cambridge one-mile telescope. This was one of the first 'pictures' of this source, the brightest in the radio sky. Resolution is 30 arcseconds. Black regions mark positions of optical wisps. [After Ryle, Elsmore, and Neville, Nature **205**, *1259 (1965).]*

Nomenclature aside, the sky (figure 1.11) was explored at several radio wavelengths and observations of individual sources were pushed to ever higher radio frequencies. Interest in the sources led to better radio techniques: higher sensitivity, better spatial and spectral resolution, and broader coverage in the radio band. The atmospheric transmission window extends from about 5 mm to 20 m wavelength and present observatories operate throughout this range. More sources were discovered, extended sources were mapped, and spectral observations gave information about the physical processes that generate radio waves.

The ability to get high resolution spectra has given celestial maps of neutral hydrogen (via the 21 cm line) and of the carbon monoxide molecule. Line radiation from hundreds of other interstellar molecules has been discovered.

The tools radio astronomers have developed for their research are also fascinating in themselves. The most obvious are the radio telescopes, large both because high sensitivity is needed to detect weak signals, and because large linear dimension is needed for good spatial resolution at long wavelengths. An equal amount of effort has gone into sensitive receivers for detecting the weak signals collected by the antennas. The same parallel development occurs in the X-ray band. A large, specially designed mirror is needed to collect the signal which then may be focused onto a variety of detectors.

Radio astronomy has not only given new significance to old objects, but has led to new discoveries – always a source of excitement:

- There are radio galaxies; distant objects bright in the radio band but faint optically. Some have enormous radio jets, usually extending in opposite directions, over distances as large as 1 Mpc, 30 times the diameter of our own galaxy.
- The quasars were first seen as 'quasi-stellar radio sources' and were singled out for attention because they are much more prominent in the radio sky than in optical photographs.

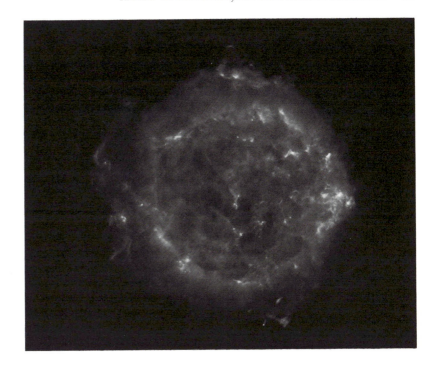

Fig. 1.10 A 6 cm map of radio emission from Cas A. This was made with the Very Large Array by P. Angerhofer and colleagues in 1983. Resolution is 0.2 arcseconds, 150 times better than that of figure 1.9. Emission is now seen to be from small clumps of material and the bright shell to be surrounded by a fainter outer region. (Courtesy of NRAO/AUI.)

Fig. 1.11 The radio sky at 73 cm wavelength. Data from Effelsberg, Germany, Jodrell Bank, England, and Parkes, Australia. The plane of the galaxy is bright with maximum towards the centre. Interesting features are the North Polar Spur, the radio galaxy Centaurus A, and the Large Magellanic Cloud. [C. G. T. Haslam, et al., Ast. & Ap. Supp. 47, 1 (1982).]

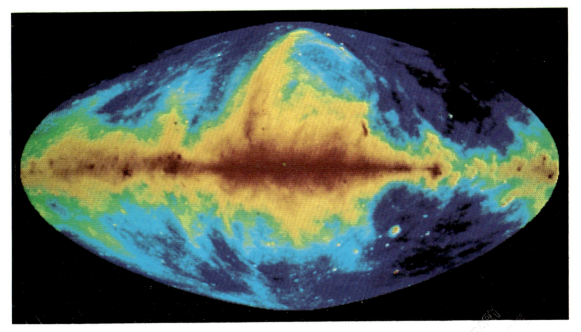

- Over 150 supernova remnants have been found in our galaxy (figure 1.10 for example). These expanding shock waves in the interstellar medium emit radio energy and are easily identified as extended sources with rather flat radio spectra.
- Neutron stars were discovered as pulsing radio sources. Before this no one dreamed that the way to find a neutron star would be to search for a regularly modulated radio signal with a period of only about one second!

1.4 Astronomy and the space age – the first major accomplishments

The atmosphere

It is not an accident that our eyes operate in the narrow waveband 4000 to 8000 Å. Not only is a large fraction of the energy of the Sun radiated in this band but the Earth's atmosphere is almost transparent throughout this 4000 Å wide waveband. Figure 1.12 shows the electromagnetic spectrum from radio waves to gamma-rays and depicts the depth to which each frequency can penetrate the atmosphere. After the transparent radio band, there are over 12 decades of the spectrum, from the far infrared to gamma-rays, where only the narrow band of 'visible' radiation reaches the Earth's surface unscathed. The 'opacity' of the atmosphere is the principal difficulty facing astronomers wishing to study radiation from the stars at wavelengths outside the visible band.

Fig. 1.12 Transmission of electromagnetic radiation by the atmosphere. The solid line shows the altitude by which half the radiation from space has been attenuated. Just below this line virtually all the radiation is absorbed. Only radio, optical, and some narrow bands of infrared radiation can reach the Earth's surface. High energy gamma-rays can be observed using balloons, but rockets or satellites are necessary for X-ray or UV detection.

Until this century, the visible part of the spectrum was all that was available for study of the heavens. Only radio astronomy was able to develop at all using ground-based instrumentation, although it is possible to undertake infrared observations from high altitude observatories through some windows less affected by water vapour.

Figure 1.12 also displays the height above sea level to which radiation of each wavelength can penetrate. All radiation from the extreme ultraviolet at 1000 Å to high-energy gamma-rays at 10^{-4} Å fails to penetrate below an altitude of approximately 30 km. It is the requirement of observing above the atmosphere that makes the study of the X-ray universe a modern one. Virtually everything in this book results from observations made during the last 30 years and most of it during the last 15 years.

The rest of this chapter is a brief review of recent accomplishments using

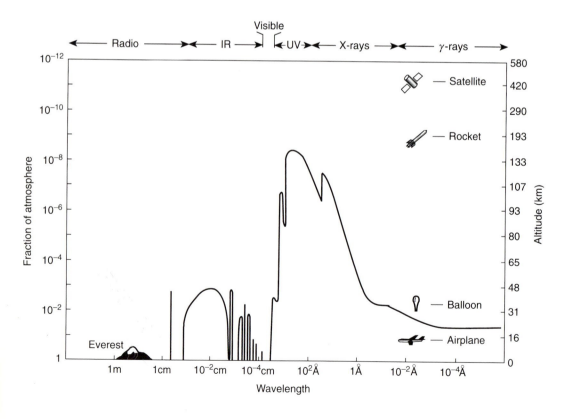

space-based information. We make no claim that this is complete. We have concentrated on fields which are connected with X-ray observations and have, for instance, not mentioned planetary flybys and lunar landings, all spectacular examples of space-age astronomy.

The ultraviolet and IUE

The ultraviolet (UV) band lies between the optical and X-ray bands, the range in wavelength being about 100 to 3000 Å. UV photons are more energetic than those of visible light but do not have the penetrating power of X-rays. The UV photons can, however, remove electrons from the outer shells of the less massive atoms that are the major constituents of the universe. Consequently UV radiation is very strongly attenuated by small amounts of matter. The Earth's atmosphere is an impenetrable barrier to all UV radiation of astronomical interest. We are familiar with the least energetic part of the UV spectrum which will produce sunburn during summer months when the path of solar radiation through the atmosphere is at a minimum. This is the near UV, attenuated by both nitrogen and ozone molecules at altitudes of 15–35 km.

The gas and dust of the interstellar medium (ISM) also strongly attenuate UV radiation from distant stars, making the band useful for probing the composition and state of the ISM. The attenuation increases with decreasing wavelength. An astronomer cannot see very far in the UV band and the horizon shrinks as the wavelength gets shorter. At a wavelength of 1000 Å distant stars in our galaxy are inaccessible, but stars closer than 3 kpc can still be observed in the galactic plane. Radiation at 912 Å wavelength or shorter removes electrons from neutral hydrogen atoms in the ISM and absorption greatly increases, preventing stars more distant than a few pc from being seen. The band from 100 to 1000 Å is called the 'extreme ultraviolet' (EUV) and, because of the strong absorption in the ISM, is largely unexplored. It is expected that the clumpy distribution of gas in the ISM will produce 'holes' where minimal absorption will make possible EUV observations of more distant objects, but targets in the EUV might be severely limited.

Why observe in the UV at all? Why bother with this spectral region which is right next to the visible band but requires the difficulty and expense of space-based instrumentation? There are good reasons. First of all, it is the only range where observations can be made of material with temperatures of 100 to 200 thousand degrees. These temperatures are characteristic of some white dwarfs, stars at the centres of planetary nebulae, and of transition regions between stellar surfaces and the million degree coronae. These must be studied if we are to understand stellar evolution and energy transport in stars. Additionally, there are elements with electron transitions in the UV that are unobservable in the visible. The most sensitive way to see silicon and carbon in stars is through UV spectra. Furthermore, as is always the case, the ability to inspect this temperature range and these elements has led to the discovery of important new phenomena.

Although the observations are not easy, one advantage UV astronomy has is that the instrumentation is not 'strange'. Familiar instruments such as reflecting telescopes, spectrometers, and photon-counting detectors can be used. Some care is required in selection of materials for filters and lenses, but the observing techniques are the familiar ones of optical astronomy. Some of the first work was done with instruments on rockets, hand-held cameras during the Gemini programme, and with small telescopes on the Spacelab mission which obtained many UV pictures of the Sun.

The band between 1000 and 3000 Å has recently been explored with instruments on several satellites, most notably Copernicus (1972 to 1980) and the International Ultraviolet Explorer or IUE, launched January 1978 and still (at the time of writing) operating! The IUE consists of a 45 cm telescope, two echelle spectrographs, UV to visible converters, and television cameras to encode the spectra for transmission to the ground. It can record spectra in the range 1150–3200 Å and is operated like a true observatory. Observations are conducted from one of two ground stations and the elliptical geosynchronous orbit keeps the spacecraft positioned approximately over Brazil, always within view of at least one of the stations. IUE is a joint project between the United States, the United Kingdom, and the European Space Agency. It has been used by over 600 observers and there are always more requests for observations than observing time available.

The scientific accomplishments are many. A few examples related to X-ray observations follow:

- The interstellar medium has been found to be hot. Absorption lines of ions such as OVI (oxygen with five electrons removed), SiIV (silicon with three electrons removed), and CIV (carbon with three electrons removed) have been recorded in spectra of many stars, implying the existence of material at temperatures of 1×10^5 to 3×10^5 K. Modern theories of the ISM must allow for vast regions of both hot and cool gas and account for the energy required for heating. Our Sun, for example, appears to be located just inside a 50 pc diameter region of hot material, perhaps heated by a long-past supernova explosion.

- The scattering of light by interstellar dust (also called extinction, or optical reddening) is well known. This phenomenon extends into the ultraviolet and the attenuation as a function of wavelength tells us about the particles which do the scattering. The attenuation increases from red to blue in the visible band and reaches a maximum at 2175 Å in the UV. This '2175 Å bump' is presently attributed to small carbon or carbonaceous particles. There is also a steep rise in extinction at the shortwave end of the IUE spectrum which is not yet understood.

- One of the most interesting developments has been the observations of strong stellar winds. Some massive early stars lose 10^{-6} to 10^{-5} $M_\odot$ yr^{-1} ($M_\odot = 1$ solar mass). The existing theories of stellar evolution, which all assumed the mass of the star remained constant, have had to be changed. The outflowing wind absorbs radiation emitted from the stellar surface, and it is, in fact, this phenomenon which accelerates material to escape velocity. The wind is 'warm' and contains ions which have strong characteristic absorption lines in the UV. The strength of each absorption line is a measure of the amount of material in the wind and, since the material is moving, the Doppler-shifted structure of the line is a measure of the velocity of the material. Lines of highly ionised carbon, nitrogen, and silicon serve as good wind indicators. Winds from the hottest main-sequence stars (O3) have terminal velocities of 3500 km s^{-1}, three times the escape velocity. Stars with highest mass-loss rates are the Wolf–Rayet stars with winds carrying over 10^{-5} $M_\odot$ yr^{-1} away from the star. These may be compared with the relatively feeble solar wind with velocity of 300 km s^{-1} and mass loss of 10^{-14} $M_\odot$ yr^{-1}.

Figure 1.13 shows a UV spectrum dominated by two broad absorption features and containing many narrow weak absorption lines. Absorption of the stellar continuum by stationary material at the base of the wind produces the narrow absorption features. The strong broad double absorption line is in the ion SiIV in the outward moving wind. It is Doppler shifted to shorter wavelengths because the material is moving towards us. The rather sharp inflection shows the maximum velocity of the bulk of the wind. The normal position of this feature is at 1393 and 1402 Å.

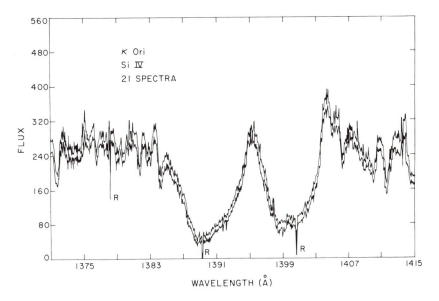

Fig. 1.13 Ultraviolet spectrum of the 2nd magnitude star κ Ori (a B supergiant and in some representations, Orion's right knee) taken with the IUE satellite and used to derive characteristics of the stellar wind. The features labelled 'R' are calibration marks. (Courtesy of A. Dupree, CfA.)

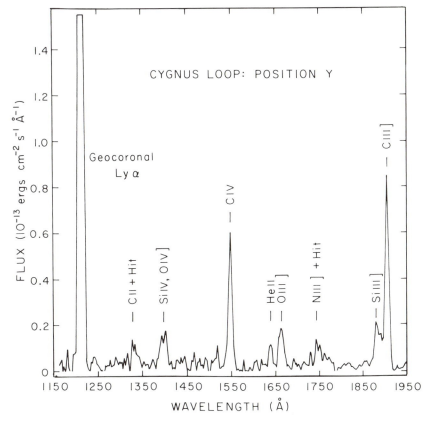

Fig. 1.14 Ultraviolet spectrum of a diffuse filament in the Cygnus Loop, a supernova remnant. Prominent emission lines from carbon, oxygen, nitrogen, and silicon are present. Carbon and silicon cannot be so clearly observed in the visible. The geocoronal radiation is from hydrogen in the vicinity of the Earth and events labelled 'hit' are from cosmic ray interactions within the detector. (Courtesy of J. Raymond, CfA.)

- IUE spectra have been useful in plasma diagnostics. Supernova remnants, HII regions (clouds of ionised hydrogen), and planetary nebulae contain diffuse hot material and many are bright optical objects. UV and optical spectra together are used to determine the composition, density, and temperature of the material. Carbon, aluminium, magnesium, and silicon can be detected only in the UV. The fact that carbon and oxygen in the Crab Nebula are not overabundant implies that the precursor star was not massive enough to produce large quantities of these elements. The strengths of carbon lines in the spectra of the Cygnus Loop (figure 1.14) have been used to derive the geometry of the filaments – they are thin sheets seen edgewise rather than string-like filaments of material. The relative intensities of all the lines in the Cygnus Loop imply that the observed gas is characteristic of material from interstellar clouds of density about 3 atoms cm^{-3} heated by a 100 km s^{-1} shock. Spectra of HII regions in the Large and Small Magellanic Clouds show less carbon than in nearby regions. The material in the Magellanic Clouds has fewer heavy elements than that within our galaxy.

Thus our understanding advances through observations of photons slightly more energetic than those of visible light.

The infrared and IRAS

Infrared radiation is not scattered strongly by interstellar dust so this waveband offers the possibility of 'seeing' the galactic centre, objects inside dense dust clouds, and other optically obscured regions. Observations, however, are difficult because infrared radiation is strongly absorbed by carbon dioxide and water vapour in the atmosphere, which starts to absorb at 0.9 microns and is opaque at wavelengths greater than 15 microns. (It is customary to state the wavelength of infrared radiation in microns. 1 micron = 10 000 Å = 10^{-4} cm.) Since the absorption occurs in molecular bands there are a few regions between 0.9 and 15 microns where the atmosphere is murky rather than opaque and some Earth-based observations are possible. Instruments on a high mountain or in an airplane are above most of the water vapour in the atmosphere, and much use is made of the telescopes on Mauna Kea, Hawaii, and of the NASA Kuiper Airborne Observatory (an airplane with small telescopes and infrared detectors).

An instrument above the atmosphere is necessary to avoid absorption bands shortward of 15 microns and the only way at all to see longward of 15 microns. In January 1983, the Infrared Astronomical Satellite (IRAS) was launched into a 900 km altitude polar orbit. This satellite, a joint project of the Netherlands, the United Kingdom, and the United States, carried a cryogenically-cooled 0.5 m aperture telescope. An array of 62 detectors and filters in the focal plane detected radiation in four wavebands centred at 12, 25, 60, and 100 microns. On November 22 the last of the liquid helium cryogen evaporated, the telescope warmed up, and observations ceased. During its 10 month life, IRAS scanned the infrared sky almost three times. The infrared cosmos was beautifully mapped without the distortion in the optical caused by interstellar absorption in the Milky Way. It is difficult to summarise these observations in a few paragraphs. A book could, and probably will be, written about the IRAS results.

Figure 1.15 shows the IRAS map of the sky. The radiation is probably mostly from dust grains heated by nearby stars or hot gas. The colour gives an indication of the temperature. Blue is warmer than red. The galactic plane is bright yellow-red and bisects the picture. There is diffuse emission and many point sources. The fainter sources do not show at all in this figure.

One component of diffuse emission, the blue background, comes from a local source. This is dust in the ecliptic plane (the orbital plane of our solar system), the same material responsible for the zodiacal light, often seen close to the horizon on a clear night. The ecliptic plane passes from lower left to upper right, through the centre, and the zodiacal dust can be seen as a broad bright-blue band with this orientation. Heated by the Sun, the dust is warmer

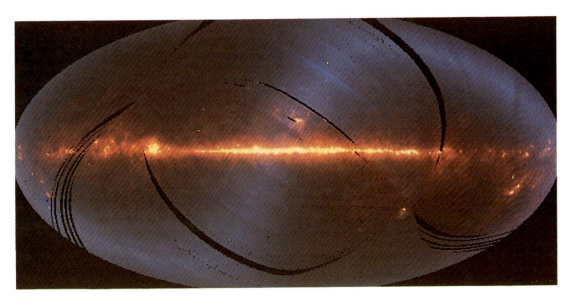

Fig. 1.15 Infrared map of the sky from IRAS – the Infrared Astronomy Satellite. Emission at wavelengths of 12, 60, and 100 microns is displayed in blue, green, and red, so colour in the map indicates temperature of the emitting dust. Dark bands are gaps in the coverage. Star-forming regions in Orion and in Ophiuchus are to the far lower right and just above the centre respectively. (Courtesy of T. Chester, IPAC.)

than the material in the galactic plane and the emission dominates the celestial background except near the galactic plane. There is also evidence for dust clouds associated with the asteroid belt and perhaps colder material in the outer solar system.

A new phenomenon, 'infrared cirrus', accounts for some diffuse emission. These highly structured, extended sources are not yet well understood and may arise from several origins. Dust in interstellar clouds is one possibility.

Solid objects within the solar system, planets, asteroids, and comets, were easily detected. Several comets were discovered by detection of their infrared radiation. The best known was Comet IRAS-Araki-Alcock which passed within 3 million miles of Earth, closer than any other comet in recent times. An extensive dust tail, not seen in visual photographs, was detected.

An excess of infrared emission has been found from the vicinity of some stars, most notably from Vega and β Pictoris, which is not from material in a strong stellar wind but from solid particles, probably remnants of the matter from which the star formed and a strong indication that planets are likely to exist in orbit about these stars.

Much of this material, suddenly revealed for inspection, is, by our standards, cold – very cold. The temperature of the dust in interstellar space, measured by IRAS, is about 30 K or –240 °C. By way of comparison, water freezes at 0 °C and the coldest temperatures recorded on the Earth's surface are about –50 °C. The dust around a star is heated by the star and 'warmed'. The material around Vega has a temperature of –190 °C and the zodiacal dust about our Sun has an apparent temperature ranging from –30 to –100 °C.

Emission from dust in star-forming regions is strong. The Orion Nebula and the ρ Oph cloud are bright IRAS sources. (See figures 4.10 and 4.11 for X-ray pictures of these.) In general most of the 12 and 25 micron point sources in the survey are stars. A long wavelength component of the hot stellar surface is being detected, not emission from surrounding dust.

Many of the 60 and 100 micron sources, however, are galaxies and the signal does come from dust. Spiral galaxies are stronger infrared sources than ellipticals and a late-type spiral can emit five times as much energy in the infrared as in the visual band. An infrared map of the nearby spiral M31 shows a bright nucleus surrounded by a ring of diameter 1.5°. This same ring shows in radio continuum maps and is a region of high density in the M31 interstellar material. Because infrared radiation can penetrate gas and dust easily, excellent images of the centre of our galaxy were also obtained.

Many infrared sources have no optical counterparts in the Palomar Sky Survey (a complete set of photographic plates mapping the northern sky and used by astronomers as a reference). A new class of object? Some of these could be Jupiter-sized bodies at a distance of 600 AU, or perhaps smaller, closer bodies associated with our solar system. A more likely possibility is that most are distant galaxies emitting 30–300 times as much infrared radiation as energy in the optical band. The reason for this is not clear and it is proposed that the energy comes from unseen Seyfert nuclei (a particular type of active galaxy) or from bursts of star formation. A very high infrared luminosity has also been observed from the optical galaxy Arp 220 which may be typical of the 'blank field' IRAS sources.

There are over 200 000 sources in the IRAS data, most of them so far unidentified. The study of these sources and of the diffuse emission will occupy astronomers for some time.

Gamma-rays

It is quite difficult to observe gamma-rays. The detectors are large, heavy, and wildly different from those useful in other wavebands. Not only are gamma-rays stopped by the atmosphere, but detectors carried by balloons and satellites have a severe background problem. The flux from the strongest cosmic sources at the top of the atmosphere is only 10^{-6} to 10^{-5} photons cm^{-2} s^{-1} with energies greater than 100 MeV, and the background produced by cosmic ray interactions in the detector and in Earth's atmosphere is 10^{5} times higher. This background must be overcome to 'see' faint astronomical sources.

Nevertheless, three spacecraft have mapped the distribution of cosmic gamma-rays at energies of approximately 50–1000 MeV. A detector on OSO-3, launched in March 1967, provided enough data so that George Clark, Gordon Garmire and William Kraushaar (MIT) could determine that the galactic plane was a bright source of gamma-rays. The second satellite, SAS-2, carried a spark chamber, designed by Carl Fichtel and colleagues (GSFC), capable of determining the direction of an incident photon to an accuracy of 1.5°. The emission from the galactic plane was shown to have a narrow distribution in latitude and to be strongest in the direction of the galactic centre. The Crab Nebula, and the Vela Pulsar were also determined to be sources of gamma-rays. The third satellite, COS-B, a cooperative ESA venture undertaken by France, Italy, the Netherlands, and West Germany, also carried a spark chamber and collected data for 7 years. Emission from the galactic plane was mapped (figure 1.16) and 25 point sources were located. Most of these are unidentified, but a few sources have secure identifications. The Compton Observatory (GRO), orbited in April 1991 carries the most sensitive detectors yet. It is rapidly adding source detections to those discussed here.

The brightest gamma-ray source, surprisingly because it is faint in other wavebands, is the Vela Pulsar. Both the Vela and the Crab pulsars emit pulsed gamma-rays at rates of 11 and 30 pulses s^{-1}. Because this emission has exactly the same period as do the radio pulsars these identifications are certain, a rare occasion in this field where the best locations have accuracies of one degree. No other radio pulsars have been so identified in spite of much searching for periodicities in the gamma-ray data.

The second brightest source is called 'Geminga' (meaning 'it is not there' in the Milanese dialect), a well-defined gamma-ray emitter in Gemini identified by Giovanni Bignami and Patrizia Caraveo (Milan) with a weak Einstein X-ray source. This source shows the soft X-ray spectrum and low optical

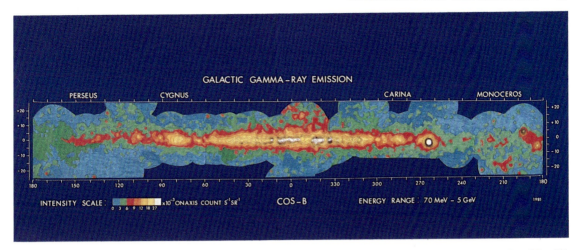

GALACTIC GAMMA-RAY EMISSION

PERSEUS   CYGNUS   CARINA   MONOCEROS

INTENSITY SCALE : [   ] x10⁻³ ONAXIS COUNT S⁻¹ SR⁻¹   COS-B   ENERGY RANGE : 70 MeV – 5 GeV
                    0 3 6 9 12 18 27

Fig. 1.16 The gamma-ray sky as mapped with the European satellite COS-B. The galactic plane is bright and the most prominent sources are the Crab Nebula and the Vela Pulsar. (Courtesy of G. Bignami, CNR, Milan.)

luminosity expected for a not-too-distant isolated neutron star. A 0.28 s X-ray period was discovered by Jules Halpern (Columbia) with ROSAT and confirmed in gamma rays detected by GRO. The identification as a neutron star is secure.

The high energy gamma-ray emission from the galactic plane is attributed to the decay of π mesons. Cosmic rays interact with interstellar material and produce π mesons which each decay to two 70 MeV gamma-rays in the rest frame. These are shifted to higher energies by the Doppler effect of the rapidly moving meson. Assuming a uniform cosmic-ray flux throughout the galaxy, the brighter gamma-ray emission from the galactic centre indicates a concentration of interactive mass in that location. In support of this, the ρ Ophiuchi dark cloud, a dense molecular cloud, smaller than the resolving power of the COS-B instrument, is identified as a gamma-ray source. The strength of this source is well predicted using the calculated mass of the ρ Oph cloud and the flux of cosmic rays in the solar neighbourhood.

The most convincing COS-B identification of an extragalactic source is the nearby quasar 3C273. Other active galactic nuclei are expected to produce high energy radiation, but the COS-B instruments were not sensitive enough to detect them.

Some high resolution spectroscopy has been accomplished with two interesting results. A variable 511 keV line has been detected from the galactic centre with a flux of 10^{-3} photons cm^{-2} s^{-1} This is a sure sign of positron annihilation. Something unusual is happening there, perhaps from material in the vicinity of a black hole. Second is the discovery of a 1.809 MeV line from the galactic plane. This line comes from the decay of radioactive aluminium atoms. About 3 $M_\odot$ of these atoms are required to produce the observed number of gamma-rays. The half-life is 10^6 years so we are seeing material which has been made by nucleosynthesis in novae and in main-sequence stars.

There has been, however, a complete and thorough survey of the sky for bright short-lived gamma-ray sources in the energy range of 0.2–1.5 MeV. The Vela Satellites, operational from 1967 to 1979 contained omnidirectional detectors to search for bursts of gamma-rays produced by clandestine tests of nuclear weapons in space. To everyone's surprise, in 1973 Ray Klebesadel, Ian Strong, and Roy Olsen at Los Alamos (LASL) reported the detection of gamma-ray bursts from astronomical sources! About two dozen events were recorded by the Vela Satellites from 1967 to 1973, and an order of magnitude more have since been detected with a variety of instruments.

A typical burst has a sudden beginning and a duration of about a second. These events can be observed with detectors on several spacecraft separated by 10^5–10^8 km (0.3–300 s light travel time). With accurate timing the directions of many sources have been located to arcminute accuracy. The origin of the bursts, however, is unknown. No steady optical counterparts have been discovered although in 1981 Bradley Schaefer (MIT) found evidence on an archival photographic plate for an optical burst from the location of a known gamma-ray burst source.

The most intense burst yet recorded is that of March 5, 1979, which was measured to come from the same direction as the supernova remnant N49 in the Large Magellanic Cloud. If this is not a coincidence, the peak rate of gamma-ray emission was 10^{44} erg s^{-1}! The gamma-ray flux during the 100 s decay of this event had a periodicity of 8 s so the origin is suspected to be a rotating neutron star within the supernova remnant. The enormous rate of radiation, however, presents formidable theoretical problems.

The distribution of gamma-ray burst directions in space is isotropic (the same in all directions) implying an origin either near the Earth or extragalactic. Although there is no general agreement yet on the mechanism of these bursts, neutron stars are considered to be the most likely sources.

Thus we have accomplished surveys and observations of at least the brightest objects in almost all wavebands. A world, or rather a universe, of data is now in hand. The rest of this book is devoted to a detailed presentation of the X-ray results. An understanding of the physical parameters which shape the cosmos will come from consideration of observations at all wavelengths. We have a nice sample of these, an appetiser which whets the appetite for future observations from radio waves to gamma-rays.

1.5 The role of chance in astronomical discovery

What is the practical value of astronomy? It is satisfying to know about the objects in the heavens and their arrangement, but it is not clear that this knowledge will lead to anything which will directly affect our lives. Perhaps the greatest benefit will come from an unexpected direction.

Astronomy is an exploration. A new instrument leads to searches through territory never before seen and things are found which are complete surprises. It is not uncommon for an astronomer to find something which cannot be explained and the data are checked, more than once, to make sure the instrument is operating properly. The finding of something which cannot be explained may lead to the discovery of a new principle of physics which will greatly influence our lives. Science, like other fields of human endeavour, is rather faddish. Most modern research is supported by the government, and research proposals are usually evaluated by a group of established scientists, the 'peer review group'. It is rare that an idea outside generally-accepted lines of thinking is supported for further development. A new discovery can blast scientific thought out of a rut and into new directions.

A classical unexpected discovery was that of Sirius-B, the white-dwarf companion of the first magnitude star, Sirius. Alvin G. Clark, testing a new telescope lens in Cambridge, Massachusetts in January 1862, was observing Sirius as it emerged from occultation by the wall of a distant building. The faint companion emerged a few seconds before the ten-thousand-times brighter primary. The discovery that the companion, known to have a mass of 1 $M_\odot$ from studies of the motion of Sirius, had a luminosity of only one ten-thousandth of the primary forced astronomers to believe that matter could exist in a state one million times as dense as that in the known stars and planets.

A more recent discovery concerned the emission-line star SS433. In 1978 optical spectroscopy revealed the presence of material ejected at one quarter the speed of light. This was not only the first large blueshift ever seen from a star-like object, but the velocity was astoundingly large. This system is still not well understood and we have devoted chapter 12 to its description.

Another current example is supernova 1987A in the Large Magellanic Cloud, the first visually bright supernova (SN) in 380 years. The precursor star, a blue supergiant called Sanduleak −69 202, has been identified. This was the first, ever, definite identification of a supernova precursor star! The early SN spectra showed the presence of hydrogen, so it is a type II (a massive star, as described in chapter 3) SN. In 1986 almost all astronomers would have said that the stars which became type II SN were red supergiants. The identification of the blue star has therefore caused a readjustment in thinking about the final stage of stellar evolution.

The following chapters will contain detailed discussions of unexpected discoveries in the X-ray band: the fact that most stars have hot, X-ray-emitting atmospheres; the existence of accretion-powered X-ray binaries – neutron stars and normal stars locked in orbits having periods of hours or days; the existence of stellar-mass black holes; hot X-ray emitting gas in clusters of galaxies which increases the amount of observable mass in the universe by 50%; and the X-ray background – a glow between the stars which is still not understood and which has cosmological implications.

2 The tools of X-ray astronomy

2.1 The proportional counter

The first instruments used for X-ray astronomy were developed originally for the detection of charged particles and gamma-rays emitted by radioactive material. The proportional counter, a close relative of the Geiger counter, has been the work-horse of cosmic X-ray observations and is still being used in modern instruments. However, the modifications necessary to adapt the simple laboratory counter to an X-ray detector capable of operating in space illustrate the technical difficulties facing the early X-ray astronomers.

First, a *window* is necessary which is thin enough to transmit X-rays but strong enough to keep the gas inside the detector from leaking into the near vacuum of space. The window must have a large area to maximise the detector sensitivity yet must be strong enough to withstand the force of the gas pressure inside the detector. Many an observation has been lost by the failure of detector windows during rocket ascent out of the atmosphere and upon first exposure to space.

Whenever an energetic particle or X-ray photon enters the detector through the thin window, it ionises (i.e. knocks out electrons from) some of the atoms of gas contained inside the counter. These electrons are attracted towards and drift to the vicinity of the central wire or anode (maintained at about +2000 volts) where they cause further ionisations by colliding with other atoms. An *avalanche* of electrons occurs close to the anode and the net result is a pulse of electrons that are collected on the central wire and detected by the associated electronics as an *event* or *count*.

In a Geiger counter (as used in the original AS&E experiment) this is a discharge which gives no more information than the fact that the event occurred. However, in the proportional counter the central wire is maintained at a slightly lower voltage so that the number of electrons collected by the anode is *proportional* to the energy of the incident X-ray photon. From this it is possible to obtain a rough idea of the X-ray spectrum of the source being detected. (This can give valuable information on the mechanism by which stars actually produce X-rays, as will be shown later.)

Most problems really are due to the hostile space environment in which these counters have to operate. Although the usual altitude of rockets and satellites place them below the principal radiation belts surrounding the Earth, there are still localised regions of charged particles and the all-pervasive cosmic rays to deal with, as described in box 2.1.

Fig. 2.1 (a) The operation of an X-ray proportional counter. An X-ray photon penetrates a thin window (usually made of metal, such as beryllium, or plastic) and enters the detection chamber, which is filled with gas. The photon ionises a number of atoms creating a small cloud of electron–ion pairs. The electrons are accelerated rapidly toward the high voltage anode, causing more collisions and new ionisations on the way, i.e. an avalanche. The final detected pulse contains typically a million electrons, amplified from the original hundred or so. A simple honeycomb collimator restricts the area of sky from which the counter can detect X-rays, in much the same way as holding up a long tube to your eye. Angular resolution is crude, but it is cheap and rugged. Although greatly improved upon in the last two decades, this basic X-ray detector was used in the Uhuru satellite (shown), and is still in use in currently operating satellites. (b) Schematic of a scintillation counter with phoswich arrangement for identifying and rejecting events caused by charged particles passing through the detector. True astronomical hard X-rays can only be seen along the direction of the honeycomb collimator. The flashes or scintillations of light produced in the crystals (C) are detected by the photomultiplier tubes (PMT).

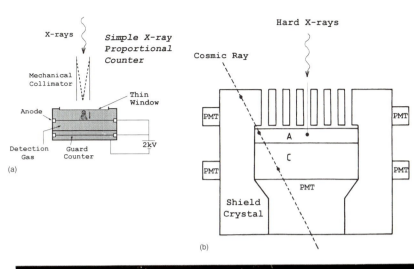

There is one region in particular, known as the *South Atlantic Anomaly*, in which the charged particle density is so high that astronomical detectors risk being severely damaged or destroyed if they are activated within that region. This is because these particles swamp the high voltage anodes causing serious breakdown problems.

Cosmic rays must be handled differently because they are always present. They are actually not *rays* but extremely energetic particles (the origin of which is still a great mystery). Because of their very high energies (vastly greater than the most energetic particles that can be accelerated on Earth) they can pass completely through the X-ray satellite detectors and the satellites themselves! In doing so, they deposit a line of charge through the counter gas which is then detected in the same way as that produced by real X-rays.

It should be clear now that it is not easy to operate X-ray proportional counters in the harsh environment of space. Nevertheless, the ingenuity of

Box 2.1 Allowed orbits for X-ray observatories

The *allowed* regions for X-ray observations above the Earth's surface are defined by two restrictions.

(i)   Although soft cosmic X-rays can be detected at altitudes above 150 km (where typical short rocket flights reached), a satellite in a circular orbit must have an altitude greater than 400 km or the drag of the tenuous upper atmosphere will soon lead to a fiery re-entry.

(ii) In regions of high charged particle flux, the background counting rate would be too high for X-ray detectors to operate.

The Earth is surrounded by the doughnut-shaped (or perhaps more like pitted-olive shaped) *van Allen radiation belts*. Here protons and electrons trapped in the Earth's magnetic field have both high energy and long lifetimes. Detectors must be below or above these belts in order to function normally. Most of the spacecraft listed in table 2.1 were launched into orbits that were below the belts. Here the particle flux is at a minimum except for the region known as the *South Atlantic Anomaly*, where the magnetic flux is anomalously low, thereby allowing particles in the inner belt to brush the top of the atmosphere.

Fig. 2.B1 The range of typical satellite orbits is shown here relative to the Earth and its magnetosphere. Normal low-Earth orbits accessible to small rockets and the Shuttle are essentially shielded by the magnetosphere from contamination and interference by the solar wind. This is not true for eccentric orbits such as that employed by EXOSAT, which was originally chosen for the purpose of allowing the Moon to occult as large a fraction of the sky as possible. In this way the Moon could have been used to accurately determine the location of a large number of X-ray sources. In fact, the contamination at high X-ray energies by the solar wind turned out to be much less than anticipated for most of the time. Hence the major advantage of EXOSAT-type orbits is the opportunity it provides for long, uninterrupted observations of a single object.

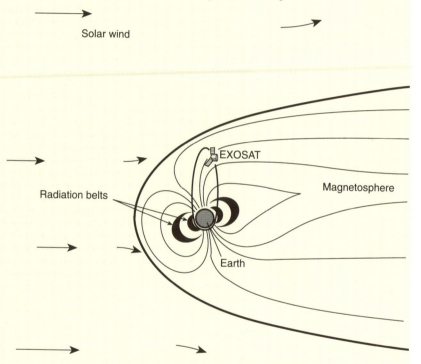

Since the primary cosmic ray flux is five times higher at the Earth's poles than the equator, a low-inclination, near-Earth orbit (which has a 95 minute period) minimises both the cosmic ray and charged particle background. Uhuru, for example, was placed in such an orbit.

However, it is also possible to operate outside the radiation belts. The Vela satellites were placed in a circular orbit with altitude approximately 17 Earth radii (which has a 4 day period). EXOSAT was in a 3 day period, eccentric orbit (a left-over from its original conception as a lunar occultation observatory) which carried the spacecraft from below the belts to an altitude of 30 Earth radii (see figure 2.B1). Since most of the time was spent near apogee, well above the belts, backgrounds were low (except during time of solar flares when the spacecraft was essentially unprotected).

instrumentalists has enabled fantastic strides to be made in the sensitivity, resolution and reliability of these devices in order to overcome these operating problems. Methods by which these *background* events are rejected are described in box 2.2.

Box 2.2 X-ray detectors in space

Figure 2.B2 shows schematically the detection of a cosmic X-ray photon (of scientific interest) by a proportional counter and the passage of a cosmic ray or energetic particle (to be eliminated from the data).

Fig. 2.B2 Operation of an X-ray proportional counter. The guard counter eliminates cosmic ray events which are detected simultaneously in both parts of the detector. The guard counter is a sealed metal chamber that is blind to real X-ray photons. The arrival time of the electrons at the anode will be different for an X-ray and a cosmic ray. Because the cosmic ray deposits charge all along its path the electrons are spread out over a much longer time interval than that of a true X-ray.

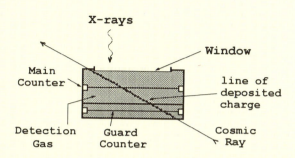

An X-ray photon can only enter the detecting volume by passing through the thin entrance window, as it would be absorbed by the metal of the detector's box. A cosmic ray, however, can pass completely through any part of the detector as well as the satellite itself! There are two basic methods by which such events are recognised and subsequently discarded.

(i)   A part of the counter's detecting volume is entirely surrounded by metal so that it is *blind* to cosmic X-rays. This region of the detector is called the *guard* counter and is only sensitive to cosmic rays. Hence, an X-ray will only be detected in the upper volume, whereas any event detected simultaneously in both volumes is almost certainly a cosmic ray and can be discarded by the on-board electronic logic. The guard counter is said to be in anti-coincidence with the main X-ray counter, or is being used as a veto system.

(ii) The rate at which electrons (or charge) arrive at the anode of the detector. An X-ray photon is *stopped* by the detector gas in a very small region of the detector and therefore yields a short (<10 ns) risetime pulse at the anode. A cosmic ray, on the other hand, produces a long trail of charge which arrives at the anode spread over a much longer period of time (>100 ns). These times can be measured electronically for every event, and only short risetime X-ray events are admitted by the electronics. This technique is called *pulse risetime discrimination*.

2.2 The scintillation counter

Proportional counters do not detect photons with energies above 20 keV efficiently. The sensitive volume of the detector is a gas and, at high energies, the gas is not massive enough to stop many of the photons. A solid state detector must then be used, the *scintillation counter*.

Another nuclear particle detector, the scintillation counter, uses crystals of sodium iodide or caesium iodide which can efficiently stop photons with energies up to several MeV. The photon energy is absorbed by an atom within the crystal and some of this energy immediately reappears as a pulse of light, or *scintillation*. A simple photomultiplier tube then detects the scintillation, thus registering the time of the event. The amount of light in the scintillation is proportional to the incident X-ray photon energy.

For rejection of events caused by cosmic rays and not real X-rays, the principal crystal can be surrounded by a second scintillating material, which is chosen so that light pulses arising from events within this material have a different risetime than that from an event in the primary detector (see figure 2.1(b)). The phototube detects light from both scintillators. The electronics then measures the shape of the pulse and rejects events which show that a charged particle has passed through the outside scintillator at the same time. This detector is called a *phoswich*, short for *phosphor sandwich*.

Scintillation counters have been the work-horses of balloon-based high energy X-ray astronomy. At altitudes of more than 50 km, balloons float above 99% of the atmosphere. The remaining 1% prevents X-rays with energies less than 15 keV from reaching the detector. Thus balloon astronomy has covered the energy range from 20 to 200 keV.

2.3 Locating X-ray sources

Lunar occultation of the Crab Nebula
Following the discovery of the existence of cosmic X-ray sources, the next major step was to identify them with objects at other wavelengths (e.g. in the visible or radio) to find out what they really were. Here the very young science of X-ray astronomy was assisted by a technique first applied by radio astronomers in the 1950s. It must be remembered that the experiment that discovered Sco X-1 actually had a very large field of view (a circle of diameter of 100°, and so it was not even clear which constellation the source was really in!) and, although this was fortuitous for that flight since Sco X-1 is the brightest X-ray source in the sky, it greatly limited the accuracy with which the source could be positioned.

Most subsequent rocket-borne detectors used honeycomb (or rectangular cell) collimators to limit the field of view to a few square degrees (as shown in figure 2.1). The collimator array had the strength to support the window and to withstand the gas pressure trying to push the window into space, whilst the walls of the honeycomb cells could not be penetrated by soft X-rays. (The use of the simple honeycomb collimator to scan the sky can be compared to observing the night sky visually through a long tube which has a diffusing filter placed over the end.)

More X-ray sources were soon discovered, including one in the constellation Taurus which was thought might be associated with the Crab Nebula (the remnant of the supernova of AD 1054; see chapter 3). In 1964 the opportunity arose to test this hypothesis when it was realised that the Crab Nebula would be occulted (or eclipsed) by the passage of the Moon in its orbit. Even though early X-ray detectors had poor spatial resolution, they had very good time resolution (the arrival of each X-ray photon could be determined to a fraction of a second), and this provided the opportunity to convert this time resolution into spatial information by letting the Moon shadow (i.e. occult) the object of interest and simply observe the time at which this happened (the Moon's position on the sky is known to very great accuracy).

This observation was undertaken by Stu Bowyer and the NRL group under Friedman's direction. The occultation was due to take place in July 1964 and would not happen again for another decade. They only had 5 months in which to prepare their experiment and modify the rocket so that it would remain pointed in the direction of the Crab Nebula. In addition the launch would have to be very precisely timed to ensure that the experiment was positioned above the atmosphere just when the occultation was due to

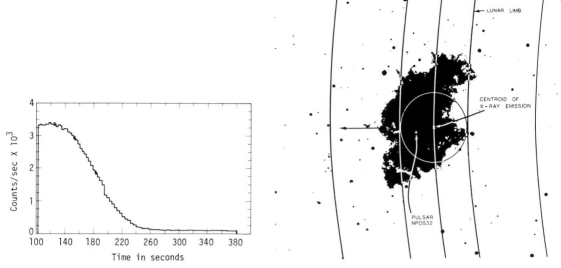

Fig. 2.2 Counting rate of an X-ray detector on November 3, 1974, during a lunar occultation of the Crab Nebula. The Moon passed in front of the Nebula at a rate of 1 arcminute (of angle) per minute (of time). The source within the Crab was gradually occulted over 120 seconds of time, thereby showing that it is diffuse and about 2 arcminutes in extent. The neutron star within the Crab, however, accounts for 5% of the X-ray flux and is a point source. When the neutron star passed behind the lunar limb at 195 seconds, a sudden drop in count rate was observed. [From Palmieri et al. Astrophys. J. 202, 494 (1975).]

start. The Moon would take only 4 minutes to cross the entire nebula but this was about as long as the useful observational length of the entire rocket flight!

The pointing of the rocket and its launch at an exact moment were major problems in those relatively early days. Although attitude control systems for accurately stabilising rockets had been developed in the late 1950s, even by 1964 they were still not reliable. (Of 14 systems tested in flight at the Goddard Space Flight Center up to this time, all but three had malfunctioned in some way!) The Aerobee rocket was also powered by liquid fuel which required careful preparation and a long countdown. To launch at an exact time for specific astronomical goals had never been tried before!

In spite of all these technical difficulties the launch occurred on time on July 7, 1964 and the experiment worked perfectly. As shown in figure 2.2 the experiment clearly identified the X-ray source with the Crab Nebula. Furthermore it showed that the source was extended, an important result that will be referred to in the next chapter. This observation gave a tremendous boost to the infant science of X-ray astronomy. Within two years of the discovery of cosmic X-ray sources one of them had been unambiguously identified with a previously known high energy object. Better still astronomers had a rough idea how far away the Crab Nebula was. At a distance of about 2 kpc (or 6500 light years) this meant that the nebula was indeed emitting at least 10 billion times the X-ray output of the Sun!

The success of the lunar occultation technique for positioning sources has been repeated for a number of X-ray objects that fortuitously lie in the region of sky through which the Moon passes. Indeed, in the early 1970s it was decided to build an X-ray satellite substantially devoted to this kind of work. In order to enable this satellite to view a much larger number of X-ray sources by lunar occultation it would be necessary to launch the satellite into a very high, elliptical orbit (see box 2.1). Unfortunately, the satellite was not launched until mid-1983, by which time the lunar occultation techique had been rendered largely redundant. Nevertheless, this unique orbit gave EXOSAT other remarkable capabilities as will be demonstrated later in this chapter.

The modulation collimator

Although the identification of the Crab Nebula was a great boost to interest in X-ray astronomy, there still remained the considerable problem of identifying and understanding the other *X-ray stars*. This included the first source

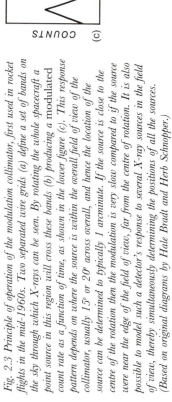

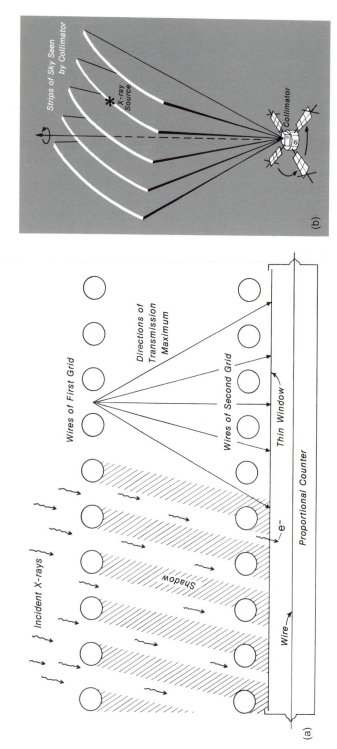

Fig. 2.3 Principle of operation of the modulation collimator, first used in rocket flights in the mid-1960s. Two separated wire grids (a) define a set of bands on the sky through which X-rays can be seen. By rotating the whole spacecraft a point source in this region will cross these bands (b) producing a modulated count rate as a function of time, as shown in the lower figure (c). This response pattern depends on where the source is within the overall field of view of the collimator, usually 15° or 20° across overall, and hence the location of the source can be determined to typically 1 arcminute. If the source is close to the centre of the rotation then the modulation is very slow compared to if the source were near the edge of the field of view, far from the centre of rotation. It is also possible to model such a detector's response to several X-ray sources in the field of view, thereby simultaneously determining the positions of all the sources. (Based on original diagrams by Hale Bradt and Herb Schnopper.)

found and still the brightest persistent X-ray star in the sky, Sco X-1. The essence of the lunar occultation technique was to acquire spatial resolution through watching how the X-ray count rate varied as the Moon passed in front of the source. Could the same thing be done artificially in some way, since an X-ray telescope capable of taking high resolution X-ray pictures was still many years away?

The solution to this problem was an invention by Minoru Oda, professor of physics at the University of Tokyo, who was visiting MIT in 1966. Instead of the simple honeycomb collimator used on all experiments until then (see figure 2.1(a)), Oda proposed an ingenious collimator consisting simply of two sets of wire grids as shown in figure 2.3.

Since the radiation from a distant star can be considered to represent a parallel beam, the first wire grid will produce a set of shadows as shown. Depending on the particular viewing direction, the second grid may or may not align with these shadows and allow the X-rays to pass through. But how can such a device be used to provide positional information? The trick is to scan the collimator across the sky. The effect of the scan is to define a set of *bands* on the sky which modulate the X-ray count rate of any steady point-like source. Hence the name *modulation collimator*, or *scanning modulation collimator* (SMC). An alternative approach was to set the collimator rotating by spinning the entire rocket or spacecraft (we have already seen that this is much easier than to try and point the experiment at a particular place in the sky).

The effect of the spin is to define a set of spinning transmission bands on the sky. This is known as a rotation modulation collimator (RMC). (Imagine viewing the Sun through a set of venetian blinds which are only just opened and then set the blinds spinning about their centre. The Sun will 'flicker' as the slits of the blinds cross it, but most importantly the frequency of the flickering will depend on the position of the Sun with respect to the spin axis.) A normal X-ray proportional counter is placed behind the collimator in order to detect the incoming modulated signal. A typical observation might produce the count rate which varies with time as shown at the bottom of figure 2.3. The modulation of the count rate from a source by this collimator clearly depends on its *position* within the overall field of view. Hence this pattern can be used to locate the X-ray source.

Of course, in order to convert such data into accurate positional information it is necessary to know where the *bands* are on the sky at any instant throughout the flight. This is obtained by means of *aspect cameras* which are co-aligned with the collimator and simply photograph the visible sky continuously during the flight. Once again we have used the good temporal resolution of the detectors to provide spatial information.

The modulation collimator was used on rocket flights in 1966 and 1967 by the AS&E group to position the bright X-ray sources Sco X-1 and Cyg X-2. The angular resolution of this device was set by the width of the grid wires divided by their separation and, for these first flights, was about an arcminute. Figure 2.4 shows the location on the sky of Sco X-1 as determined by earlier, simple experiments and the modulation collimator flight of March 8, 1966.

It was rather a shock to see that the final accurate position for Sco X-1 was outside of some of the earlier *error boxes* which indicated the grave difficulties of identifying X-ray sources without arcminute locations! There are two small *error boxes* for Sco X-1 from the modulation collimator experiment because of an ambiguity in locating the source with respect to the spin axis of the collimator (if you look at figure 2.3 you will see that an X-ray source positioned two *bands* to the left of the spin axis and in the centre of the band will give the same count rate as the source marked already, hence the

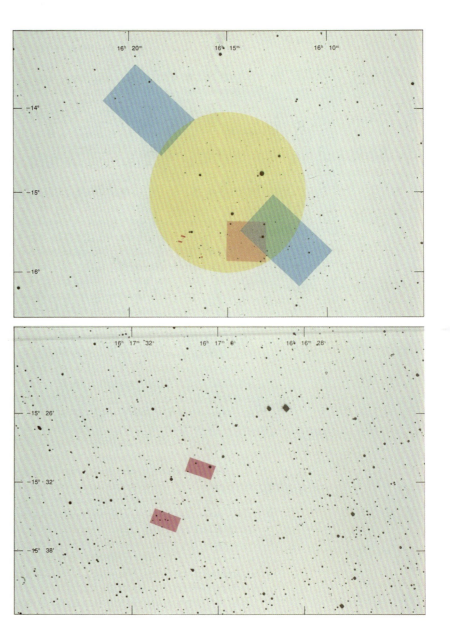

Fig. 2.4 The X-ray location of Sco X-1 on the sky as determined by a variety of instrumentation. The upper photograph covers a large area of sky, more than 15 square degrees. The large rectangles and circles superposed represent the accuracy of location of Sco X-1 from several different rocket flights using conventional collimated proportional counters. Not only are there very many stars within the error boxes, they do not even agree with one another! The location derived by the modulation collimator technique is represented by the two small rectangles in the bottom left of the large circle. This area of sky is blown up in the lower photograph, and the optical counterpart is in fact the brightest star in the upper box. (There are two boxes because the modulation collimator in fact exhibits a 180° ambiguity in the count rate modulation, as is evident from examining figure 2.3. If the source were actually at the same value of r, but at an angle θ + 180°, the modulation produced would be exactly the same. This ambiguity can be removed in satellite versions of the modulation collimator by making further observations with a slightly different centre of the field of view.) (Figure produced by Jeff McClintock, CfA.)

ambiguity). Nevertheless it did not take optical astronomers very long to discover that the 13th magnitude star in the upper rectangle (the brightest star in fact) was extremely peculiar, unlike anything ever seen before. It was immediately hailed as the optical counterpart of Sco X-1 and was clearly completely different from the Crab Nebula. Cyg X-2 was identified with a rather similar star, and both of these objects will be described in detail in chapter 7.

The modulation collimator has been described in some detail because of its subsequent extensive use on satellite missions in which many X-ray sources could be accurately positioned. Undoubtedly one of the most important and successful (in terms of optical identifications) was the SAS-3 (third Small Astronomy Satellite) launched in 1975 and operated by MIT.

However, the British Ariel V and American HEAO-1 satellites also used the RMC and SMC as did the more recent Japanese mission Hakucho (built by Oda's group in Tokyo).

2.4 Scanning the X-ray sky

Exploration by rocket

The period from the time of the discovery of Sco X-1 to the launch of the first X-ray astronomy satellite in December 1970 was a time of exploration and surprises. The sky was searched for sources of high energy radiation using rocket-borne proportional counters and scintillation detectors carried by balloons. The counter itself was cheap, easy to build, and naturally rugged. The weakest part was the window which, to maximise sensitivity, was made as large as could be accommodated by the dimensions of the rocket payload. Diligent research was done to find and develop the thinnest low-Z (atomic weight) material which would survive the rigours of launch and ascent and still manage to contain the gas filling of the counter for the duration of the observation. Ingenious schemes were invented to reduce the sensitivity of detectors to cosmic rays without reducing their ability to detect soft X-rays. At first, large areas of sky were surveyed using the spin and precession of the payload to scan the detector field of view over the heavens above the rocket-horizon. Mechanical collimators made from slats or honeycomb restricted the field of view to a few square degrees at any instant. Later, as attitude control systems became more reliable, small regions were surveyed with greater precision and sensitivity.

This was an exciting way to do science. After a year of planning, building, and calibration came a sometimes frantic week in the field when detector, payload, and rocket were mated for the first time (see figures 2.5 – 2.6). After final assembly there was often a delay of several days due to unfavourable weather or range conditions. The launch itself was a thrilling moment. As the flame of the rocket accelerates up into darkness (solar X-rays scattered from the upper atmosphere and collimator usually necessitated a night launch), the dominant emotion of the watchers was relief. 'We finally got it off and we won't have to struggle with that particular payload again.' This was tempered with some apprehension over the detector windows surviving ascent and exposure to vacuum, and with perhaps fervent hope that the launch time had been calculated correctly.

If preparations had been done properly, and if the experimenters were not unlucky, they were rewarded with 5 minutes of trouble-free flight above the atmosphere. Since most of the sky was unexplored, each successful flight produced new and exciting data. There was elation over success and deep gloom over failure. Failure, however, was not crushing since the attempt could usually be repeated in about a year. Those who build satellite-borne instruments have sometimes invested 10 years of their professional lives, or more. The stakes and risks are higher now.

By the end of 1970, about 50 cosmic X-ray sources had been discovered. Most of these were bright sources in the galactic plane; the *galactic bulge* sources which are now recognised as accretion-powered binaries. As described above, two of these had been precisely located and identified with star-like optical objects. Some of these X-ray sources were recognised to be highly variable and two X-ray transients or *X-ray novae* had been observed to flare up and then disappear. The diffuse X-ray background was discovered and determined to be uniform to 5–10%. Six supernova remnants had been identified as bright, sometimes extended, sources and X-rays had been detected coming from the Large Magellanic Cloud, the active galaxies M87 and NGC 5128, the Coma Cluster and the quasar 3C273.

Fig. 2.5 A Terrier-Sandhawk rocket, with payload attached, on the launcher at the Kauai Test Range, May 11, 1970, the day before launch. Sandia Corporation engineers have just mounted the second stage rocket and it is supported by a strap while the launcher is in a horizontal position. Two tanks of gas for the final proportional counter checks are strapped to the top of the launcher. These solid fuel rockets will each burn for only a few seconds but will produce an acceleration of 20–30 g and will propel the payload to an altitude of 306 km. It will spend 6 minutes above the atmosphere detecting X-rays and will fall (with parachute and flotation bag) into the ocean 10 minutes after launch. (Figure courtesy of Lawrence Livermore Laboratory and Sandia Laboratories.)

All this from a total observing time of a few hours. It boggled the mind to think what might be done with a detector on board a satellite... .

Clearly the next major step would come from placing the same instrumentation on board a satellite, thereby changing the typical observing times per object from seconds to days. NASA recognised the scientific importance of such a mission and contracted Giacconi's AS&E group to construct the payload. SAS-1 was launched off the coast of Kenya on Kenyan Independence Day, December 12, 1970 and, in honour of the date, the satellite was named Uhuru, the Swahili word for 'freedom'.

Fig. 2.6 The payload has been recovered from the water and has just been returned to the scientists and engineers who built the detector. They are all smiling because everything has worked. Things of interest on the payload, proceeding from front to back, are: two deployed attitude control gas jets in the nose; the white paint on the nose cone has been charred black by heating during ascent; a square hole shows where the star-field camera has been removed to recover and develop the film (used to determine pointing direction); one of the telemetry antennas has been bent, probably during the process of dragging the payload from the water onto the recovery boat; the proportional counter collimator is intact. (Figure courtesy of Lawrence Livermore Laboratory and Sandia Laboratories.)

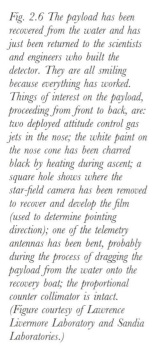

Uhuru

The Uhuru payload, at 64 kg, weighed no more than a typical rocket experiment. But, with the advantage of the spin-stabilised spacecraft, the two sets of conventional proportional counters with simple honeycomb collimators (operating as in figure 2.1) were to undertake the first X-ray survey. Even though spinning rapidly (each scan of the sky took only 12 minutes) Uhuru would scan many times over the same region thereby greatly increasing its sensitivity to weak sources. In addition, Uhuru was able to incorporate both of the background rejection techniques described earlier. The net result of this was that Uhuru was able to detect X-ray sources 10 times fainter than the faintest detectable on earlier rocket flights. The 4U (fourth Uhuru) catalogue contains 339 sources.

Satellites of the 1970s

After this first important X-ray survey had been completed by Uhuru there was clearly an enormous amount of follow-up work to be done on all these X-ray sources, since little was actually known about them. Fortunately a number of missions involving European as well as American astronomers had been planned and the 1970s saw an explosive growth in observations and interest in X-ray astronomy.

Table 2.1 lists all the X-ray astronomy satellites since the field began.

2.5 X-ray imaging

X-ray optics

For visible light, producing an imaging telescope with normal incidence reflecting mirrors is not difficult. Indeed, very large apertures (in excess of 6 m) are possible and are now being constructed in optical astronomy for operation before the end of this century. However, below about 1000 Å and particularly in the X-ray part of the spectrum, there is a severe difficulty in that X-rays do not reflect at normal incidence, they scatter! This has required a different approach in order to produce a truly imaging X-ray telescope. X-rays can in fact be reflected if they strike a smooth metal surface at a shallow or *grazing* angle as shown in Fig. 2.7.

The application of this principle to create X-ray imaging optics was developed by the German physicist Hans Wolter in the early 1950s as part of an effort to produce an X-ray microscope. However, such a microscope was never completed because it was found to be impossible to polish the mirror surfaces to a high enough accuracy. When this work came to the attention of Giacconi he realised that this limitation would not be important if the optics were inverted and used as a telescope. Figure 2.7 shows how such a set of mirrors at grazing incidence can be used to produce a true X-ray image, and the photographs show those actually constructed for use on HEAO-2, the Einstein X-ray Observatory.

Position sensitive detectors

It is all very well producing X-ray images at the focal plane of such X-ray telescopes, but the simple proportional counters described earlier have positional resolution given only by the field of view of the honeycomb collimator. How then are these images going to be recorded and transmitted back to Earth? There are two main types of *position sensitive detectors* currently in use (see figure 2.8).

The basic principle of the imaging proportional counter (IPC) is the same as in a conventional proportional counter. An X-ray photon passes through the thin *window* of the detector and produces a small electron cloud when it encounters the atoms of the counter gas. However, the anode of an IPC is

Table 2.1 *X-ray astronomy satellites 1969–94*

Satellite	Country	Launch	Demise	Type
Vela 5A,B	USA	May 1969	Jun 1979	Scanning, small scintillation counter, gamma-ray
Uhuru	USA	Dec 1970	Jan 1975	Scanning, conventional collimator
OSO-7	USA	Sep 1971	May 1973	Scanning, proportional counters
Copernicus	USA/UK	Aug 1972	Feb 1981	Pointed, X-ray telescopes (non-imaging)
ANS	Netherlands	Aug 1974	Jul 1976	Pointed proportional counters, Bragg crystal
Ariel-V	UK	Oct 1974	Mar 1980	Scanning, RMC + large proportional counters
SAS-3	USA	May 1975	Apr 1980	Scanning, RMC
OSO-8	USA	Jun 1975	Oct 1978	Scanning
HEAO-1	USA	Aug 1977	Jan 1979	Scanning + short pointings (non-imaging)
Einstein	USA	Nov 1978	Apr 1981	Pointed, true images + spectra
Hakucho	Japan	Feb 1979	Apr 1985	Scanning, RMC
Tenma	Japan	Feb 1983	~ 1985	
EXOSAT	ESA	May 1983	Apr 1986	Pointed, true images + spectra
Spartan 101	USA	Jun 1985	Jun 1985	Proportional counter
Ginga	Japan	Feb 1987	Oct 1991	Pointed, large proportional counters
Kvant	USSR	Jun 1987	—	Pointed, GSPC, coded mask, scintillation counter
Granat	USSR	Dec 1989	—	Pointed, coded masks, all-sky monitor
ROSAT	Germany/UK/USA	Jun 1990	—	Scanning, pointed X-ray telescopes
Astro-1	USA	Dec 1990	Dec 1990	Pointed, BBXRT collector, Si detector
ASCA	Japan	Feb 1993	—	Pointed, collector, CCD, IGSPC

(a)

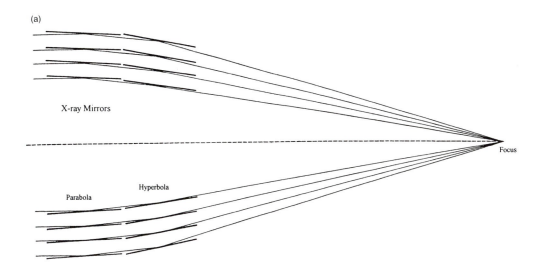

Fig. 2.7 X-ray reflection at grazing incidence and the X-ray optics first pioneered by Wolter. (a) X-rays incident on a mirror at large angles are scattered, not focused. This limits the grazing incidence technique to energies below 10 keV, with most X-ray telescopes concentrating on the 0.1–2 keV band, i.e. soft X-rays. (b) The photograph shows how this work was put into practice on the Einstein Observatory's set of nested X-ray mirrors. Nesting enables the effective collecting area to be maximised. A single grazing incidence mirror collects only a small fraction of the light that would be collected by a normal incidence telescope mirror. The X-rays enter the telescope through the concentric annuli and are focused by the mirrors behind. (Photograph by Leon van Speybroeck of CfA.)

much more sophisticated than the simple wire shown earlier in figure 2.1 as it has to *read out* the X-ray image it is detecting. Two systems for accomplishing this have already been tried in satellite experiments. The first method involves a system of *crossed* wires, the relative amounts of charge collected by different wires giving the *x-y* location. In the second method the amplification avalanche is restricted by the screening grid to a much smaller section of the IPC. This prevents significant spreading or blurring of the X-ray image. Seen from above, the anode is actually a resistive disc of material which enables the X-ray image to be recorded. (See box 2.3 for a technical description of how the Einstein IPC operates.)

The micro-channel plate (or MCP), on the other hand, is a completely different kind of device. Figure 2.8(b) shows how a single glass tube can function as an electron multiplier and an MCP is simply a very large number of such tubes. Single tubes are fused together by means of glass fibre technology in order to create MCPs. They have extensive military applications and are perhaps best

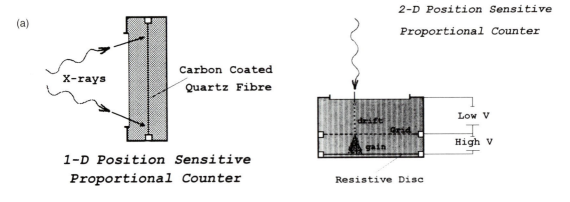

(a)

X-rays

Carbon Coated Quartz Fibre

2-D Position Sensitive Proportional Counter

drift

Grid

gain

Low V

High V

1-D Position Sensitive Proportional Counter

Resistive Disc

(b)   *Channel Electron Multiplier*

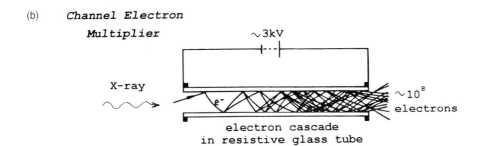

∼3kV

X-ray

e⁻

∼10⁸ electrons

electron cascade in resistive glass tube

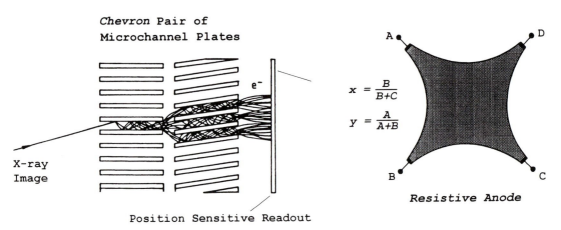

Chevron Pair of Microchannel Plates

e⁻

$$x = \frac{B}{B+C}$$

$$y = \frac{A}{A+B}$$

X-ray Image

A   D

B   C

Position Sensitive Readout

Resistive Anode

Fig. 2.8 Imaging systems at work: (a) An imaging proportional counter (IPC); the principle of operation is best demonstrated in the 1-D case. The anode wire is not a pure conductor (as in figure 2.1) but is a semi-conductor, a carbon coated quartz fibre. This has the result that the risetime of the electron pulse produced by the incident X-ray depends on where along the wire the electrons were collected. In this way the position of the X-ray can be inferred. In the 2-D case the anode is now a resistive disc with four output terminals so that both x and y coordinates can be measured. An extra grid is introduced here in order to control the avalanche region. (b) The micro-channel plate (MCP); this is simply a disc with thousands of tiny pores, each of which is a resistive glass tube. By applying a large potential across the tube, an electron released by an incident X-ray is accelerated down the tube, knocking further electrons out of the wall as it goes. Each incident X-ray produces a 10 000 electron pulse at the end. By combining a pair of such plates (termed a chevron *pair) then a gain of 10⁷ to 10⁸ can be achieved, with the electron image being read out by for example a resistive disc again.*

Box 2.3 The Einstein imaging proportional counter (IPC)

The Einstein IPC used a stack of three active grids: two cathode planes and an anode plane, to measure the position of X-ray events. The wires in the cathode grids were mutually perpendicular, and signals were induced on them by X-ray initiated charge avalanches at the anode. This geometry is shown in figure 2.B3.

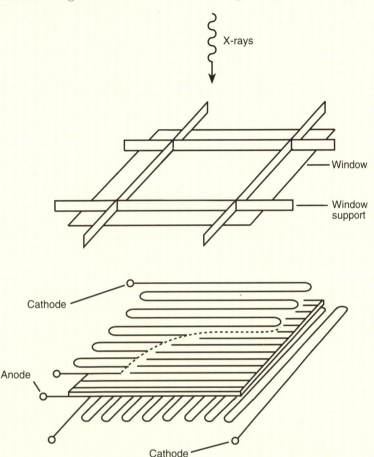

Fig. 2.B3 Schematic diagram of the Einstein IPC showing the grids used to measure the position of X-ray induced charge avalanches. (Material in this box is taken from the thesis of Dan Fabricant of CfA.)

The anode plane was wound with 12 micron gold-plated tungsten wires spaced one millimetre apart and operated at a potential of ~4000 volts relative to the cathode so as to obtain sufficient proportional gas gain for a good signal-to-noise ratio in the position-determining process. The counter was operated at this potential and produced about 105 ion pairs at the anode for each initial X-ray photoionisation.

The determination of position for each cathode plane was achieved through discrimination. The cathode grids were made from a continuous nichrome resistive wire wound around pegs which established a constant 1 mm spacing in one dimension. The risetime of the induced signal increased with the distance it travelled through the cathode wire. This is because the high frequency components of the pulse (recall how to construct a square wave by superposing a series of sine waves, see box 8.2) were removed by the RC time constant of the wire resistance and the coupling capacitances (which act as a filter) of the cathode to ground and the other grids. A preamplifier was placed at each end of this wire and a 1-D position was determined by comparing the relative risetimes of the two preamp outputs. The orthogonal cathode provided the other dimension.

The entrance window of the IPC was made of thin polypropylene which had been coated with an even thinner layer of carbon (in order to reduce the amount of unwanted UV light that was transmitted, e.g. from a bright star). This window was supported by a tungsten mesh and a cross-ribbed structure (the infamous *ribs* which cast the shadows seen in some of the X-ray pictures). Because the window was not completely gas tight, the IPC was operated as a flow counter, and was continuously resupplied with gas during the mission.

known for their use in *night-vision binoculars*. A typical MCP of 25 mm diameter might have 3 million individual micro-channels, each 20 microns across. It is capable of producing a gain of about 10 000 (i.e. a single electron entering a micro-channel will be amplified to yield 10 000 electrons at the other end). The spacing of the channels is so small that the incident image is reproduced as an *electron image* which can be recorded by a variety of techniques.

Here are all the elements needed to put together a true X-ray telescope.

Box 2.4 Sensitivity gain of an imaging system

For most X-ray detectors the major *background* contribution to the counting rate comes from charged particles in the Earth's upper atmosphere. These are *independent* of the particular direction the detector is looking in, and the size of the detector's field of view on the sky (as mentioned earlier, the charged particles and cosmic rays can enter the detector from any angle). It is therefore clear why an imaging system gives such a large gain in sensitivity to point sources of X-ray emission. The telescope focuses the X-rays onto a small area of the detector, whereas the background events are spread uniformly over the detector. This does not happen in a simple scanning system, as illustrated in figure 2.B4.

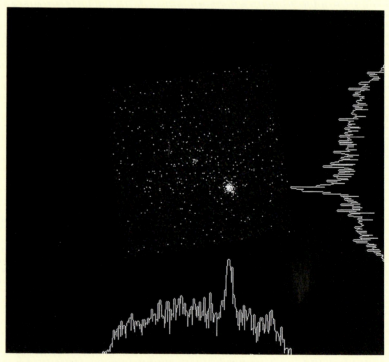

Fig. 2.B4 Appearance of the data from an imaging X-ray system and a simple scanning proportional counter. We have used here the projection of an Einstein X-ray image onto the x and y axes to simulate the effect of a scanning detector. Notice that the brighter source shows clearly in both scans, but the weaker source at the field centre does not show well in either scan.

For typical parameters of the detectors involved, the gain of an imaging system over a scanning one (e.g. Einstein compared to HEAO-1) is about a factor of 1000.

Such a device gives an enormous gain in sensitivity compared to scanning instruments which have no intrinsic spatial resolution (see box 2.4). Because of the technological developments required and the high cost of accurately pointed spacecraft, there have only been three X-ray satellites launched (Einstein, EXOSAT and ROSAT) containing truly imaging telescopes for studies of faint cosmic X-ray sources.

2.6 X-ray spectroscopy

The proportional counter acquired its name because the number of electrons produced in the avalanche after the X-ray photon interacted in the detector was proportional to the energy of that photon. Although this does yield spectral information, it is very crude because of the limited number of electrons produced in the first step of this process. To improve upon this situation, two instruments were incorporated in the Einstein Observatory of which the principal aim was X-ray spectroscopy.

One of these was entirely novel. Called the *solid state spectrometer*, or SSS, it worked in a similar way to the proportional counter in that the incoming X-ray ionises atoms of material in the detector. However, instead of then triggering an avalanche as in the proportional counter (in order to get a detectable signal!), the SSS collects these electrons and measures the pulse directly (see box 2.5). This works because the detector is solid state and is cooled to very low temperatures (80 K or −193 °C) so that the effect of heat in the material is greatly reduced. Because each X-ray which deposits energy in the detector initially generates many more ion pairs than in the gas of a proportional counter, the resolution of the SSS is about three times better, yet with virtually no loss of efficiency.

As always, though, there are swings and roundabouts. An SSS can only be used at the focus of a telescope because it is almost impossible to build one with a large collecting area (due to the cooling requirement). It also has no spatial resolution. The need to cool it is the major limitation on the lifetime in space of such a device. The Einstein SSS worked very well for the first 11 months of the mission, until the cryogen ran out! Within hours the SSS had warmed up, which rendered it useless and the device took no part in the remainder of the mission.

The other spectrometer used in the Einstein Observatory was the focal plane crystal spectrometer (or FPCS). The operation of the FPCS is also explained in box 2.5. Crystals of this kind were used to look at the X-ray spectrum of the Sun in the 1960s and 1970s with great success. Because their

Box 2.5 X-ray spectroscopic devices

Figure 2.B5 shows the principle of operation of the two spectroscopic devices that were part of the Einstein Observatory.

An X-ray photon produces an ion pair in the detecting volume (the depletion layer) of the *solid state spectrometer* which is collected in the field set up between the reversed bias junctions and hence produces a charge pulse. Unlike the proportional counter, *no* amplification of charge occurs, and so very low noise electronic amplifiers are needed for the initial processing of the signal. Also the whole system must be cooled to 80 K in order to avoid thermal excitation into the conduction band. The great advantage of this system is that it requires very little energy to produce each ion pair (only about 3 eV) and so each incident X-ray photon produces far more ion pairs than in a proportional counter (which requires 30 eV of energy). This results in higher spectral resolution (about 5% full width half maximum at 6 keV) but with no loss of sensitivity.

The *focal plane crystal spectrometer* however, utilises a totally different piece of physics. X-rays (of wavelength λ) incident upon the crystal are only reflected when the Bragg condition is satisfied:

$$n\lambda = 2d \sin\theta$$

(a)

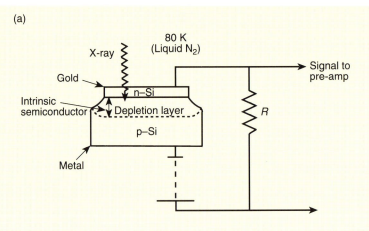

(b)

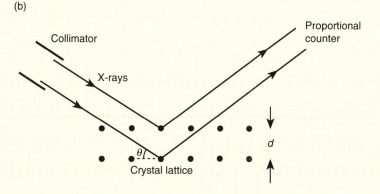

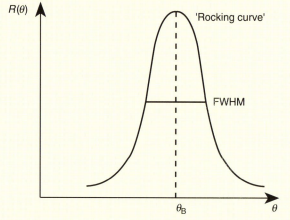

Fig. 2.B5 Principle of operation of the two spectroscopic devices on the Einstein Observatory, the solid state spectrometer (SSS) and the focal plane crystal spectrometer (FPCS). (a) The SSS is essentially a reverse biased junction that creates a depletion layer as its detecting volume. An X-ray photon creates ion pairs in this volume in much the same way as in a proportional counter, but much more easily, and there is no avalanche. To reduce thermal noise the whole system must be cooled to 80 K and the small number of electrons requires very low noise amplifiers. (b) The FPCS is based on Bragg reflection at different layers of atoms in the crystal lattice structure. The resulting different path lengths travelled cause constructive interference when the Bragg condition is satisfied, creating a sharp maximum (the rocking curve*) for a given photon energy. Unfortunately, the reflection efficiency is very low.*

where $2d$ values for useful crystals are in the range $1\sim25$ Å and n is integer. Different wavelengths are sampled simply by changing the value of θ. Resolution is extremely high ($\lambda/d\lambda \sim 300$) but the efficiency of reflection is *very* low (typically 0.1% or less), and so the technique can only be applied to very bright X-ray sources.

Developments in both medium and high resolution X-ray spectroscopy will be mentioned in chapter 17.

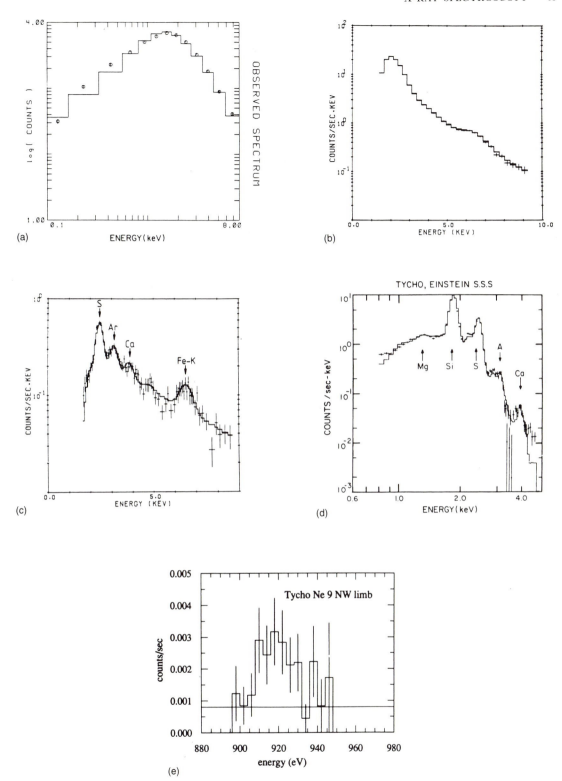

Fig. 2.9 X-ray spectrum of Tycho's supernova remnant as measured by a variety of instruments operating in different ways. Shown in order of increasing spectral resolution they are from (a) the Einstein Observatory IPC, (b) the EXOSAT Medium Energy array, (c) the EXOSAT GSPC, (d) the Einstein SSS, and (e) the Einstein FPCS.

resolution is set by the crystal's lattice spacing, the FPCS has a resolving power almost 100 times greater than the SSS. However, the big drawback here is efficiency. This is typically a thousand times lower than a proportional counter! Hence it works well when used to observe the Sun simply because the Sun is so close, but requires extremely long observation times on even the brightest cosmic sources. Nevertheless, the FPCS has made some major discoveries, particularly in its observations of supernova remnants (chapter 3).

Figure 2.9 shows the X-ray spectrum of a supernova remnant obtained with a variety of instruments so as to illustrate the gain in information that occurs as the spectral resolution is improved.

2.7 Advanced spectroscopic techniques

CCDs

Charge coupled devices (or CCDs) have been used in ground-based optical astronomy for over a decade now with revolutionary results. Their substantial gain in efficiency over previous devices (mostly a combination of image tubes and photographic plates) has been directly responsible for many of the important discoveries of the 1980s. They typically consist of an array of 1000 by 1000 individual picture elements (pixels) which store the photoelectrons produced as each photon arrives until they are read out at the end of an exposure. Larger CCDs, some approaching 4000 by 4000 pixels in size, are becoming available as we enter the 1990s. Each pixel is analogous to an individual photomultiplier, but with higher efficiency and *ruler-flat* linearity.

These devices are now being developed for use as X-ray detectors. It has been found that they can be used, not only for imaging, but for spectroscopy too as is being demonstrated by their use on ASCA. Pictures of extended objects can be obtained along with spectral information from each pixel. Thus it becomes possible to combine the spatial resolution of the Einstein high resolution imager (HRI), with the spectral resolution of the SSS.

High efficiency gratings

In addition to the SSS and FPCS, there was a further spectroscopic option on the Einstein Observatory which we have not so far mentioned because it enjoyed very limited use during the mission. This is the *objective grating spectrometer*, or OGS. Working like a conventional diffraction grating, this consists of a gold transmission grating that can be inserted into the X-ray beam of the telescope (see figure 2.12). The diffracted spectrum that is so produced is detected by the HRI as a pair of extended *lobes* either side of the main X-ray target. The OGS was not extensively used because the efficiency of the transmission grating was very low (only very bright sources gave useful spectra). However, it was employed on EXOSAT, producing some useful results, although the overall efficiency was still low, and very long integration times were required.

In the mid-1980s a completely new approach to the use of diffraction gratings for X-ray astronomy was developed. Rather than use transmission gratings (which are difficult and expensive to manufacture), a conventional reflection grating was used, but at low angles in order to overcome the natural difficulty of reflecting X-rays (as in X-ray imaging; see section 2.5). Figure 2.10 illustrates the geometry of such an arrangement. Laboratory testing of this technique is currently under way, and it is expected to achieve resolving powers greater than 1000 simultaneously with very high efficiency. Such an instrument is planned to be part of ESA's XMM spacecraft in the late 1990s (see chapter 17).

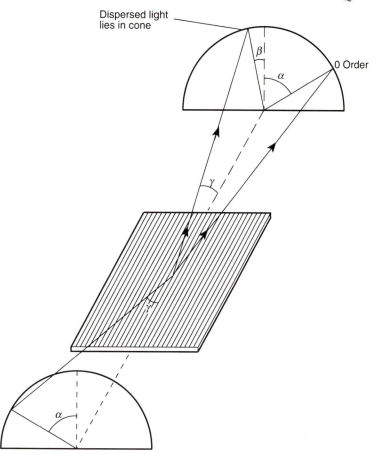

Fig. 2.10 A grazing incidence diffraction grating. Such reflection gratings can be constructed much more efficiently than their transmission counterparts, and they are made to work at X-ray wavelengths by using them at grazing incidence giving conical diffraction. This novel method has not yet been flown in space but offers potential for very high spectral resolution and high efficiency on future missions. (Diagram by Webster Cash, University of Colorado.)

X-ray calorimetry

A remarkable and completely new device (at least in space astronomy) is being designed and tested by Steve Holt and the GSFC X-ray group. This is the *quantum calorimeter* in which individual X-ray photons are absorbed by a crystal which is maintained at a temperature very close to absolute zero (< 0.1 K). The energy of the X-ray causes the temperature of the crystal to

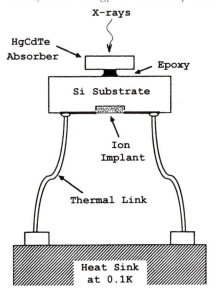

Fig. 2.11 The X-ray quantum calorimeter. This completely new device detects an X-ray photon by the heat it deposits in a very cold (<0.1 K) absorber made of mercury, cadmium and tellurium. The amount of heat deposited depends directly on the energy of the photon. Detection efficiency is virtually 100% and a spectral resolution 50 times better than the SSS is believed possible with this technology. (Diagram by Steve Holt, GSFC.)

increase, and this is measured. The higher the energy of the photon, the greater the temperature increase, hence the device is a spectrometer. An energy resolution of 3 eV (50 times better than the SSS!) is the performance target, with 17 eV already (1989) having been achieved in the laboratory.

The large, future X-ray astronomy missions described in chapter 17 should exploit these developments, and greatly enhance the field of X-ray astronomical spectroscopy.

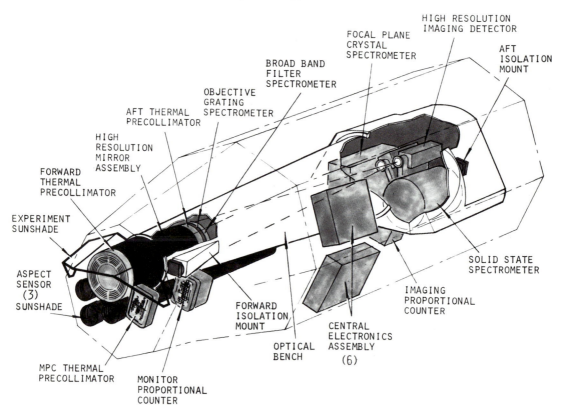

Fig. 2.12 Cut-away schematic of the Einstein X-ray Observatory. The X-rays from the main telescope mirrors are focused onto one of several detectors that can be positioned at the focal plane using a turntable mechanism. The monitor proportional counter is not part of the main telescope assembly but observes along the same axis. The objective grating was located just below the main mirrors and could be commanded to enter the X-ray path, the diffracted spectrum being read out by one of the HRIs. The spacecraft dimensions were 6.7 m long by 2.4 m diameter, with a weight of 3175 kg. (Diagram produced by Riccardo Giacconi, STScI.)

2.8 The Einstein X-ray Observatory

Launched in November 1978, HEAO-2 (later to become known as the Einstein X-ray Observatory) was the culmination of Giacconi's programme of development in X-ray astronomy which had started with his discovery of Sco X-1. The Einstein spacecraft and its instrumentation is shown schematically in figure 2.12.

In order to be able to use a variety of detectors at the focal plane of the X-ray telescope mirrors, Einstein employed an ingenious turntable system by which different detectors could be used according to the type of object that was being looked at. Since X-ray imaging was the principal aim of the mission, Einstein included both an IPC and an MCP device (called the *high resolution imager* or HRI). Such devices were being flown on a satellite for the first time, hence two of each were provided as a precaution against the failure of either type of instrument. Table 2.2 summarises the instruments that were flown on Einstein. Of these instruments, only the MPC was not part of the turntable mechanism at the telescope's focal plane. Instead, the MPC was riding *piggy-back* on the telescope but pointing in the same direction so that it could monitor variability in the brighter X-ray sources.

With this new X-ray observatory a dramatic gain in sensitivity was

Table 2.2 X-ray detectors on Einstein

Name	Acronym	Aim	Spatial resolution	Spectral resolution
Imaging proportional counter	IPC	Imaging	1 arcmin	typical PC
High resolution imager	HRI	Imaging	4 arcsecs	none
Solid state spectrometer	SSS	Spectra	5 arcmins	$3 \times$ PC
Focal plane crystal spectrometer	FPCS	Spectra	$\sim$ arcmins	$300 \times$ PC
Monitor proportional counter	MPC	Monitor for variable sources	3/4 degree	typical PC

achieved over what had gone before. This is illustrated in figure 2.13 where Einstein's sensitivity is compared with earlier X-ray instruments. The figure also extends into the optical and radio parts of the spectrum, so as to compare sensitivities in those regions too. These are put into perspective by plotting the spectra of known objects, such as the Crab Nebula and the nearest quasar, 3C273.

Box 2.6 Flux and luminosity

Fluxes quoted are measured at the top of the Earth's atmosphere. To give an intuitive feeling for the X-ray brightness of a source, fluxes are sometimes quoted in units of photons $cm^{-2} s^{-1}$. To be precise we should use ergs $cm^{-2} s^{-1}$ and specify the exact energy range covered. The counting rate of an X-ray detector, C, is equal to the product of photon flux, F, detector efficiency, η, and detector area, A, integrated over the energy range of the detector.

$$C = A \int F(E) \, \eta(E) \, dE.$$

Since detector efficiencies usually range from 0.1 to 1.0, and detector areas from 100 to 1000 cm^2, the counting rate of a modern X-ray detector is very roughly 100 times the photon flux quoted.

The observed flux is a measure of the brightness of a source. The intrinsic luminosity, L, is related to the flux, F, through the square of the distance to the source, d. Thus $L = 4\pi d^2 F$. As a matter of interest, the most luminous X-ray source known is the quasar PKS 2126–158, at a redshift of 3.27 and with an X-ray luminosity of 5×10^{47} erg s^{-1}. The least luminous extra-terrestrial X-ray source detected is the Moon with X-ray luminosity 7×10^{11} erg s^{-1}, a range of physical processes which produce X-ray emission varying by 36 orders of magnitude!

2.9 The European X-ray Astronomy Observatory, EXOSAT

EXOSAT has already been mentioned in this chapter under the application of the lunar occultation technique for positioning X-ray sources. A satellite dedicated to this technique was first proposed in the late 1960s for launch in 1974–75. Had it been launched then, it would have made a major contribution to the field by identifying many of the brightest X-ray objects. However, the project was subject to many delays, and then SAS-3 together with Einstein achieved most of the major goals of the lunar occultation project using the modulation collimator and imaging techniques. Clearly the EXOSAT mission had to be modified in the light of these changing circumstances.

The principal modification was to add two sets of imaging telescopes and detectors of a kind and performance similar to those on Einstein. A large array of proportional counters was kept (such an array had been intended as

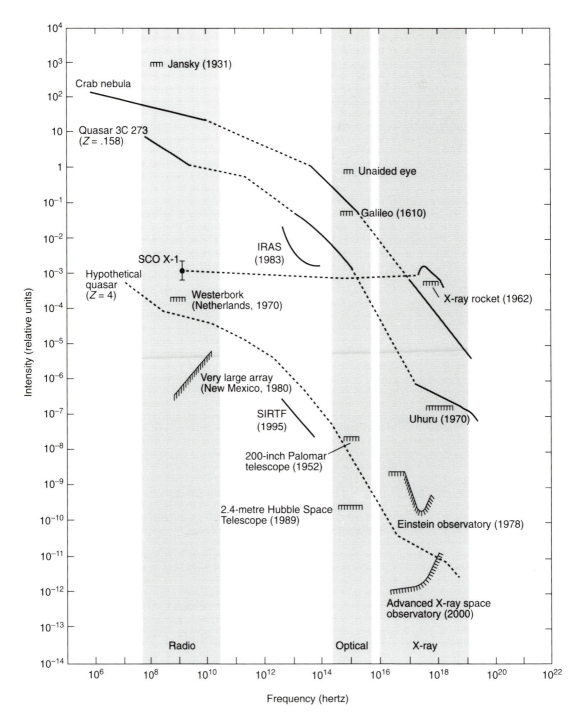

Fig. 2.13 Comparison of the sensitivity of the Einstein Observatory with a variety of instruments in X-ray and other wavebands. Typical astronomical spectra are shown so that the relative contribution of flux in different wavebands can be estimated. The gain in sensitivity from the early rocket flights of the 1960s to Einstein, is comparable to that from Galileo's first telescope to the Palomar 200 inch! But it has been achieved in 25 years rather than over 300 years. This demonstrates most vividly perhaps the reason why X-ray astronomy has progressed so rapidly during the last decade or so. (Diagram based on an original by Riccardo Giacconi, STScI.)

the main detector for use in lunar occultations) to study very rapid time variability in the brighter X-ray sources. It was also decided to keep the highly elliptical orbit that had originally been necessary to provide many more occultation opportunities. The usual low-Earth orbit (altitude 200–300 miles) has an orbital period of about 100 minutes. Although such low orbits provide protection from the solar wind there are disadvantages:

(i)   with the orbit so close to the Earth, a large fraction of the sky is obscured at any instant. Except at the poles of the orbit this means that the observations will be repeatedly interrupted for part of every 100 minutes;

(ii) all detectors have to be turned off when passing through the *South Atlantic Anomaly* where high charged particle densities are found. This causes another interruption on every orbit.

These two effects combine to produce an average viewing efficiency on any object of about 40–45%. An uninterrupted observation of any X-ray source for longer than 40 minutes is almost impossible from conventional low orbits, except for sources at the poles of the orbit (see box 2.1).

The EXOSAT occultation orbit, on the other hand, enables uninterrupted observations to be made for as long as *3 days*! The highly eccentric orbit extended to a distance of 200 000 km from the Earth, and took 4 days to complete one revolution of the Earth. Only during *perigee* (i.e. when passing closest to the Earth) was the satellite inoperative as it rapidly crossed the radiation belts and was not in contact with its tracking station at Villafranca, near Madrid in Spain. Such extended observing facilities have already provided fascinating new results. The major disadvantage of this orbit is the absence of protection from the *storms* of the solar wind. Gigantic flares on the Sun can inject huge amounts of energetic particles out into the solar system. We notice them on Earth as causing telecommunications problems and also, from high latitudes, as the spectacular *aurora borealis*. A large flare sometimes resulted in the need to close EXOSAT down, perhaps for as long as several days. Fortunately this was a rare occurrence as 1983–85 was a time of solar minimum in the 11 year cycle of solar activity!

Figure 2.14 shows an exploded view of the EXOSAT spacecraft and a detail of one of the X-ray telescopes. A summary of the instruments flown on EXOSAT is given in table 2.3. As is clear from the figure, only the position sensitive detector (PSD) and the channel multiplier array (CMA) form part of the X-ray telescopes, the other instruments have simple collimators and no intrinsic spatial resolution of their own. The gas scintillation proportional counter (GSPC) is a new type of detector that has not been described yet. It is actually very similar in operation to a conventional proportional counter. However, instead of detecting the electron cloud produced when the X-ray photon enters the counter, the GSPC detects the optical flash or *scintillation* that occurs when the gas atoms recombine (i.e. rejoin with an electron). This process is analogous to that of the crystal-based scintillation counter used for detecting very hard X-rays (see section 2.2). Again, the avalanche is dispensed with and so the intrinsic energy resolution is very similar to that of the SSS but without the need for cooling to very low temperatures. Future devices based on this principle are expected to combine this spectral resolution with spatial resolution comparable to the IPC.

Unfortunately, the PSDs experienced significant breakdown problems during early operation and played only a small role in the EXOSAT mission. However, all other detectors worked well and results from them will be displayed in later chapters. In particular, although the highly eccentric orbit led to a higher background flux in the imaging telescopes than had been anticipated, the background in the medium energy (ME) array was much lower and hence its sensitivity higher. Also, the capability for very long, uninter-

Fig. 2.14 The EXOSAT
spacecraft and its suite of X-ray
instrumentation (top). There were
two low energy telescopes similar
in design and operation to that of
Einstein. These were co-aligned
with the medium energy (ME)
array of proportional counters,
very much larger and more
sophisticated than Einstein's
MPC. Finally there was a gas
scintillation proportional counter
(shown in detail below), flown
here for the first time on a satellite
mission. (Diagram courtesy of
ESA's Astrophysics Division.)

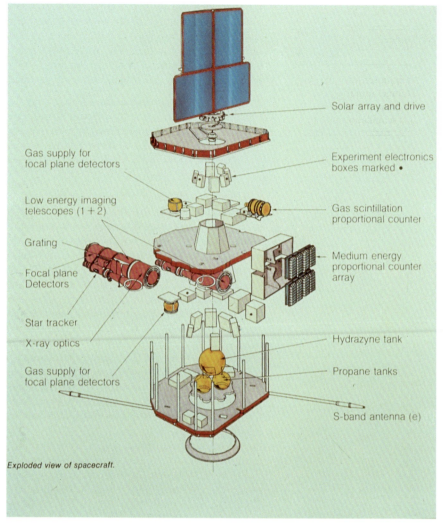

Exploded view of spacecraft.

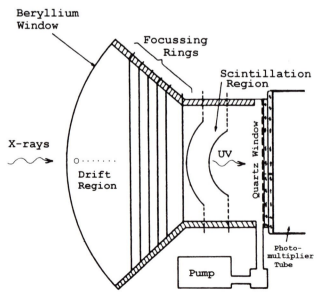

Table 2.3 EXOSAT instrument complement

Name	Acronym	(Einstein equivalent)	Type	Spatial resolution
Position sensitive detector	PSD	(IPC)	Imaging	3 arcmins
Channel multiplier array	CMA	(HRI)	Imaging	18 arcsecs
Medium energy array	ME	(MPC)	PC	45 arcmins
Gas scintillation proportional counter	GSPC	(SSS)	Spectra	45 arcmins

rupted observations (never before possible) led to some remarkable discoveries (see chapters 7–10).

2.10 ROSAT, The first deep all-sky survey

Box 2.4 has already shown how much more sensitive an imaging X-ray telescope is compared to a simple scanning proportional counter, but the only surveys of the entire sky have been performed with non-imaging devices, such as those on Uhuru, Ariel V and HEAO-1. Although this has enabled the brightest X-ray sources to be mapped out, there are entire classes of astronomical objects which were too weak to be seen until they were studied by Einstein and EXOSAT. Unfortunately, given the pointed nature of these missions, it was necessary for astronomers to try and guess beforehand which were the best objects to study. Sometimes this worked, somtimes it did not.

The best solution would be to employ an imaging X-ray observatory which was capable of surveying the entire sky with a sensitivity comparable to that of individual pointed observations with, say, Einstein. This is the goal of ROSAT, or Roentgen-Satellit, a German-led mission also involving the UK and US, which is designed to perform the first deep survey of the whole sky in soft X-rays (0.1–2 keV) and the EUV (extreme ultraviolet; 60–300 Å). The X-ray telescope is similar in design to that of Einstein (employing nested mirrors in a Wolter type I configuration similar to that of figure 2.7), but is larger and hence more sensitive. Following the initial check-out of the spacecraft and its instrumentation after launch, the mission was divided into two phases.

The first six months was the survey phase, in which the telescope (which has a 2° field of view and 1 arcminute spatial resolution) is scanned slowly (at 4° per minute) and continuously across the sky. This means that X-ray sources traversed the field of the detector in about 30 seconds, with the same pattern repeating 90 minutes later on the next scan. Because the detection of each X-ray photon is accurately timed and located by the imaging detector (a position sensitive proportional counter), a sophisticated computer program builds up a narrow strip image (2° wide around the sky) using the 16 scans for each day. Eventually, after six months, the satellite had scanned the entire sky in this way, with a minimum exposure time for each point in the sky of 600 seconds (rising to 4000 seconds for the poles of the orbit). This gave a source detection sensitivity comparable to that of the average Einstein pointing, but for the whole sky! Once this was completed, the pointed phase of ROSAT began in which it operated in a similar mode to Einstein. Only now it has the benefit of having surveyed the entire sky for sources, and thus is in a better position to choose the best targets for further study.

Launched in June 1990, ROSAT has completed the all-sky survey and the first results from both the X-ray and EUV instruments are coming in at the time of writing, and will be included in subsequent chapters. Indeed, the preliminary all-sky map of EUV sources has already been produced and is shown in figure 2.15.

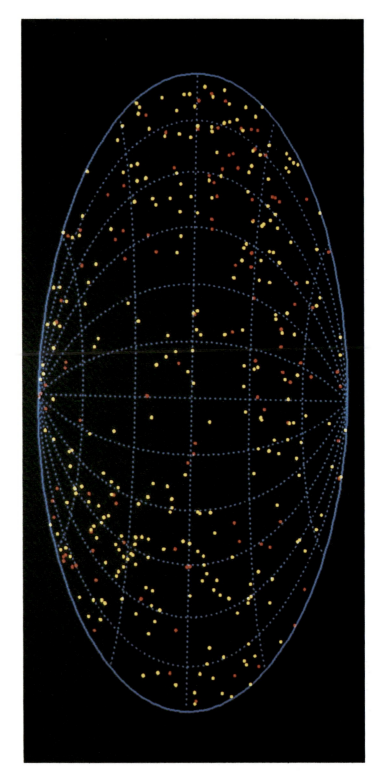

Fig. 2.15 Map of the whole sky at EUV wavelengths obtained for the first time by the UK's wide field camera on the ROSAT satellite, launched in mid-1990. The colour gives an indication of the temperature or spectrum of each object. Red objects are cooler and very soft, whereas yellow ones are hotter (courtesy Mike Watson, University of Leicester).

The techniques of X-ray astronomy are developing rapidly for exploitation in the next generation of satellites in the 1990s. Involvement is becoming ever more international as the costs of these increasingly more complex missions rise. Japan, the US, Germany, the UK and the rest of Europe are all involved in satellites due for launch in the next few years. Our last chapter describes these future developments in detail.

3 Supernova remnants

3.1 Introduction

On May 1, 1006, a new star appeared in the heavens, and, within a matter of days, it became the brightest star observed in all of recorded history. According to records kept by Chinese and Arabic scholars at that time, this star seemed 'glittering in aspect, and dazzling to the eyes.' 'The sky was shining because of its light.' 'Its form was like the half Moon, with Pointed rays shining so brightly that one could see things clearly.'

This nearby supernova (a very bright 'new star') was an awe-inspiring event. The new star was probably visible for three months during daylight, and only after three years did it fade below naked-eye visibility at night. It captured the interest of more people than just the professional astronomers of the day; understandably, these events were considered portents of natural disaster and political change.

The study of supernovae and their remnants is now one of astronomy's most interesting topics. Although much has been learned in recent years, the basic nature of the explosion and the kinds of stars that explode are not well understood. At least seven times in the last 2000 years such explosions have occurred in our galaxy and near enough to Earth to appear as bright new stars. These are listed in table 3.1. The last one was discovered on October 9, 1604, a few years before the first astronomical use of the telescope by Galileo, so none of the progenitor stars of these nearby supernovae could be identified.

On February 23, 1987, to the great joy of present-day astronomers, a supernova appeared in the Large Magellanic Cloud (LMC) – the first bright one in 383 years! Although not as bright as the historical supernovae, SN 1987A was near enough to identify the progenitor star – a B3 supergiant (a bright blue star, see box 4.3). This is the very first time that past observations have been available which reveal the character of the star just before the explosion.

Even more exciting was the detection of a burst of neutrinos which marked the exact time of the event! This was a remarkable confirmation of theories about processes occurring deep in the core of a collapsing star, a region hidden from conventional astronomical observations. More discoveries are expected in the next few years as the expanding debris, blown outward by the explosion, disperses to reveal the inner layers of the exploded star.

Aside from the LMC supernova, astronomers, routinely using telescopes having 10 000 times the light gathering power of Galileo's instrument, find 10–20 supernovae a year in extragalactic nebulae. Figure 3.1 shows such a

Table 3.1 Nearby historical supernovae

Date	Observers	Optical magnitude	Type	Location	Remnant	Radio character
185	Chinese	−8	I?	Centaurus	RCW 86	shell
1006	Many	−9.5	I	Lupus	SNR 1006	shell
1054	Many	−4	II?	Taurus	Crab Nebula	bright amorphous
1181	Chinese	0	?	Cassiopeia	3C58	faint amorphous
1572	T. Brahe	−4	I	Cassiopeia	Tycho	shell
1604	J. Kepler	−3	I	Ophiuchus	Kepler	shell
1667?	Nobody	?	II?	Cassiopeia	Cas A	bright shell
1987	Everybody	3	II	Dorado	SN 1987A	to be determined

Fig. 3.1 Discovery of an extragalactic supernova: Supernova 1972E in NGC 5253, a nearby elliptical galaxy in the Centaurus group. The supernova reached a peak brightness of 8th magnitude; only once every 30 years is a supernova of this brightness observed. This galaxy was also the site of an earlier bright supernova in 1895. (Photo from Hale Observatories.)

discovery. Although the supernova was one of the brighter specimens, it was not bright enough to be visible to the naked eye. These extragalactic supernovae are very distant and it has been impossible to detect and locate the stars which explode to make them. Only the progenitor star of SN 1987A has been identified. The nature of other presupernova stars and the explosion mechanism must be deduced from the study of events after the explosion. Such information includes optical spectra and light curves of the supernovae themselves and the structure of the remnants which are hundreds or thousands of years old. These observations are compared with models and with theoretical calculations which tax the capability of the largest computers.

Supernova explosions are the most energetic stellar events known. The

energy deposited in the debris and immediate surroundings is typically 10^{51} ergs. This is as much energy as the Sun will radiate at all wavelengths in its 10 billion year lifetime! Although the optical energy radiated during the first year is only 1% of this, the brighter historical supernovae at their peak were visible during the day and all could be seen at night for months or even years. Most of the energy appears as energy of motion. The outer layers of the star are thrown into space with initial velocity typically 10 000–15 000 km s^{-1}. As the remnant expands, this energy heats both the ejected and the surrounding material to temperatures well above a million degrees, and X-rays are radiated profusely.

Not surprisingly supernovae and their remnants were the only class of object predicted to be X-ray emitters before cosmic X-ray sources were discovered in 1962. Indeed, the Crab Nebula, a famous supernova remnant, was the first X-ray source to be identified with a previously known object. In order to understand observations of the remnants at X-ray and other wavelengths, it is necessary first to consider the mysteries of the late stages of stellar evolution. What leads up to these cataclysmic events?

3.2 The supernova explosion

Historically astronomers have recognised two types of supernova which are distinguished by their optical spectra. The change of brightness with time, or the 'light curve' also carries important information. Type I events all display similar light curves, suggesting a common type of progenitor star and explosion mechanism. The intensity rises quickly to maximum, reaching a luminosity of more than 10^9 suns in about two weeks. An initial rapid decay is followed by a long slow decline in brightness. The luminosity falls exponentially with a characteristic time of about 55 days until the supernova fades to invisibility. The spectra show a variety of lines. Most are difficult to identify and all are Doppler-broadened indicating velocities up to 15 000 km s^{-1}. Surprisingly for astronomical spectra, there is a marked absence of hydrogen lines.

Type II supernovae rise more slowly to a maximum and are generally 2 magnitudes less luminous than those of Type I. The maximum is broader and the decay on less regular timescales. A great deal of individuality characterises the decay of Type II supernovae, implying a marked variation in the progenitor stars or explosion characteristics. The spectra are dominated by broad emission lines of hydrogen suggesting that the explosions occurred in young stars which had hydrogen-rich outer envelopes. SN 1987A was an example of a Type II supernova with a most peculiar light curve (figures 3.2 and 3.3).

The distribution of Type II events observed in distant galaxies supports this classification scheme. Type II outbursts occur in the arms of spiral galaxies, regions known to contain bright young (massive) stars and dense clouds of gas and dust characteristic of recent star formation. They rarely occur in elliptical galaxies where the stellar population is older. In contrast, the rarer Type I events occur in all kinds of galaxies and show no preference for spiral arms. This suggests that the progenitor stars of Type I supernovae are billions of years old and consequently not very massive.

Evolution to the explosion

Regardless of whether the supernova is of Type I or Type II there is agreement that the cause is a catastrophic release of energy at the end of the life of a star. Throughout its evolution a star generates internal energy by thermonuclear reactions. The consequent high internal pressure and temperature support the outer layers and resist the star's tendency to collapse under its own weight. At first hydrogen is fused in the core to form helium, then the helium is fused to form carbon and oxygen and, if the star is massive enough, fusion

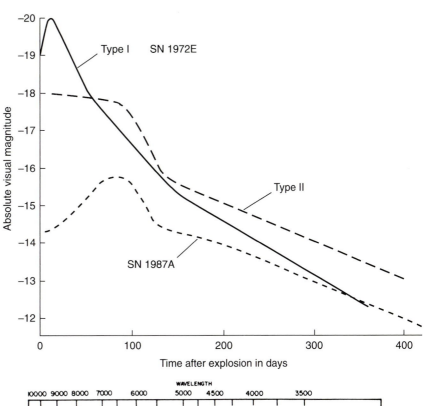

Fig. 3.2 Typical light curves of the two types of supernovae and that of 1987A. Maximum of the type I events is brighter and of shorter duration than the type II. The type I light curve shown here is that of SN 1972E which was followed for over 700 days, an unusually long time. The decay of type II events varies greatly and 1987A is peculiar even for a type II SN. (Courtesy R. Kirshner, CfA.)

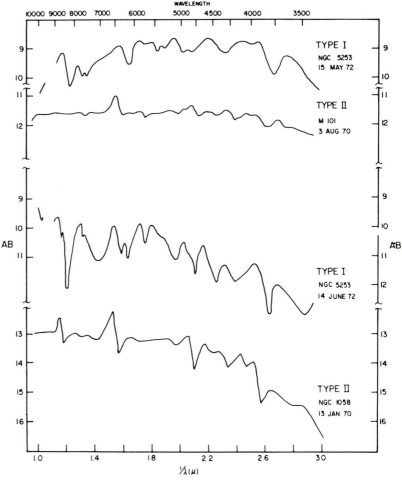

Fig. 3.3 Optical spectra of supernovae taken at maximum light and about a month after maximum. Type I spectra show broad absorption lines attributed to ions of various elements. Type II spectra are dominated by the 6563 Å emission line of hydrogen. This line is emitted from a thick high-velocity shell and is blueshifted and broadened in these spectra. (Courtesy R. Kirshner, CfA.)

continues until the core is mostly iron. The nuclear binding energy of iron is the maximum of all the elements so further energy release through fusion is impossible. Throughout this process gravity has inexorably compressed the core of the star, thereby creating the higher densities and temperatures which enable thermonuclear burning of the heavier elements. The internal energy generated eventually reaches the surface and escapes as radiation. When the fuel is exhausted, collapse continues until halted by other forces.

A star the size of the Sun is stabilised before thermonuclear burning is complete by an effect called degenerate electron pressure. At this point the core consists of carbon and oxygen and is a million times denser than ordinary matter. The material has been compressed until all empty space has been squeezed out of the atoms. At this point the electrons resist being crowded closer together and collapse is halted. The surface of the star is now 'white hot' and it is called a white dwarf – a star with the mass of the Sun and the size of the Earth. A star with mass less than $1.44\ M_\odot$ (a mass called the 'Chandrasekhar limit') cannot evolve beyond this stage. Examples of white dwarfs abound; Sirius B, only 1.3 pc distant, is the closest and probably the best known.

Type II explosions: gravitational collapse of the stellar core

The evolution of a star starting its life with more than $10\ M_\odot$ is different. It consumes nuclear fuel very rapidly. A typical luminosity is 10^4 times that of the Sun. When its resources are exhausted, the end is marked by a great catastrophe – a Type II supernova explosion. Gravity has been the dominant force in the life of the star and the energy released in the explosion is gravitational also.

At the beginning hydrogen in the core is fused to helium. When this is depleted the star contracts. It contracts until the temperature is high enough to fuse helium into carbon and oxygen. At the same time the pressure increases in the layer of hydrogen surrounding the core. This results in initiation of nuclear burning in this outer layer, creating a shell of helium surrounding the core of heavier elements. At the end, after a life perhaps as short as ten million years, the star is stratified into onion-like layers. The central core is now iron, the end point of the fusion process. This is surrounded by a shell of silicon and sulphur, then a shell of neon and oxygen, then carbon, then helium, and the outermost shell is hydrogen. The density in the core is that of a white dwarf whereas the outer layer is very tenuous. Most stars are probably red giants or supergiants at this phase, surrounded with glowing shells of hydrogen as large as the inner part of our solar system. Unfortunately, we cannot see through these hydrogen shells to observe the inner part of the stars where the final stages of evolution rapidly occur.

The iron core can no longer generate nuclear energy. It is stabilised by electron pressure and is at the Chandrasekhar limit. The mass of the core, however, is continuously increasing as the adjacent layer of silicon is fused into more iron. The core is compressed more and the internal temperature increases. This causes some of the iron to decompose into lighter nuclei. As a result energy is absorbed, reducing the pressure and causing the core to shrink. Free protons are created which combine with electrons to make neutrons (and neutrinos). Some of the electrons which have been supporting the core thus disappear causing further pressure drop. The process now runs away, and gravity overwhelms the electron pressure. Within milliseconds the core collapses until nuclear density is reached – a solar mass within a radius of 10 km.

This process is difficult to believe. The mass is slightly larger than that of the Sun. The gravitational field is enormous and the rapidity of the collapse surprising. Yet on February 23, 1987 the neutrinos created in such a collapse were actually observed as the first signal from SN 1987A.

The energy explosively released by the infalling matter creates a shock wave. This shock propagates outward through the still-infalling outer layers until it reaches the outermost part of the star. The shock ejects this last material outwards.

Box 3.1 Shock propagation in a collapsing star – not as simple as expected

Once formed, the neutron star resists further compression with a force greater than any yet used in the struggle against gravity. The nuclear force itself prevents the neutrons from being more tightly packed together and the collapse is halted. The momentum of infall, however, is so great that the neutron star is compressed to a density perhaps as much as 50% greater than its normal density. The neutron star rebounds from this compression and the outward push it gives to material at the boundary propagates rapidly away from the core as a shock wave.

Since the material in which the shock is moving is infalling with high velocity, outward progress is slow. Furthermore, the strong shock raises the temperature of the material to the point where iron nuclei are broken up into lighter elements, cooling and slowing the shock. In many computer simulations the shock 'stalls' and never reaches the outer boundary of the star. In some cases neutrinos escaping from the neutron star can be absorbed by compressed material behind the shock giving it the additional energy needed to break through this layer of rapidly infalling dissociating material. The calculations are difficult and are done only on the largest computers. Only 1% of the energy released in the collapse appears in the expanding ejecta of a supernova. It is not surprising that small details of the calculation have great observational consequences.

We observe light from the expanding hot material as a Type II supernova. A rapidly-cooling hot neutron star is left at the centre of the explosion and, if this remnant of the core is spinning rapidly, it is later observable as a pulsar at the centre of the remnant. Total energy generated by the collapse and subsequent nuclear changes is in excess of 10^{53} ergs. This is largely carried off by neutrinos – emitted as electrons and protons combine to form neutrons. The neutrinos, because of their low reaction cross section, both escape easily from the collapsing star and are extremely difficult to detect. The light from the supernova itself carries only 10^{49} ergs of energy. The kinetic energy of the expanding debris is typically 10^{51} ergs. Rotational energy of the neutron star can also be appreciable; if spinning at 30 cycles s^{-1}, this is 2×10^{49} ergs.

SN 1987A

SN 1987A was promptly classified as Type II. The optical spectrum clearly showed the presence of hydrogen. After some initial confusion, the progenitor star was identified as the 12th magnitude star Sk −69 202, a B3 supergiant and well documented in previous photographic plates. This was a young star of about 20 $M_\odot$ but not the red supergiant progenitor expected!

There is no doubt that the event originated in gravitational collapse of the core of the star. This was established by the simultaneous detection of a burst of neutrinos by two Earth-based detectors. It is rather interesting that these detectors were not made for astronomical research. They were built to search for extremely-rare proton decays predicted by some theories of fundamental particles. They are roughly 10 m cubes of pure water surrounded by photomultiplier tubes and buried deep in the Earth to reduce the background from cosmic rays. The detector at Kamioka, Japan, recorded 11 neutrino-induced events in 13 seconds. The Irvine-Michigan-Brookhaven detector at Cleveland recorded 8 events in 6 seconds, a 'burst' in a system having a background of 0.2 events per day. Thus we know that collapse occurred at 0735 UT (Universal Time) February 23, 1987, 18 hours before Ian Shelton from the University of Toronto started the exposure of the discovery photograph.

The detection of these neutrinos was a triumph of both theory and observational science. The measurements yield neutrino flux and energy, which is approximately equal to the total energy release in the collapse. The total energy derived from these data is 3×10^{53} ergs, exactly the gravitational binding energy of a neutron star. The fact that the neutrinos did not arrive simultaneously, but over an interval of several seconds, implies many interactions in the collapsing core before escape. The timing of the observed events is now being used to study the motion of material during formation of the neutron star.

The optical light curve observed is unusual, but explained by the nature of the progenitor star. A prediscovery photograph by Robert McNaught (Siding Spring Observatory) shows appreciable brightening only 3 hours after core collapse. The star brightened rapidly reaching a maximum of magnitude 4.5 in 4 days. After a small decline in brightness, there was a slow rise to maximum light of magnitude 3.0, 70 days after discovery. A 'typical' Type II supernova in the LMC would have reached magnitude 0–1 at maximum brightness. The fact that the progenitor lacked the extensive envelope which characterises the enormous red supergiants accounts for the very rapid initial rise and for the fact that the maximum brightness was a factor of 10 less than expected.

Light from the supernova now is in a gradual exponential decline. The energy comes from radioactive decay of ^{56}Ni created in the explosion. The ^{56}Ni decays with half life 7 days to ^{56}Co which in turn decays with half life of 77 days into stable ^{56}Fe. There is also a possible contribution from the rapidly spinning neutron star at the centre.

T. Dotani (ISAS) and colleagues of the Ginga Large-Area Counter team have reported the detection of X-rays starting 130 days after stellar collapse (figure 3.4(a)). These were originally gamma rays from the decay of ^{56}Ni and ^{56}Co. They have been downgraded in energy by many collisions with electrons in the expanding shell. In spite of continuous monitoring, no X-rays were seen before this time. At early times the shell was too thick for X-rays to escape.

X-rays were also detected by Rashid Sunyaev (SRI, USSR) and colleagues with a high energy detector on the Soviet Mir-Kvant spacecraft (figure 3.4(b)). The spectrum was measured up to 200 keV. It is unusually 'hard' and fits the scattering calculations nicely.

Type I explosions: white dwarf ignition in a binary system

The cause of the Type I events has been more difficult to determine. The similarity of optical characteristics of the Type I supernovae would seem to require similar progenitor stars. White dwarfs, at the Chandrasekhar limit, would provide a constant starting point. Such stars, however, are stable and it is necessary to drive the star over the limit to get an explosion, probably by adding more mass. Thus Type I progenitors could be white dwarfs in binary systems with mass transfer (box 3.2). Material is added to the white dwarf until the mass is high enough for gravity to overcome the resistance of electron pressure. The resulting collapse raises the temperature until the carbon and oxygen in the core start to fuse. This creates an explosive wave, a deflagration (an explosion which proceeds without initiation by a shock wave), which propagates through the core in seconds. The resulting nuclear fusion reactions create about a solar mass

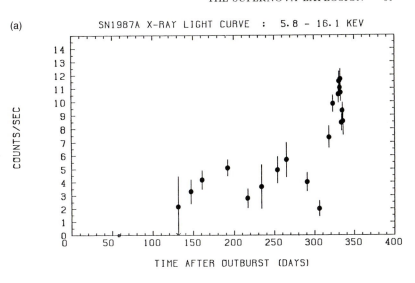

Fig. 3.4(a) The time history of
X-rays from SN 1987A observed
by Ginga. Fluctuations in the
intensity are probably real. The
X-ray luminosity in this energy
band is 10^{37} erg s^{-1}. (Courtesy of
the Ginga LAC team.)

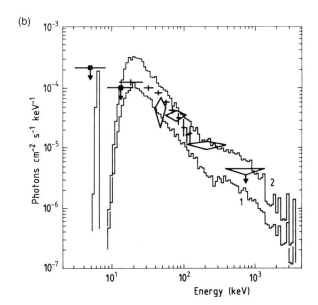

Fig. 3.4(b) High energy X-rays
from SN 1987A at an age of
170 days. Crosses and diamonds
show data points from two
instruments on board Mir-Kvant.
Lower and upper histograms
(labelled 1 and 2) show
calculation of spectra at 180 and
240 days after the event. The
envelope becomes more transparent
as it expands. (Courtesy of J.
Trümper, MPE.)

of radioactive ^{56}Ni and release 10^{52} ergs of energy, completely disrupting
the star. The delayed release of energy from the ^{56}Ni causes the expanding debris to glow and explains the slow decay characteristic of Type I
light curves. This model thus accounts for the basic properties of Type
I supernovae. The major problem to its acceptance is the lack of detection
of the predicted large mass of diffuse iron in the central regions of the remnants.

Type Ib explosions

Some Type I supernovae, although having spectra with no trace of hydrogen
and light curves of exactly the right shape, show anomalous characteristics.
They are underluminous, fainter than typical at maximum light by a factor
of 4, and at late times the spectra are dominated by lines of oxygen. It has
been proposed by Robert Kirshner (Harvard), Craig Wheeler (University of
Texas), and others that these be classified as Type Ib. The progenitors are
thought to be massive stars that have lost all their hydrogen envelope.
Wolf–Rayet stars, for example, are possible progenitors since they have shed

Box 3.2 Mass transfer in a binary system – a question of evolution

Consider a binary system consisting of two stars of moderate mass (say 1 $M_\odot$ and 3 $M_\odot$) in a reasonably close orbit. Both stars are formed at the same time. The time in which a star completes its hydrogen-burning phase on the main sequence is inversely proportional to the square of the star's mass. The more massive star therefore evolves first. Once the central hydrogen has been exhausted, the helium core shrinks until the density is sufficiently high for helium burning to begin. Then, once again, the energy generated by thermonuclear processes will be sufficient to support the star against the force of gravity. The higher temperature, however, causes the outer layers of the star to expand. This material, instead of forming the usual red giant envelope, is either captured by the less massive star or flung into space. After this first stage of mass transfer is complete the second star is left as the more massive star and only a carbon-oxygen white dwarf remains of the first. The second star now evolves rapidly and starts to enter its own red giant phase, at which time it transfers matter back onto the white dwarf. The mass of the white dwarf now increases until it passes the Chandrasekhar limit. At this point gravity initiates the collapse which leads to the nuclear detonation of the interior of the white dwarf.

their outer hydrogen in a massive stellar wind. Time will tell if a classification scheme which adds the Ib class to the usual I and II is feasible.

3.3 Evolution of supernova remnants

The following sections discuss how observations of the remnants, which are hundreds or thousands of years old, can yield information about the explosion mechanism and the progenitor stars. As in many astronomical pursuits, it is difficult to extract precise numbers from the data. Nevertheless, the new X-ray observations have revealed gross properties of the remnants which are most useful and often definitive.

Most remnants have been discovered as spatially extended sources in radio surveys. They are distinguished from HII regions (clouds of ionised hydrogen usually associated with luminous hot stars) by the relatively flat shape of their radio spectra. Many remnants carry the names assigned after the radio discovery. Cas A and Tau A (the Crab Nebula) are the brightest radio sources in these two constellations and Vela XYZ is formed of three radio-bright regions. Catalogues of such nonthermal extended radio sources now contain about 150 objects in our galaxy and about 30 objects in the Magellanic Clouds. Most of these sources appear to be approximately circular in shape and are limb brightened as if we are looking at emission from a large hollow shell which is transparent to its own radiation.

This is the shape expected for the expanding debris from a supernova explosion. It is simplest to imagine that material is ejected uniformly in all directions and that the progenitor star was embedded in a uniform medium. The shell of ejected material expands rapidly and sweeps up the surrounding medium like a snowplough. A low density region is left in the interior. The material in the shell soon becomes of low enough density for the shell to be transparent or 'thin'. During this phase the mass of swept-up material is negligible compared to the mass of the ejecta and the expansion proceeds at a uniform velocity. This 'free expansion' is the first phase in the life of a supernova remnant. The total ejected mass might be 1 $M_\odot$ and the density of the surrounding medium 0.3 atoms cm^{-3}. If so, this phase will last until the radius is 3 pc when the swept up mass becomes equal to that of the ejecta. If the initial velocity is 15 000 km s^{-1}, the age of the remnant at this time will be 200 years.

In the assumed uniform interstellar medium the speed of sound is about 10 km s^{-1}. This is the speed at which small fluctuations in pressure and temperature propagate. It is much less than the velocity of the stellar ejecta. A shock wave consequently forms at the leading edge of the ejecta and travels just before it into the interstellar material. An atom or ion of interstellar hydrogen far from the star will suddenly find itself violently thrown into its neighbours as the shock passes. The pressure increases, the temperature becomes 10^7 to 10^8 K, and electrons are separated from atoms. All material is propelled outward, in the direction of motion of the shock, at a velocity somewhat less than that of the shock itself.

As time passes the expansion slows and the remnant enters the second phase, an adiabatic expansion. This is sometimes known as the Sedov–Taylor or 'blast wave' phase. The mass of swept up material is now large compared to the original mass of the ejecta. The energy radiated by the material in the shell is still small compared with its internal energy. Hence the rate of expansion is determined only by the initial energy deposited by the explosion, E_0, and the density of the interstellar medium. As it expands, the remnant sweeps up cold interstellar material and becomes cooler as its mass increases. Box 3.3 gives more detail.

Box 3.3 Adiabatic expansion of supernova remnants – a useful model for calculations

The average remnant, between the ages of about 100 and 1000 years, accretes mass rapidly through the 'snowplough' effect as it expands; yet the loss of energy through radiation is small. As the size increases, this material is accumulated in a 'shell' just behind the shock. Although the temperature of material in this shell decreases with time, the thermal energy within the remnant is always 72% of E_0, the initial energy release. This model is appealing to astrophysicists because it predicts a precise behaviour of the remnant with time, dependent only on E_0 and the density of the surrounding medium, n. Furthermore, if the distance is known, the age, t, and E_0 can be derived from the measured X-ray flux and spectrum (which directly yield X-ray luminosity, L_x, and temperature, T).

The radius, R, of the blast wave is

$$R = 14 \, (E_0/n)^{1/5} \, t^{2/5} \, \text{pc},$$

where E_0, n, and t are in units of 10^{51} ergs, cm^{-3}, and 10^4 years respectively. The shock temperature is also given as

$$T = 1.0 \times 10^{10} \, (E_0/n) R^{-3} \, \text{K}.$$

The only other necessary formula relates L_x to the calculated (and tabulated) emissivity, P, of the hot gas. Both are dependent on temperature and X-ray photon energy, E. The density distribution of gas in the shock also requires a dimensionless factor, $q(T)$, which has value approximately 0.6 over the range of interest here.

As the material behind the shock cools, the rate of radiation of energy actually increases. This somewhat paradoxical result is due to the fact that, after the temperature drops to about 2×10^5 K, some electrons have recombined with the carbon and oxygen ions and the gas is then able to radiate by the very efficient process of ultraviolet line emission. The remnant has now reached phase 3, the 'radiative' phase. During the 10^5 year duration of this phase, most of the internal energy is radiated away. The shell coasts through interstellar space becoming fainter and fainter until it is indistinguishable from the surrounding medium. This phase is also called the 'constant-momentum' phase even though momentum is of course conserved through-

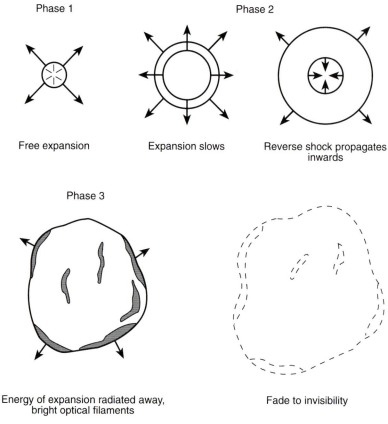

Fig. 3.5 Evolution of a supernova remnant. The young remnant at first expands freely into the surrounding space. As the size increases, it incorporates an ever increasing amount of interstellar material until it eventually cools and is no longer distinguishable from the background.

Phase 1

Free expansion

Phase 2

Expansion slows

Reverse shock propagates inwards

Phase 3

Energy of expansion radiated away, bright optical filaments

Fade to invisibility

out the life of the remnant. Figure 3.5 shows the evolution of a supernova remnant.

X-rays are emitted most profusely during phase 1 and 2 from the hot material behind the shock (table 3.2). During phase 3 the bright optical filaments characteristic of older remnants form. Radio energy is radiated from the vicinity of the shock and from the cooling filaments and is easily detectable throughout the life of the remnant. When a remnant is observed in these different bands, different processes produce the radiation detected. All these processes are stimulated by, and are most prominent in, the vicinity of the shock wave.

The X-rays are thermal. They are generated, as explained in box 3.4, by fast electrons colliding with positive ions or by electron–ion recombination. The temperature of this plasma is a million degrees or higher. The optical emission comes from material with a temperature of ten thousand degrees and consists of discrete optical and ultraviolet lines; all have wavelengths characteristic of the radiating atoms. The radio emission is synchrotron emission which occurs when high energy electrons move through a moderately strong magnetic field. Thus optical and X-ray emission do not come from the same material. Remnants which exhibit both (which most of them do) must consist of material spanning a large temperature range and probably pervaded by high energy electrons.

Many remnants are also bright infrared sources. Dust immersed in the hot gas is heated until it radiates in the IR. Maps of the brighter remnants indeed show that the origin of most of the IR is within the X-ray emitting shell. (Fig. 3.9(b)). Total energy radiated in the IR is sometimes greater than in the X-ray band. This is expected to somewhat hasten the evolution of these remnants.

Table 3.2 Well-studied shell-like supernova remnants

Name	Age (yr)	Distance (kpc)	Diameter (pc)	Angular diameter (arcmin)	Phase	Einstein IPC rate (count s^{-1})	X-ray luminosity (10^{35} erg s^{-1})
Cas A	300	3	3.5	4	1–2	61	38
Kepler	380	5?	4.4?	3	1–2	7	10
Tycho	410	3	7.0	8	1–2	22	6
SN 1006	980	1	9	30	1–2	11	1
IC 443	3000?	1.5	22	50	2–3	12	1
Pup A	4000?	2	26	45	2	250	73
Vela XYZ	10 000?	0.5	48	330	2–3	500	5
Cyg Loop	20 000?	0.8	40	170	2–3	620	11

The detailed arrangement of the matter within remnants will be discussed in the following sections. A glance at any of the supernova remnant pictures, however, reveals that none of them shows the precise spherical shape assumed in the above discussion. Bright regions exist as shreds, patches, or filaments reflecting sometimes large asymmetries in the distribution of ejected material or in the surrounding medium. Mathematical models require a simple geometry in order to derive a solution. Nature seems always to be more complicated.

In this chapter the characteristics of several remnants will be described as well as the attempts to classify them as the products of Type I or II supernovae. Each of these objects, however, has its own personality. It might at first appear that 10 remnants imply 10 different kinds of supernova explosions instead of two. Nevertheless, allowing for differences in age and in the surrounding environment, it is possible to see that many of the remnants are compatible with one of the two explosion mechanisms previously discussed. We hope that the reader is not bewildered by the rich array of sizes, shapes, and luminosities of these remnants. A variety have been included to give a feeling for the quality of the available data, the successes and limitations of the available models, and the value of instruments proposed for future observations.

Box 3.4 Three astrophysical mechanisms for generating X-rays

There are three dominant physical processes expected to produce X-rays in an astronomical setting. The spectral signature of each is unique and is one of the first clues an observer has concerning the nature of an unknown X-ray source. If the spectrum can be measured with high resolution over a broad energy band, then usually both the emission process and the physical conditions within the source can be deduced. All three mechanisms are encountered in the study of supernova remnants.

Thermal emission from a hot gas

Consider a hot gas of low enough density that it can be described as 'thin' and is transparent to its own radiation. This is not difficult to achieve for X-rays. At temperatures above 10^5 K, atoms are ionised, and a gas consists of positive ions and negative electrons. The thermal energy is shared among these particles and is transferred rapidly from one particle to another through collisions. Indeed 'thermal equilibrium' means that the average energy of all particles is the same and is determined only by the temperature. When an electron passes close to a positive ion, the strong electric forces cause its trajectory to change. The acceleration of the electron in such a collision causes it to radiate electromagnetic energy, and this radiation is called 'bremsstrahlung' (literally, 'braking radiation').

Electrons in thermal equilibrium have a well-determined distribution of velocities (called Maxwellian after the physicist James Clerk Maxwell) and the radiation from such electron-ion collisions is a continuum with a characteristic shape that is determined only by the temperature. This is 'thermal bremsstrahlung'. The higher the temperature, the faster the motion of the electrons and the higher the energy of the photons in the bremsstrahlung radiation. For temperatures above one million degrees, these photons are predominantly X-rays.

The thermal bremsstrahlung spectrum falls off exponentially at high energies and is characterised by the temperature, T. The intensity, I, of the radiation at energy, E, is given by,

$$I(E,T) = A\ G(E,T)\ Z^{2}n_{e}n_{i}\ (kT)^{-1/2}\ e^{-E/kT}.$$

where k is Boltzmann's constant and G is the 'Gaunt factor', a slowly varying function with value increasing as E decreases. The form of this spectrum can be seen in figures 3.B1 and 3.B3. Note that the intensity is proportional to the square of the charge of the positive ions, Z, and the product of the electron density, n_{e}, and the positive ion density, n_{i}. A is a constant.

In a hot gas X-ray line emission is also an important source of radiation. The elements heavier than hydrogen are not completely ionised except at very high temperatures. When a fast electron strikes an ion with bound electrons, it often transfers energy to that ion, causing a transition to a higher energy level. The ion is left in an excited state which lasts only briefly. The ion decays rapidly to its ground state by radiating photons of energy characteristic of the spacing of energy levels through which the excited electron passes. This radiation appears as spectral lines with energies determined by the radiating ion species.

Radiation from a thermal gas is thus a blend of thermal bremsstrahlung and line radiation (other processes also make small contributions but X-ray diagnostics rely on these two). For a gas of 'cosmic' composition (which for every 10 000 atoms of hydrogen, contains 800 atoms of helium and 16 atoms of carbon, oxygen, and heavier elements), at temperatures below 1×10^{6} K, most of the energy is radiated as ultraviolet lines. At 2×10^{6} K half of the energy is radiated as soft X-rays, at 1×10^{7} K all the energy is radiated as X-rays, half in lines and half as thermal bremsstrahlung. At 5×10^{7} K almost all of the ions have been stripped of their bound electrons and almost all of the energy is radiated in the X-ray continuum.

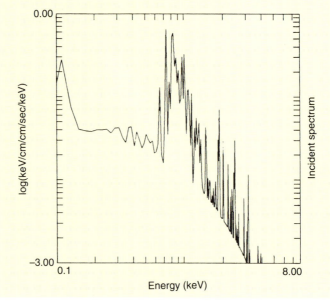

Fig. 3.B1 The calculated spectrum of X-rays from a thin, hot gas at a temperature of 10^{7} K. Lines from ions of oxygen, iron, and other 'heavy' elements are superposed on a thermal bremsstrahlung continuum. The rapid falloff of intensity at high energies indicates that there are few electrons with very high energies.

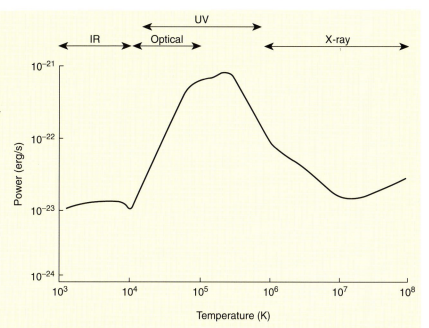

Fig. 3.B2 The calculated rate of radiation of 1 cm³ of hot gas, with density 1 ion cm⁻³, as a function of temperature. The form in which the radiation appears is indicated at the top. The highest rate of cooling is at temperatures close to 10⁵ K. (From the work of D. Cox and E. Daltabuit, University of Wisconsin).

Thus by measuring X-ray spectra, the shape of the continuum and/or the presence of lines can identify the origin as a hot gas. The temperature of the gas can be calculated from the particular lines present and from the shape of the high-energy end of the bremsstrahlung continuum. The strength and energies of the lines also reveal the elemental composition of the gas.

Synchrotron radiation from relativistic electrons

A fast electron traversing a region containing a magnetic field will change direction because the field exerts a force perpendicular to the direction of motion. Because the velocity vector changes, the electron is accelerated and consequently emits electromagnetic energy. This is called magnetic bremsstrahlung or synchrotron radiation (after radiation observed from particle accelerators by that name). The frequency of the radiation depends only on the electron energy, the magnetic field strength, B, and the direction of motion relative to the field.

In an astrophysical setting, the magnetic field can be somewhat aligned but particle velocities are expected to be isotropic; so the observed spectrum depends only on B and the energy spectrum of the electrons. The usual spectral form assumed for the electrons is a power law and, if this is so, then the spectrum of the resulting synchrotron radiation is a power law also. Indeed, when an observed spectrum is a power law over a reasonably large energy range, it is usually taken as a strong indication that the source is emitting synchrotron radiation. If the magnetic field is aligned, the radiation will be polarised, and observed polarisation is usually proof of synchrotron emission. The form of the power law spectrum is simply

$$I(E) = A\,E^{-\alpha},$$

where A is a constant and α is the 'spectral index'. The larger the value of α, the softer the spectrum. This spectrum is illustrated in Fig 3.B3 for $\alpha = 1$.

As a matter of interest, the power radiated as synchrotron radiation is proportional to $B^2\,E^2$, and the average photon energy to $B\,E^2$. Two examples of synchrotron radiation are: (1) radio emission from shell-like supernova remnants where the magnetic field strength is about 7×10^{-5} gauss and the electrons which radiate radio waves each have an energy of about 1 GeV; (2)

the central region of the Crab Nebula where the magnetic field is 10 times stronger than in other SNR. Radiation is emitted over most of the electromagnetic spectrum. The electrons which produce synchrotron X-rays have energies of about 10^4 GeV, a few ergs each! Thus synchrotron X-rays indicate the existence of very energetic electrons.

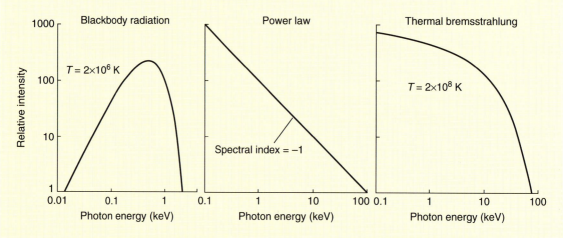

Fig. 3.B3 Three basic spectral forms expected from astrophysical processes. At the left is the blackbody spectrum expected from a dense object at a temperature of 2×10^6 K. At the centre is a power-law spectrum expected from synchrotron radiation produced in a region containing a magnetic field and high energy electrons. At the right is thermal bremsstrahlung from a thin, very hot gas. The different shapes are signatures of the physical production processes.

Blackbody radiation from star-like objects

A 'black' surface completely absorbs any radiation incident upon it. Reflectivity is zero. It is a law of nature that the surface must not only absorb but also emit radiation. The spectrum radiated is a well-defined continuum with peak emission at an energy dependent only on the temperature, T. The higher the temperature, the more energetic the photons. A familiar example is the electric heating element on a stove or hot plate. As it is heated it first glows a deep red and then, as the temperature increases, becomes orange and almost yellow.

The form of the spectrum is given by

$$I(E,T) = 2\, E^3 \left[h^2 c^2 \left(e^{E/kT} - 1 \right) \right]^{-1}$$

where h is Planck's constant, and c is the speed of light.

The stars radiate as blackbodies with temperatures from 2500 K (red dwarf) to 40 000 K (O star). Although strongly modified by the stellar atmosphere, the spectra retain the overall gross shape imposed by the blackbody emission process. A newly formed neutron star is expected to have a hot surface. If the surface temperature is one million degrees or higher, it will emit blackbody radiation with photons in the X-ray range. There is great interest in searching for such a neutron star at the centre of the expanding debris of SN 1987A.

3.4 Young shell-like remnants

Chances of learning the nature of the progenitor star are best when the remnant is young. At this time its characteristics are determined by the nature of the star and the explosion. The observed material will be predominantly ejecta. A measure of the mass and composition of the remnant will then give the mass of the progenitor star and might reveal heavy elements made in the explosion. The energy released can be calculated using the observed mass of material and the velocity of expansion. Any asymmetries in spatial distribution or composition must be caused by the explosion itself. The youngest remnants known in our galaxy are those listed in table 3.1. Only SN 1987A is young enough to be expanding freely. The others are probably in a transition between phase 1 and phase 2 so the interaction with surrounding material, although not yet dominant, cannot be ignored.

If there were no surrounding material, the stellar ejecta would expand freely with velocity of 10 000 to 20 000 km s[-1]. The interstellar gas, however, forms a barrier that is increasingly difficult for the expanding ejecta to push aside. As the rapidly moving material ploughs into the surrounding medium, some of its kinetic energy is transferred to this medium. The interstellar gas is swept up and moves out with the ejecta. Two shock waves form (figures 3.6 and 3.7). One propagates into the gas ahead of the ejecta, and a second, the 'reverse' shock, propagates back into the ejected gas. The boundary between circumstellar material and ejecta is called the 'contact discontinuity'. As seen by an outside observer, both shocks initially travel outward with the radii of the two shocks differing by about 25%. After the accreted mass becomes greater than the ejecta mass, the reverse shock propagates back to the centre.

Fig. 3.6 A shock wave formed by a small explosion. A fraction of a second after the detonation of a dynamite cap the gaseous products of the explosion are black and are surrounded by an outgoing shock. Supersonic particles travelling ahead of the shock are also clearly seen. The same phenomena are visible in supernova remnants, even though the energy release is 10^{41} times greater. (Picture by Harold Edgerton, MIT.)

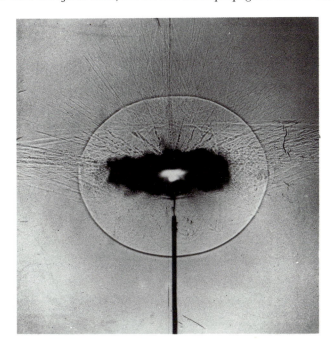

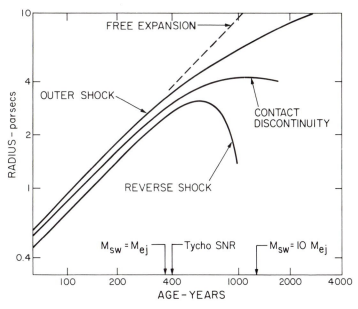

Fig. 3.7 Position of the shock waves in a young remnant as it expands into a uniform surrounding medium. Numbers have been picked to model Tycho's remnant; swept-up mass equals ejected mass after 370 years and after 1300 years is 10 times the ejected mass. (Based on the work of R. Chevalier, University of Virginia.)

Only between the two shock waves is the material hot. Here the interstellar gas has been heated and compressed by the expanding ejecta. The ejecta have in turn been slowed and compressed by the pressure of the interstellar gas. At large distances the interstellar gas is cool and does not know about the expanding debris from the stellar explosion. In the central region the material is not hot and is freely expanding. The information that there will be resistance ahead has not yet been received. Only the shocked material is hot enough to emit X-rays and this forms the bright shells of young remnants.

The geometry, however, is complicated by irregularities in both the ejecta and in the surrounding medium. Supernova shock fronts are not exactly spherical. Furthermore, the ejected material pushing into interstellar gas is compressed and unstable. It is expected to break into clumps. This is a 'Rayleigh–Taylor' instability, the same effect which causes a large blob of water falling through air to break into small droplets; or the breakup of a bubble of oil floating to the top of a container of water.

The remnant of Tycho's supernova

The Danish astronomer Tycho Brahe first saw the supernova of 1572 on the night of November 11. He was outside, after dinner, planning a night of observing after a stretch of bad weather. When he spotted the new star close to maximum brightness he could not believe his eyes. He was a seasoned observer and in his experience the stars were fixed features in the arrangement of the heavens. This was also the belief of everyone else at the time. He asked others with him to verify that the new star was really there before going to his observatory and starting the first of many careful observations.

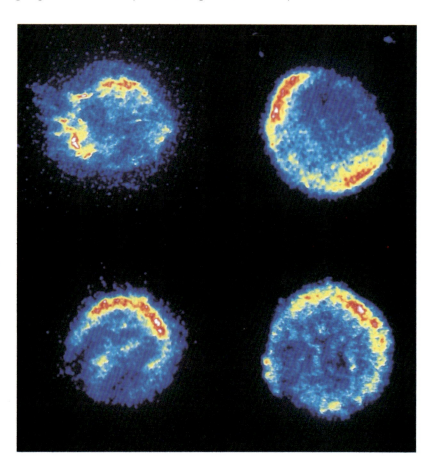

Fig. 3.8 Einstein X-ray pictures of four young shell-like remnants (false colours are used to show surface brightness). At the upper left is Cas A, the youngest and the most luminous. A faint outer shell of shocked ISM surrounds a bright incomplete ring of ejecta. A jet-like feature in the east may indicate high-velocity ejecta. Tycho's remnant, at the lower right also has an outer shell of shocked ISM but it lies closer to the bright ring of clumpy ejecta and is harder to distinguish. At bottom left is the remnant of Kepler's supernova, at top right, the remnant of SN 1006 which, as seen from the Earth, subtends the same angle as the Moon – about 30 arcminutes.

Tycho measured the brightness of the new star from maximum until it faded below naked-eye visibility 15 months later. He obtained the position by measuring the angular distance between the new star and the bright stars in Cassiopeia. His position is within 2 arcminutes of the centre of the remnant we see with modern instruments! Tycho's light curve is clearly that of a Type I supernova. Consequently the remnant is expected to have been formed by a low-mass star.

The remnant of this supernova was discovered in the radio band 378 years after the supernova had faded to invisibility by Robert Hanbury Brown and Cyril Hazard in 1952. Extremely faint optical wisps at this location were found on a photograph taken by Walter Baade in 1949. The X-rays were first recorded with a rocket-borne detector by Herbert Friedman, Ted Byram and Talbot Chubb (NRL) in 1965 . The X-ray source was identified as the remnant by Paul Gorenstein and colleagues (AS&E) in 1968. Detailed radio and X-ray maps of the remnant have been made with radio interferometers and with the Einstein telescope and are shown in figures 3.8 and 3.9.

These observations of Tycho's remnant show both the shock in the interstellar medium and the shocked ejecta. The leading shock is quite clear in the radio picture. The outer edge of the remnant is sharp and smooth. High energy electrons moving in a region of relatively strong magnetic field pro-

Fig. 3.9(a) Radio emission from the remnant of Tycho's supernova at a wavelength of 20 cm taken with the Cambridge 5 km radio telescope. Note the remarkable similarity between this and the Einstein X-ray picture (figure 3.8). Most of the emission comes from a shell of the same diameter as seen in X-rays, but the brightest regions are located at different places. This picture clearly shows the shock in the interstellar medium. (Courtesy D. Green and S. Gull, MRAO.)

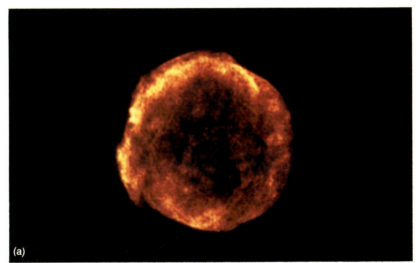

(a)

Fig. 3.9(b) Infrared emission from the remnant of Tycho's supernova. These data were obtained by the Infrared Astronomical Satellite at a wavelength of 60 microns. Emission is probably from dust grains heated by the hot X-ray-emitting gas. (Courtesy R. Braun, NRAO.)

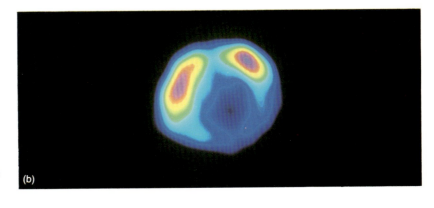

(b)

duce this radiation. This field is thought to be an interstellar field compressed and amplified by the outer shock.

The ejecta show most clearly in X-rays. The brightest part of the remnant is a shell of material, broken into clumps, following the outer shock, and 80% of the X-ray emission comes from this region. The composition of the material is revealed by the X-ray spectrum (figure 3.10). This is dominated by strong emission lines from medium-weight elements, perhaps made within the star during the explosion. The abundances of silicon and sulphur appear to be greater than in 'solar' material. (The abundance of iron, however, does not appear to be enhanced.) Thus the X-ray picture can be interpreted as a direct observation of material from the disrupted star. The deviations from symmetry, other than the small scale clumping, are not understood and may indicate an asymmetry in the explosion itself. The mass of the X-ray emitting matter can be derived from these data through detailed comparison with a model for the distribution of the material.

The comparison of observations with mathematical models is, however, not trivial. Expected features are not always clear and easily discernible. The interstellar shock, for example, is not well separated from the ejecta ring. Some of the ejecta clumps have almost overtaken the outer shock.

Fig. 3.10 X-ray spectra from the Einstein SSS spectrometer show prominent emission lines from silicon and sulphur, probably made in the supernova event. There are also weaker lines from magnesium, argon, and calcium. Observation of these lines proves that the emission is thermal radiation from a gas heated to a temperature of several million degrees. (Courtesy A. Szymkowiak, GSFC.)

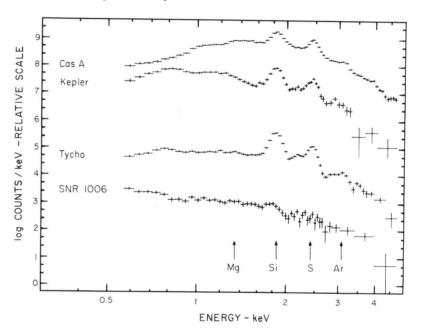

Furthermore, since we are viewing a projection of a three-dimensional object, there is always an element of uncertainty in knowing the actual structure. In any case, the presence of lines in the X-ray spectrum identifies the emission process as thermal radiation from a hot gas (or more correctly a plasma since it consists of ions and electrons) and enables calculation of the temperature and composition of the material. Knowing the temperature and the elements, we can derive the mass of material from the observed X-ray luminosity.

The mass of ejecta calculated in this way for Tycho's remnant is between 1 and 2 $M_\odot$. This is also about the calculated mass swept up from the interstellar medium, but (and this 'but' is the bane of many astronomical calculations) the numbers are not precise. One basic problem is that the mass calculation depends on the 5/2 power of the distance to the remnant. This is easily uncertain by 30%, introducing a factor of 2 uncertainty in the mass. A further complication is that the time required for electrons and ions in the shocked interstellar gas and ejecta to reach equilibrium is larger than the lifetime of the young remnant. The strength of line emission from a given mass of material depends on the equilibrium state which, in this situation, is not well known (box 3.5).

The measured mass of ejecta, therefore, is about the same as the expected 1.4 $M_\odot$ of the exploding-white-dwarf model. Although the lack of any strong X-ray lines attributable to iron is difficult to understand, the apparent absence of a neutron star in the interior argues against gravitational collapse.

Box 3.5 Ionisation equilibrium of material in remnants – another unknown variable

Shock waves in the shells of supernova remnants are thought to be 'collisionless'. Energy is transmitted through the magnetic field to the surrounding material, which has been ionised by ultraviolet radiation from the supernova itself. To an observer riding on the outward-moving shock, the kinetic energy of incoming material resides in the positive ions which carry almost all the mass. This kinetic energy is thermalised by the shock, and immediately after passage of the shock, the energy is found in rapid motion of the positive ions which remain temporarily in a state (characterised by the number of bound electrons) corresponding to a much lower temperature than that characterising the motion. After a time, the free electrons, through collisions, share the energy of motion and come into thermal equilibrium with the positive ions. As more time passes, the now fast-moving electrons collide with and remove more electrons from the heavier positive ions and, eventually, the state of ionisation is increased to that appropriate to the electron temperature. Thus the three energy-indicators of the gas, positive ion velocity, electron velocity, and ion state, all eventually achieve the same temperature. The time required for this to come about depends on density and temperature and is calculated to be typically hundreds of years.

Information about the state of the gas is obtained from the X-ray spectrum. The high energy continuum shows the velocity of free electrons, and the emission line intensity and energy give the nature of the ions. If these are characterised by different temperatures, analysis of the results is difficult. Such is the case for the remnant of Tycho's supernova. The electrons and the ion states are not yet in equilibrium and this increases the uncertainty in the calculated amount of material.

So, although the Einstein X-ray data now show the structure of this young remnant in great detail, more information is needed before the proposed models for supernova explosions can either be definitely confirmed or eliminated from consideration.

Results for the other young remnants shown in figure 3.8 are similar. Kepler's 1604 supernova has been classified as Type I from its light curve. However, at maximum light, it was a magnitude fainter than Tycho's supernova. Consequently it was not observed for as long and the classification is not as firm. The supernova of 1006 is thought to have been Type I because it was so bright at maximum. The morphology and spectrum of the Kepler and Tycho remnants are so very similar that it is easy to believe that the progenitor stars were almost identical. The remnant of SN 1006, allowing for its greater age and its location in a less-dense region of the galactic plane, is also quite like Tycho's remnant. Optical emission from these three remnants is very faint; a few wisps are seen in deep Hα photographs, but nothing bright.

As a final thought, let us consider the physical nature of the remnant of Tycho's supernova. Although one of the smaller remnants, its 7 pc diameter is very large compared to the distance scale we are used to. If the Sun were at the centre, our nearest celestial neighbours, the bright stars Alpha Centauri, Sirius, and Procyon, and 12 faint red stars would all lie within the outer boundary of the remnant. The X-ray brightest clumps of material have a density of only 3 atoms cm^{-3}. Even though 4 $M_\odot$ of material are within the outer boundary, the remnant is completely transparent to radiation at all wavelengths. It is expanding into an almost pure hydrogen gas with density only 0.4 atoms cm^{-3}, a particle density 2×10^{-20} that of the air we breathe. The remnant is a very tenuous thing, truly a ghostly inhabitant of interstellar space.

Cassiopeia A

Cas A is commonly regarded as the remnant of a Type II supernova. The motion of apparent optical fragments observed by Karl Kamper and Sidney van den Bergh (Dominion Astrophysical Observatory), if run backwards in time, implies an origin at about 1670, a time when there was a keen interest in astronomy in Europe. It is perplexing that this event, which occurred in a north circumpolar location, was not observed. Since Type II supernovae are dimmer at maximum than Type I, a Type II event is considered more likely. There are other unique features. The radio luminosity is very high; Cas A is the brightest radio source in the sky. The interior of the remnant is full of bright optical knots, some highly enriched in oxygen. A high resolution radio map (see figure 1.10) is also very clumpy. The remnant seems composed of bright knots, more so than the other young remnants.

The X-ray morphology and spectrum (figures 3.8, 3.10, 3.11), on the other hand, look quite similar to the young remnants of Type I supernovae. The mass of X-ray emitting material has been calculated as 10–15 $M_\odot$, implying a Type II origin. If this were the case, we would expect a bright young neutron

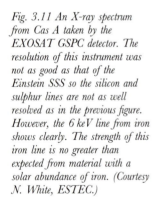

Fig. 3.11 An X-ray spectrum from Cas A taken by the EXOSAT GSPC detector. The resolution of this instrument was not as good as that of the Einstein SSS so the silicon and sulphur lines are not as well resolved as in the previous figure. However, the 6 keV line from iron shows clearly. The strength of this iron line is no greater than expected from material with a solar abundance of iron. (Courtesy N. White, ESTEC.)

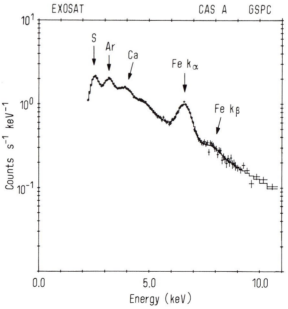

star at the centre but there is no evidence for such an object. So neither of the two proposed models fits. Perhaps gravitational collapse leading to a black hole is the explanation. More data might help, but at this time Cas A seems to be a unique remnant requiring a special set of circumstances for its formation.

3.5 Four older remnants

After a few hundred years of expansion, the nature of the remnant changes. Most of the material is now swept-up circumstellar matter. Consequently the form of the remnant and the X-ray spectrum are determined by the surrounding material rather than by the characteristics of the explosion. Most of the known remnants are in this state. Phase 1, the free expansion, lasts only 300–1000 years and most remnants are detectable until the end of phase 3, at an age of 30 000–100 000 years. The old remnants typically appear as large radio shells containing an array of optical filaments. Most are reasonably luminous X-ray sources. The X-ray morphology is sometimes similar to the radio structure and sometimes not. It is important to remember that gas at high enough temperature to emit X-rays is too hot to have the bound electrons necessary for optical emission. A shell visible in both these energy bands must consist of many interacting regions, each with gas at different temperature.

The gas within a large remnant is thought to be in approximate pressure equilibrium. The force at the boundary of a high density cloud of cool gas can be balanced by that of hot gas at low density. This force or pressure is important in the formation of filaments. A cloud of hot gas will cool slowly to the point where electrons are captured by protons and hydrogen atoms are formed. Energy is then easily radiated as ultraviolet lines and the cooling rate increases enormously. This is the start of phase 3. Pressure within the cooling cloud drops and the cloud is compressed further by the external pressure of surrounding hot material. The increase in density increases the rate of radiation so the process proceeds even more rapidly until a dense, usually elongated, cloud remains – the optical filament. Any magnetic field within the cloud during this compression is frozen into the ionised material and is compressed also. Since radio emission is expected from regions of relatively high magnetic field, the filaments are radio-bright. High resolution radio observations of IC 443 by R. Duin and Harry van der Laan (Leiden) in 1975 demonstrated that old remnants show a high degree of correlation between radio and optical emission.

Although the surrounding material is referred to as interstellar, it is quite likely that much of it was originally part of the progenitor star. Early stars in particular have strong stellar winds. A typical O star might lose mass at a rate of 10^{-6} $M_\odot$ yr^{-1} and the terminal velocity of the wind, as determined from the shape of lines in the ultraviolet spectrum (see figure 1.13), is about 2500 km s^{-1}. (This is 10^{10} times more energetic than the solar wind where 10^{-14} $M_\odot$ yr^{-1} is lost at a velocity of 300 km s^{-1}.) This material of course comes from the outermost layer of the star, which is hydrogen. After a few million years the star will be surrounded by a bubble of hydrogen. The original interstellar medium and its inhomogeneities have been pushed away. It is not surprising that the star can do this. If the wind just described continues for 5 million years, 5 $M_\odot$ will have been lost. The total energy loss in the wind will have been 3×10^{50} ergs, comparable to the kinetic energy of debris from the supernova explosion itself! Chapter 5 will discuss other manifestations of such strong stellar winds.

Imagine the shock wave from an expanding supernova travelling through such a circumstellar bubble. Because the density of material is low, the shocked hydrogen is not a strong X-ray source. If this shock then encounters a region of high density (a cloud with 10–100 atoms cm^{-3}, either within the bubble or outside it), the shock will propagate into the cloud. Material within the cloud will be compressed still further, and some material will be 'evaporated' from the outer surface of the cloud. The increase in density caused by this evaporated material will increase the X-ray emission. Optical filaments will first be formed in the interior of the shocked cloud. Using this hypothesis, the differing appearance of old remnants can be explained.

These remnants are enormous. If centred on our sun, a 40 pc diameter remnant would surround thousands of stars, including the bright stars Alpha Centauri, Sirius, Procyon, Aldebaran, Castor, Pollux, Fomalhaut, and Regulus. The diameter is large enough to be an appreciable fraction of the 'thickness' of the layer of gas, dust, and bright stars which form the galactic disc. The older remnants are thus expected to show not only the density structure of the local interstellar medium (ISM) but also large-scale density variations of the Milky Way. The radiating material ranges from X-ray bright clumps of hot gas with densities of a few atoms cm^{-3} to optically bright filaments with densities of a few hundred atoms cm^{-3}.

The Cygnus Loop and the Vela Remnant

Here are two remnants with shape determined by the local circumstellar environment. Both are fairly close to Earth and soft X-rays (having energy below 0.3 keV) are easily transmitted over the path to Earth. Temperatures are low and most of the X-ray emission from these remnants is soft.

X-rays from the Cygnus Loop were first detected by Friedman, Byram, and Chubb in April 1965. The source was identified as the supernova remnant in November 1968 by Rod Grader, Richard Hill, and Peter Stoering (LLL) who noted that the source was extended as would be expected if emission came from throughout the volume of the 2.5° diameter remnant. Over the next 10 years various observers, with great enthusiasm, attempted to map the X-rays from the remnant. They used well-collimated detectors on rockets and the rather small detectors on the Copernicus and ANS satellites. They determined that X-rays came from the vicinity of optically-bright regions but the maps were unimpressive. In July 1977, Saul Rappaport and colleagues (MIT) used a rocket-borne X-ray telescope to produce an excellent map with resolution 10 arcminutes. The main features seen in the Einstein result (figure 3.12) are all clearly there. This was a remarkable result considering that only 5 minutes of data were obtained during the rocket flight.

The Cygnus Loop really looks like a bubble. The brightest filaments form a symmetrical ring with a faint southern extension. The brightest X-ray emission comes from this ring and emission from the interior is faint and relatively uniform. The shock has encountered dense clouds only at the outer boundaries of the remnant. The progenitor star has apparently swept the surrounding area free of clouds out to a distance of 18 pc where they are now illuminated by energy from the shock wave. The observed symmetry indicates that the remnant is in phase 2 and that the blast-wave model might be used to calculate physical conditions. This model requires an interior temperature of 3 million degrees, an initial energy release of 3×10^{50} ergs, 100 $M_{\odot}$ within the shell, and an age of 18 000 years. The average interstellar density needed is 0.16 atoms cm^{-3} but the optical filaments are probably formed from compressed interstellar clouds having original densities of about 10 atoms cm^{-3}. The 'breakout' in the south may indicate a warm low-density region in the ISM where the shock has propagated with higher velocity.

X-rays from the Vela Remnant (or Vela XYZ, a complex of three radio sources) were first detected in May 1968 by Grader and colleagues as a very soft source in the southern sky. This source was identified as a superposition of both the Vela Remnant and the nearby remnant Puppis A in May 1970 by Tom Palmieri and colleagues (LLL). Crude maps of the emission were obtained with rocket-borne detectors and with the SAS-3 satellite. Not until the Einstein map, however, was the structure of the emission from this region clearly visible. The Einstein picture shown in the figure is a mosaic of 40 IPC

Fig. 3.12 Einstein X-ray pictures of four older supernova remnants. At upper left is the Cygnus Loop. This remnant is remarkable because it shows, except for a 'breakout' in the south, the circular symmetry expected from a blast wave propagating through the ISM. At the upper right is the Vela Remnant. Forty IPC fields were used, yet coverage is not complete. Brightest emission is from clouds of interior material. At lower left is Puppis A, very bright in soft X-rays, and showing filamentary structure over much of the remnant. Most X-rays come from a bright knot on the east limb. At lower right is IC 443, a remnant which shows almost no limb brightening and with morphology determined by two nearby dense interstellar clouds. Note that the rectangular grid-like structure is an instrumental effect due to the detector's window supports. (Courtesy W. Ku, Columbia University; F. R. Harnden Jr., CfA; R. Petre, MIT.)

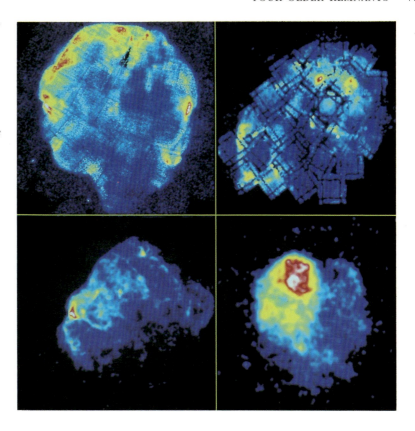

fields. The grid-like structure superposed is an artefact due to shadowing by detector window supports.

The Vela Remnant has a different appearance than the Cygnus Loop. It is filled with optical filaments and the brightest X-rays come from great irregular patches of gas inside the boundary. The outer shock is visible in a few places but is not bright in X-rays. This remnant is perhaps the product of a less massive star. The explosion has occurred before the stellar wind could remove lumpy interstellar material from the near vicinity. The denser regions now show as bright optical filaments surrounded by hot X-ray emitting gas. Some of this gas is probably that evaporated from the filament-forming clouds. The pulsar within the Vela Remnant, although clearly visible and showing that the remnant originated in gravitational collapse, is not the dominant source of emission.

Puppis A

This remnant is the brightest X-ray source in the sky in the range 0.5 to 1.0 keV. It is particularly well suited for high resolution measurements. The Einstein HRI map, compiled by Robert Petre and colleagues (MIT), (figure 3.12) shows two bright knots and an overall filamentary structure. The radio morphology (figure 3.13) is quite similar to that in X-rays and there are a few faint optical filaments in the vicinity of the northeast maximum of radio emission. As a matter of interest, Pup A is immediately adjacent to the Vela Remnant and may obscure the edge of this larger, cooler remnant.

The density of material in interstellar space varies greatly from region to region and particularly with distance from the galactic plane. The average density and number of clouds are maximum close to the plane. The appearance of Pup A is best understood in terms of variations in density of the surrounding medium (with again only very small influence from a progenitor stellar wind). This remnant is younger than the Vela Remnant and the X-rays are observed to come from filamentary regions which (because they emit X-rays) must be too hot to emit optical lines. The observed large-scale variations in X-ray brightness can be explained by density variations of a factor of 4 in the emitting material. There are two dense clouds, a small one in the north and a larger one on the east limb. Most of the emission comes from the eastern condensation where the density is high. The density on the west limb where the X-ray emission is too faint to appear above background is at least one hundred times lower.

Because of its X-ray brightness, the east limb of Pup A was selected as one of the prime targets for the Einstein crystal spectrometer, the FPCS. The results were gratifying to see (figure 3.14). Working with a spectral resolution a hundred times greater than anything used before, Frank Winkler (Middlebury

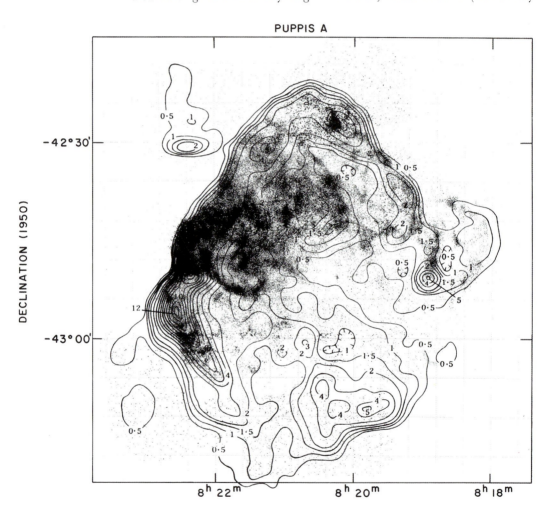

PUPPIS A

DECLINATION (1950)

RIGHT ASCENSION (1950)

Fig. 3.13 Radio and X-ray emission from Puppis A. Contours of constant radio surface brightness have been overlaid on a grey-scale X-ray picture. Note the general, but not small-scale, similarity. Outline of the remnant is the same in both wavebands and maximum emission comes from bright regions on the northwest limb, but regions of maximum brightness do not exactly correspond. (Courtesy R. Petre, MIT.)

Table 3.3 Properties of material in Pup A

Eastern cloud:		Temperature	$2-5 \times 10^6$ K
		Density	25 atoms cm^{-3}
		Mass	1.4 $M_\odot$
Density of western limb			< 0.2 atoms cm^{-3}
Mass of total remnant			100 $M_\odot$

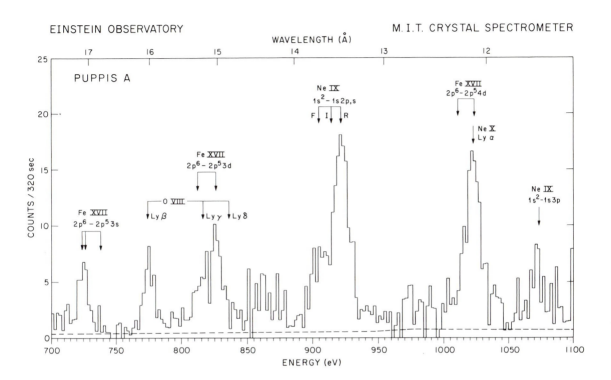

Fig. 3.14 X-ray spectrum of the bright knot in Puppis A from the Einstein FPCS spectrometer. Emission lines from highly ionised oxygen, neon, and iron were clearly detected. This figure is noteworthy as the first non-solar, high-resolution, X-ray spectrum to show sharp features. (Courtesy F. Winkler and colleagues, MIT.)

College), Claude Canizares (MIT), and colleagues detected emission lines of highly ionised oxygen, neon, and iron. These data enabled detailed calculations to be made of the elemental abundances and temperature of the gas (table 3.3). Results indicate that oxygen and neon are present with a concentration 3–5 times greater than normal relative to iron. This is assumed to be material from the progenitor star, ejected during the explosion and now mixed with about 100 $M_\odot$ of X-ray emitting gas. This remnant is apparently the result of a type II supernova since a massive star is required to produce the 3 M$_\odot$ of oxygen and neon observed. These elements are expected in the outer layers of such an evolved star. Puppis A is one of a class of 'oxygen-rich' supernova remnants (which includes Cas A). The lack of an observable pulsar, if Pup A is a type II remnant, is again a puzzle. Although there is much theoretical and observational work to be done, these FPCS observations clearly show the powerful role that high resolution X-ray spectroscopy will play in the future.

IC 443

Here is a unique middle-aged remnant with properties best explained through interactions with two nearby high-density regions; a neighbouring HI cloud and an overlying molecular cloud. The remnant has an apparently complete optical shell but the appearance is dominated by a network of bright optical filaments filling the northeast quadrant. These filaments probably arise from the interaction of the shock with a large-scale density enhancement in the surrounding medium – the HI cloud. The molecular cloud lies in the line of sight covering the middle, the southern, and the western portions of IC 443.

The X-ray morphology of IC 443 is different from that of Pup A and the Cygnus Loop which show definite shell-like structure (figure 3.12). The soft X-ray luminosity is also an order of magnitude smaller. The clouds explain these differences. The decreasing surface brightness towards the west and southwest is due to absorption of X-rays in the overlying molecular cloud. The X-ray bright region in the northeast is not only clear from the shadow of the molecular cloud but is also enhanced by interaction with the HI cloud (figure 3.15). There is no limb brightening where the shock has interacted with either cloud. In a high-density medium, evolution is rapid. That part of the shell within dense regions is now in phase 3 where the shock has cooled below the temperature required for X-ray emission.

The observed X-ray temperature and surface brightness of IC 443 can be used to derive physical information directly, even though the non-uniform

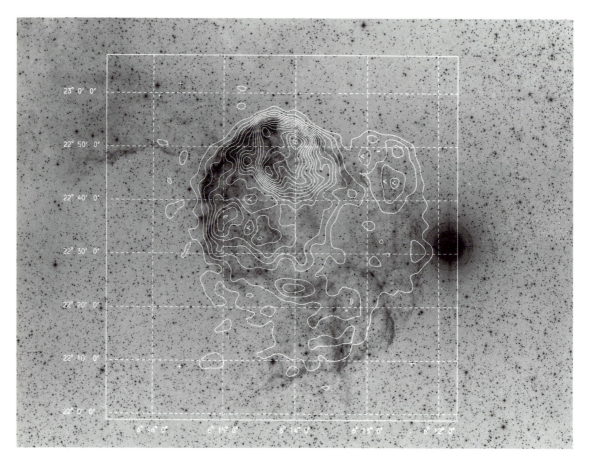

Fig. 3.15 Optical and X-ray emission from IC 443. Contours of constant X-ray surface brightness have been overlaid on the Palomar Sky Survey picture. Interaction of the remnant with a large dense cloud to the northeast causes the bright emission from this sector. Figure 3.12 also shows the X-ray data.

medium makes the blast-wave model difficult to apply. Derived characteristics are listed in table 3.4. Information about the filaments came from optical spectra. The surrounding, cool, swept-up material is hydrogen and is visible only to a radio telescope at 21 cm.

Thus older remnants can give information about their origin in several ways. The arrangement of surrounding material can in some cases show the influence of the presupernova stellar wind which depends strongly on the mass of the star. High resolution spectroscopy (both X-ray and optical) can

Table 3.4 Properties of material in IC 443

X-ray emitting gas:	Temperature	12×10^6 K
	Density	0.6 atoms cm^{-3}
	Mass	11 $M_\odot$
	Internal energy	1×10^{51} erg
Optical filaments:	Density	500 atoms cm^{-3}
	Mass	2 $M_\odot$
Mass of swept-up cool gas:		100 $M_\odot$

detect enriched material in middle-aged remnants even though the ejecta can no longer be spatially distinguished from shocked circumstellar material. Finally, before most of the energy is radiated away during phase 3, the thermal energy of the material, derived from the X-ray luminosity, is a measure of the energy released in the explosion.

3.6 Remnants with neutron stars

The Crab Nebula
The Crab Nebula is dear to the heart of most astrophysists. It contains the first-discovered and only undisputed example of a neutron star formed in a supernova explosion. The evidence for this is overwhelming. First and most important: the supernova was observed in 1054 and its location is the same as that of the remnant. Second: the observed expansion of the remnant, run backwards in time, converges at the date of the supernova. Third: the pulsar is located exactly in the middle of the remnant. Fourth: the characteristic age of the pulsar is close to the age of the remnant. Fifth: the energy lost by the pulsar is high enough to support the high luminosity of the Crab Nebula at all wavelengths. Here then is a remnant known to have originated in gravitational collapse and a neutron star known to have been formed in a supernova explosion! With a period of 33 milliseconds, this pulsar was, for many years, the fastest known pulsar. Figure 3.16 shows the Crab Nebula at four different wavelengths.

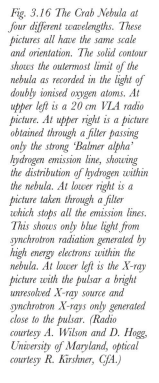

Fig. 3.16 The Crab Nebula at four different wavelengths. These pictures all have the same scale and orientation. The solid contour shows the outermost limit of the nebula as recorded in the light of doubly ionised oxygen atoms. At upper left is a 20 cm VLA radio picture. At upper right is a picture obtained through a filter passing only the strong 'Balmer alpha' hydrogen emission line, showing the distribution of hydrogen within the nebula. At lower right is a picture taken through a filter which stops all the emission lines. This shows only blue light from synchrotron radiation generated by high energy electrons within the nebula. At lower left is the X-ray picture with the pulsar a bright unresolved X-ray source and synchrotron X-rays only generated close to the pulsar. (Radio courtesy A. Wilson and D. Hogg, University of Maryland, optical courtesy R. Kirshner, CfA.)

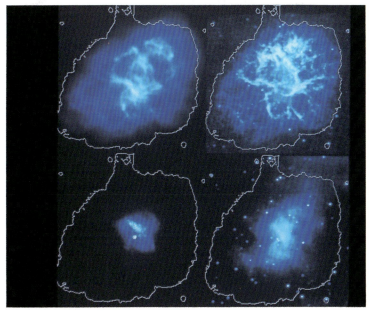

Until the discovery of radio pulsars by Jocelyn Bell and Anthony Hewish (Cambridge University) in 1967, 'neutron stars' had existed only in the minds of theoretical physicists. First proposed as an end state of stellar evolution in 1939 by Robert Oppenheimer and George Volkoff, they are now commonly accepted as the only explanation for radio pulsars.

The neutron star is, to us, a bizarre state of matter. A mass about equal to that of the Sun is confined to a region 15–30 km in diameter. The material is almost all neutrons. The density is that of the atomic nucleus, about 10^{14} g cm^{-3}, perhaps 10^{15} g cm^{-3} at the centre. A 10 km cube of rock (about the size of Mt Everest) compressed to a 10 cm cube (about the size of a grapefruit) would have this density.

The gravitational field close to the surface is enormous. The tidal forces will pulverise to dust any solid object which approaches closer than a few thousand kilometres. We know these stars exist and we know the masses of a few which are in binary systems. There is no straightforward measure of the radii or of the internal structure. We would dearly like to know more.

Neutron stars are thought to have very powerful dipole magnetic fields (like bar magnets) which are not aligned with the axes of rotation (figure 3.17). These fields are simply the fields of the progenitor stars compressed and concentrated by the collapse of the stellar interiors. As a neutron star rotates, electrons are torn from the surface and accelerated by large-scale electric fields. Then, held in the firm grip of the intense magnetic field, they are swept around by the rotation and flung outward. Far from the surface, the particles and field are forced to stop corotating because they cannot go faster than the speed of light. At this point some of the considerable energy acquired by the particles is radiated. Thus a pulse of radiation is observed when one of the magnetic poles rotates through the observer's line of sight. Radio waves (and higher frequencies) are produced by the 'synchrotron' process as the electrons move across magnetic field lines some distance from

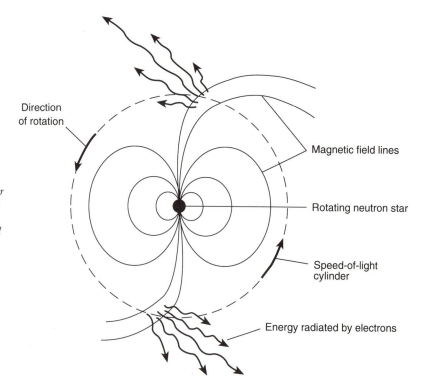

Fig. 3.17 A spinning neutron star with dipole magnetic field aligned perpendicular to the axis of rotation. Electrons are constrained to follow the field lines. At the poles, electrons move outward until they reach a point where they must travel faster than the speed of light to corotate with the star. At about this radius they radiate and an external observer sees a pulse of radiation every time a magnetic pole swings by his or her line of sight.

Direction of rotation

Magnetic field lines

Rotating neutron star

Speed-of-light cylinder

Energy radiated by electrons

the neutron star. The width of this beam of radiation is unknown. If it is narrow, like a lighthouse beam, there follows the interesting consequence that most pulsars are unobservable because we happen not to be illuminated by the rotating beam.

The extreme regularity of the observed pulsations requires that pulsars be associated with a massive, but rapidly spinning solid body. Long-term radio monitoring, however, soon showed that pulsars were slowing down. For example, the period of the Crab pulsar is steadily increasing at a rate of 15 microseconds per year. This loss of rotational energy powers the Crab Nebula. Because of this the nebula has a different appearance than most other supernova remnants.

The discovery of the Crab pulsar by David Staelin (MIT) and Edward Reifenstein (NRAO) in 1968 solved a long standing problem of astrophysics. The pulsar was the powerhouse supplying the prodigious energy radiated by the nebula. The total energy output of the Crab Nebula from radio to X-rays is an incredible 10^{38} erg s^{-1}, or about 100 000 times the energy output of our sun. And this, remember, is over 900 years after the supernova explosion!

Because of the central injection of energy by the pulsar, the Crab is not shell-like but 'filled-in' in appearance. Hence the term 'plerions' (from the Greek word *pleres*, meaning 'full') to describe Crab-type remnants.

Irrespective of pulsar orientation, the intense low frequency electromagnetic radiation of the pulsar (a giant rotating bar magnet) has a braking effect which slows the rotation. A large fraction of the energy in this low-frequency radiation is transferred quickly into high energy electrons and magnetic field. These electrons form a 'relativistic wind' flowing outward from the pulsar. The exact mechanism by which electrons are accelerated to high energies is, alas, not understood. In any case, the energy density is high in this cloud of relativistic particles surrounding the Crab pulsar. When the relativistic electrons cross magnetic field lines they radiate synchrotron radiation, forming a bright diffuse nebula. The radiation extends throughout the electromagnetic spectrum – from radio waves to gamma rays. This 'synchrotron nebula' dominates the appearance of the Crab at X-ray wavelengths. (Radiation from this nebula is quite separate from the pulsed radiation which originates close to the neutron star and which appears as a point source to all detectors.)

The Crab Nebula was first positively identified as a strong X-ray source during a lunar occultation in 1964. The emitting region was not only precisely located but was established as being extended. After precise calculations, a rocket was launched by Stuart Bowyer and colleagues (NRL) and the detector was pointed at the nebula. The relative motion of the Earth, the Moon, and the rocket then caused the nebula to pass behind the limb of the Moon; all during the 5 minute flight of the rocket!

The next series of lunar occultations occurred in 1974. During these the Crab Nebula was observed by several rocket and balloon-borne payloads and the shape and size of the X-ray nebula determined. To everyone's surprise, the synchrotron nebula was not centred on the pulsar. The brightest X-ray emission was centred 10 arcseconds northwest of the pulsar. The pulsar itself appeared clearly as a pulsing point source. The Einstein observations in 1979 produced the images shown in figures 3.16 and 3.18, confirming past occultation results and showing detail clearly.

Why is the X-ray nebula off-centre? An explanation offered by Martin Rees (IOA) and James Gunn (Princeton) is that the bright region is a shock in the relativistic wind from the pulsar. Particles and field flow from the pulsar at a velocity close to the speed of light. These are stopped by the network of filaments at the boundary of the nebula, so there must be an abrupt transition in velocity somewhere inside. The energy dissipation associated with such a shock would be a possible source of electron acceleration, known to be

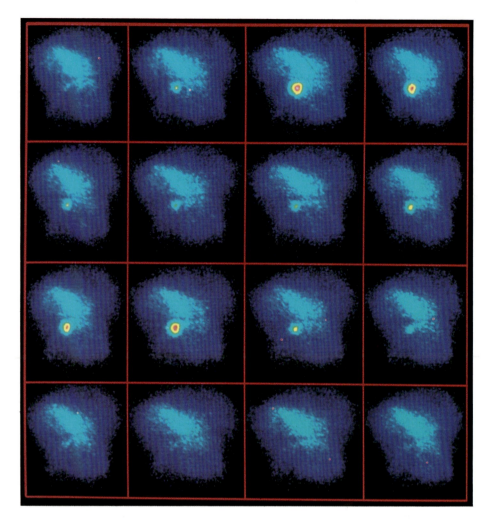

Fig. 3.18 X-rays from the Crab Nebula and its pulsar as a function of pulsar phase. A 5 hour exposure with the Einstein HRI has been sorted into 16 equal phase bins using the known pulsar period. The pulsar is brightest during the primary pulse, is faintly visible during the 'interpulse' time, brightens again during the secondary pulse, and fades into invisibility at the end of the cycle. (Courtesy F. R. Harnden Jr. CfA.)

necessary to maintain the nebula as we see it. The lifetime of electrons responsible for the diffuse X-ray nebula is only a few years. Since we observe an almost steady state 930 years after the explosion, an internal source of energetic electrons is needed. The off-centre X-ray nebula shows that all electron acceleration does not take place close to the pulsar and that the process is not symmetrical. Perhaps the pulsar is moving to the northeast and this compresses the magnetic field in this direction.

The Vela supernova remnant

The Vela Remnant and its pulsar, with a period of 0.089 s, are quite different in appearance to the Crab Nebula. The Vela Remnant is considerably older and is also probably the nearest known supernova remnant to Earth. It is 7° in diameter and filled with a beautiful network of optical filaments. At only 500 pc distance there is little absorption of X-rays in the interstellar medium. Figure 3.12 shows a map made of the low-energy data (photon energy < 1 keV) from 40 separate Einstein pointings. The interior of the remnant is filled with wisps of X-ray-emitting plasma characterised by a temperature of two million degrees. The pulsar is visible as a bright unresolved source 1° distant from the apparent centre of the remnant. The high-energy (photon energy 1–4 keV) data have been assembled into the picture shown in figure 3.19 where the pulsar stands out more clearly.

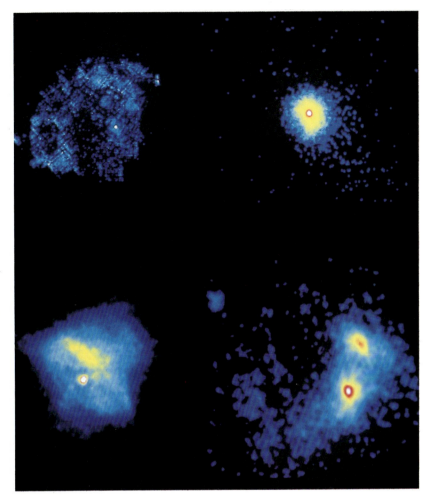

Fig. 3.19 X-ray pictures of three diverse remnants with internal neutron stars. The Crab Nebula, at the lower left, is the brightest diffuse X-ray source observed by Einstein. The Crab Pulsar is unresolved and accounts for only 4% of the X-ray counts. The Vela Remnant, at the upper left, is older and larger. The Vela Pulsar appears as a small bright source and the interior of the remnant is filled with clouds of hot X-ray emitting gas. At upper right, a single high resolution picture shows a field only 10 arcminutes in extent which shows the synchrotron nebulosity surrounding the Vela Pulsar. At the lower right, MSH 15–52 has two X-ray bright regions: the central pulsar and surrounding nebulosity and, to the northwest, a bright patch on the rim caused by interaction of the expanding shock wave with a dense region of the interstellar medium. (Picture by G. Dodge and F. Seward, CfA.)

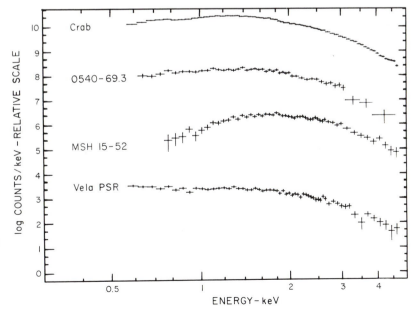

Fig. 3.20 X-ray spectra of remnants with neutron stars and synchrotron nebulae obtained with the Einstein SSS. These featureless spectra are characteristic of continuum radiation emitted by high energy electrons moving through a magnetic field. The spectra are also 'hard' in that the radiation is strong at high energies. Note the different shape compared to the thermal spectra shown in figure 3.10. This illustrates the value of spectral measurements in determining the radiation mechanism. (Courtesy A. Szymkowiak, GSFC.)

The high-resolution X-ray image in figure 3.19 shows that the pulsar is surrounded by a small diffuse nebula 4 arcminutes in diameter. A close inspection of the low-resolution image to the left reveals that this nebula extends even further to the south. This extended nebula is clearest when the lowest energy X-rays (seen in figure 3.12 and which constitute most of the thermal emission) are excluded from the image.

Thus the Vela Pulsar (like the Crab Pulsar) is surrounded with nebular emission. The X-ray luminosity of this nebula, however, is only 0.0002 that of the Crab synchrotron nebula. Nor is there evidence for a shock standing off from the pulsar. The electrons are apparently energised close to the pulsar and diffuse outward with no subsequent acceleration. Far from the pulsar the magnetic field is low and electron lifetime long. The emission here may indicate past activity more than the present energy deposition from the pulsar. The X-ray luminosity of the pulsar itself is only 0.001 that of the Crab Pulsar. Nevertheless there is plenty of energy available from the pulsar to power the nebular emission.

Compared to the Crab Pulsar, the Vela Pulsar is weak at optical and X-ray wavelengths. It could not be detected if it were not so close. It is surprisingly bright, however, at gamma-ray energies. The gamma-ray luminosity is 1/4 that of the Crab Pulsar but, since the Vela Pulsar is 4 times closer, it is observed to be brighter in gamma-rays than the Crab Pulsar.

The Vela Pulsar was discovered in the radio band by Michael Large, Arthur Vaughan, and Bernard Mills (University of Sydney) in 1968, one month before the Crab Pulsar. Diligent radio searches, however, failed to find pulsars within any other of the 140 known galactic remnants, even those few considered to be plerionic in character. This was perplexing. Half the remnants were expected to have originated in gravitational collapse and young neutron stars were thought to be bright. Even if 'beaming' were to make 80% of them undetectable, about 10 pulsars should have been easily observable. Why were only two young neutron stars detected?

X-ray observations have helped to understand this puzzle. In some cases, the pulsar emission is easier to see above background in the X-ray band than in the radio band. Several rather faint point-like X-ray sources within supernova remnants were discovered with Einstein. Four of these exhibit regular pulsations, a certain identification of a neutron star. Apparently not all young pulsars are bright.

Four Einstein discoveries

The first discovery was accidental. Observers Phil Gregory and Greg Fahlman (University of British Columbia) pointed the Einstein telescope at a variable radio source in Cassiopeia. A bright unexpected source appeared in the corner of the field. It had the shape of a half circle 30 arcminutes in diameter with a bright point at the centre (figure 3.22). This source had been catalogued as a radio source, CTB 109. But since it is only 3° from Cas A, the brightest radio source in the sky, it had not been mapped or even identified as a supernova remnant. Follow-up observations showed that X-rays from the central source are strongly pulsed with period of 6.98 seconds. The source is probably in a binary system. (Chapter 7 contains a full discussion of X-ray sources in binary systems.) A faint optical counterpart for the central object has been found with a blue magnitude of 23.5, showing that the companion star is not massive. This then is probably an accretion-powered X-ray source and information concerning the spindown age of the neutron star cannot be obtained, although the existence of the remnant indicates it is relatively young. Here also is a binary system not disrupted by the supernova explosion.

The unusual form of the remnant is caused by a neighbouring molecular cloud. In the dense material of the cloud to the west, the shell has evolved rapidly and is now past the point of X-ray or even radio emission (although at an earlier phase this may have been the brightest part of the remnant).

The second discovery, by one of us (FDS) and Rick Harnden (CfA), was in the remnant MSH 15–52 in the constellation Circinus, far in the southern sky. At radio frequencies this remnant is undistinguished; a patchy, rather diffuse shell with diameter 30 arcminutes. The appearance in X-rays is shown in figure 3.19 and is quite different. X-ray emission is concentrated in two places: at the centre, and along the shell in the northwest. A high resolution image shows an unresolved source at the centre which is pulsing 7 times each second. This new pulsar, PSR 1509–58, is surrounded by an X-ray nebula extending 10 arcminutes southeast and northwest. The X-ray spectrum of the nebular emission (figure 3.20) is a hard continuum – characteristic of synchrotron radiation.

This nebula is not bright. It has 10^{-2} the X-ray luminosity and 2×10^{-5} the surface brightness of the Crab. This nebula is faint because energy input from the pulsar is only $1/25$ that of the Crab Pulsar. Since the surface brightness is too low to detect this synchrotron nebula at radio or optical wavelengths, the discovery was made in the X-ray band.

The pulsar was confirmed in the radio band by Richard Manchester (CSIRO), Ian Tuohy (Mt. Stromlo), and Nick D'Amico (Palermo). Observations spanning two years time showed the pulsar to be spinning down steadily with no observed 'glitches' (transient increases in spin rate with sudden onset which appear irregularly in observations of the Crab and Vela Pulsars.) This pulsar exhibits the highest period derivative, $\dot{P}$, of any radio pulsar. There is no doubt that this is an isolated neutron star.

This system presents a puzzle. The pulsar seems to be much younger than the shell of the remnant. The appearance of the shell is that of a 'middle aged' supernova remnant, about 10 000 years old. The characteristic pulsar age is only 1550 years. (Box 3.6 describes how this is calculated.) The bright spot on the rim is full of optical filaments and is catalogued as an Hα emission region, RCW 89. There is one knot which is particularly bright in X-rays, optical lines, and infrared emission from iron. The supernova shock wave has probably run into a dense interstellar cloud here. If the interstellar medium is highly nonuniform, this might explain the unusual appearance of the shell. This is one possible resolution of the problem.

The third discovery reaches out to the Large Magellanic Cloud (LMC). A neutron star has been found by Seward, Harnden, and David Helfand (Columbia) within the remnant 0540–69.3 (a radio source named by using its coordinates). The properties of this remnant/pulsar combination are so much like the Crab Nebula that it is finally clear that the Crab is not unique. A further piece of good fortune is that the distance to the remnant is known to be 55 kpc (because it is in the LMC), and it is a rare remnant for which the distance is so accurately determined. It is difficult, however, to measure structure that would be easily resolved in a remnant 10 times closer.

The pulsar pulses 20 times per second. This is $2/3$ the rate of the Crab Pulsar, yet the new pulsar has a pulsed X-ray luminosity 2.5 times higher than that of the Crab Pulsar. The rate of loss of rotational energy is $1/3$ that of the Crab Pulsar. Strong optical pulsations have been found by J. Middleditch (LASL) and C. Pennypacker (Berkeley). This is only the third isolated pulsar which has been detected optically and it is in the Large Magellanic Cloud!

The most striking optical feature of the remnant is a bright ring 8 arcseconds in diameter which strongly emits a line characteristic of ionised oxygen. This is $1/2$ the linear size of the Crab, and inside this ring, surrounding the pulsar, Gary Chanan (Columbia), Helfand, and Steve Reynolds (NRAO)

Box 3.6 Calculation of pulsar characteristics

A rotating neutron star with a magnetic field must radiate electromagnetic energy at the rotation frequency. There is a torque associated with this radiation which acts to slow the rotation. Thus rotational energy is changed to radiation and, as time advances, the star slows down.

If the field is a dipole field (like a bar magnet) this torque is at a maximum. Several quantities can be simply calculated from the observed rotational period, P (measured in seconds), and the spin-down rate (or period derivative), $\dot{P}$. It is also necessary to know the rotational moment of inertia, I, which is calculated from the assumed equation of state of nuclear material, and for a 1 $M_\odot$ neutron star usually has a value close to 10^{45} g cm^2 s^{-2}. R is the neutron star radius and is about 10 km.

The rotational energy and the rate of loss of this energy are:

$$E = 2\pi\, I\,/P^2 \text{ ergs and } \dot{E} = -4\pi\, I\, \dot{P}/P^3 \text{ erg s}^{-1}.$$

$\dot{E}$ is negative because the pulsar is losing energy.

If the period is now appreciably larger than the period at birth, the age of the pulsar should be close to the 'characteristic age',

$$A = P\,/\,(2\dot{P})\text{ s},$$

and the magnetic field at the pole is,

$$B = (3c^3 IP\dot{P}/8\pi^2 R^6)^{1/2}\text{ gauss}.$$

The magnetic field is, of course, strongest at the poles and is very strong indeed for most pulsars. The field of the Crab Pulsar, for example, is calculated to be 5×10^{12} gauss at the poles. This is stronger than any field in our experience. The natural magnetic field at the Earth's surface is strongest at the south magnetic pole where it is 0.7 gauss. The strongest steady field we have been able to make in a laboratory is 3×10^5 gauss. Materials available on Earth do not have the strength to contain the fields generated by these stars.

found a diffuse blue continuum which is probably optical synchrotron radiation. Both faint optical and radio emission indicate the existence of an outer shell, five times larger than the bright ring, a feature long sought but not found around the Crab Nebula. Thus this distant remnant contains a fast pulsar, bright over a large range of wavelengths, and diffuse features quite similar to those of the Crab Nebula. We must await results from a high resolution southern-hemisphere radio telescope to see more detail in the outer part of this remnant.

The fourth discovery is within the remnant CTB 80 which has been an enigma for years. A bright radio core anchors three faint arms which extend 30 arcminutes to the north, east, and southwest (figure 3.21). Although there has been speculation that this was the result of a supernova occurring in 1408, the large size has always seemed incompatible with such a recent origin.

A faint point-like X-ray source was detected by Einstein in the center of CTB 80. An accurate position was determined but the source was too weak to find pulsations. Armed with this information, Richard Strom (Dwingeloo), who has long been interested in this remnant, discovered an unresolved radio source at this position. Because the radio spectrum was unusually steep, like most radio pulsars, he proposed that this was indeed an isolated neutron star. Soon after this, Shrinivas Kulkarni (CIT) and co-workers detected radio pulsations with a period of 0.039 second. The characteristic age of the pulsar is 1.1×10^5 years. This is compatible with the large size of CTB 80 and, based on present knowledge, incompatible with a supernova observed in 1408.

*Fig. 3.21 A 49 cm radio map of
CTB 80, a large irregular
remnant. This figure is about 1
degree on a side. A 39 millisecond
pulsar has recently been discovered
in the bright central region.
(Courtesy R. Strom, Netherlands
Foundation for Radio Astronomy,
Dwingeloo.)*

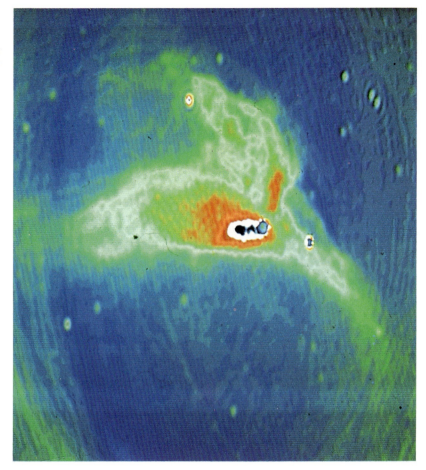

Fig. 3.21 A 49 cm radio map of CTB 80, a large irregular remnant. This figure is about 1 degree on a side. A 39 millisecond pulsar has recently been discovered in the bright central region. (Courtesy R. Strom, Netherlands Foundation for Radio Astronomy, Dwingeloo.)

A radio discovery

The most sensitive radio pulsar searches have been done with the Arecibo (Puerto Rico) telescope. Because it has been built into a natural bowl-shaped feature, it can only observe a 40° wide band of the sky which the Earth's rotation causes to pass overhead. Alek Wolszczan (Arecibo) and co-workers have observed supernova remnants which happen to be located in this band and have found a young pulsar within the shell of the remnant W44. The pulsar has a period of 267 milliseconds, an age of 20 000 years and is losing rotational energy at a rate only one one-thousandth that of the Crab Pulsar. The age of the pulsar is about the same as that of W44.

W44 is unusual. Although the radio picture shows the expected shell-like structure, the interior is filled with hot X-ray emitting gas and no X-rays are observed from the radio shell. Einstein was unable to distinguish any X-rays from the vicinity of the pulsar over this background of thermal emission. Apparently X-ray emission from the pulsar was too faint to see. Perhaps ROSAT will find it.

Other likely neutron stars

The Einstein high resolution X-ray pictures show unresolved sources within several other remnants, listed in table 3.5 and illustrated in figure 3.22. These are point-like X-ray sources with the location appropriate for the remnant of the core of the progenitor star. Regular pulsations, however, have not been detected. In most cases not enough photons were collected to enable a search for pulsations to be undertaken. This is not the case for SS433, an accretion-

Table 3.5 Supernova remnants containing (and likely to contain) neutron stars

Remnant	Neutron star	Period (s)	Age (yr)	Rotational energy loss (erg s^{-1})	Comments
Crab Nebula	PSR 0531+21	0.033	1240	5×10^{38}	remnant of SN 1054
0540−69.3	PSR 0540−69	0.050	1670	1.5×10^{38}	in Large Magellanic Cloud
MSH 15−52	PSR 1509−58	0.150	1550	2×10^{37}	
Vela XYZ	PSR 0833−45	0.089	1.13×10^4	7×10^{36}	
CTB 80	PSR 1951+32	0.0395	1.05×10^5	4×10^{36}	large radio lobes
W44	PSR 1853+01	0.267	2.0×10^4	4×10^{35}	
CTB 109	1E2259.0+5836	6.98			
W50	SS 443				binary system
3C58	1E0201.8+6435				remnant of SN 1181
Kes 73	1E1838.6−0459				
RCW 103	1E1613.7−5055				
Kes 75					synchrotron nebula only
G21.5−0.9					synchrotron nebula only
PKS 1209−52	1E1207.3−5209				

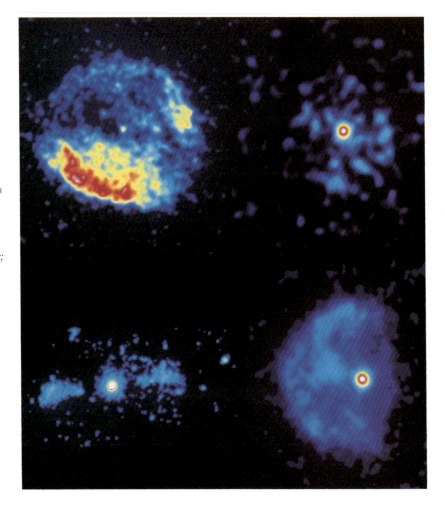

Fig. 3.22 Einstein X-ray pictures of four remnants with internal unresolved X-ray sources. Clockwise from top left: RCW 103, 8 arcminutes in diameter; Kes 73, 4 arcminutes in diameter; CTB 109, 30 arcminutes in diameter; and W50 with point source SS433, 80 arcminutes east-west extent. SS433 and the source within CTB 109 are rather bright and are probably binary systems. RCW 103 has an X-ray bright shell like the remnants in figure 3.8, but the internal source is at the limit of detectability. At the other extreme, the internal source in Kes 73 is moderately bright, but the surrounding diffuse emission is faint and its shape not well determined. (Picture by G. Dodge and F. Seward, CfA.)

powered binary which is rather bright in X-rays. This non-pulsing source is the subject of chapter 12, and the surrounding region, W50, is an example of a remnant with characteristics probably largely determined by the central object.

The young pulsars within the remnants shown in figure 3.19 are all isolated (not in a binary system) and are all surrounded by diffuse X-ray synchrotron nebulae. Four isolated radio pulsars (PSR 0656+14, 0950+08, 1055–52, and 1929+10) have also been detected as X-ray sources by France Cordova (LANL), Andrew Cheng (Rutgers) and Helfand. Only one shows evidence for X-ray pulsations at the radio frequencies. The luminosity of this X-ray emission depends on $\dot{E}$, the rate of loss of rotational energy; the higher $\dot{E}$, the more luminous the pulsar and surrounding nebula. Some of the rotational energy lost by all isolated pulsars is converted to magnetic fields and relativistic particles. Radiation from the resulting nebula, which may be too small to be resolved from the neutron star, is isotropic. Even if radiation from the pulsar itself is beamed and not directed towards us, emission from the nebula will always be detectable.

There are two remnants, Kes 75 and G21.5–0.9 where this is happening. Both have radio shells a few arcminutes in diameter with X-ray emission concentrated at the centre. High resolution X-ray observations by Robert Becker (University of California, Davis) and Helfand show small, clearly diffuse nebulae with no point-like component. Even the unpulsed radiation from the Vela Pulsar could be such a synchrotron nebula. Although appearing point-like to us, it could be 10 000 AU in diameter. Thus the presence of an energetic pulsar within a remnant is, in principle, always detectable.

The converse of this is also true. The absence of a synchrotron nebula means there is no energetic pulsar. Thus the shell-like remnants such as Tycho's remnant and the Cygnus Loop do not contain energetic isolated neutron stars. There is, however, a loophole in this argument. Diffuse emission from the shell can obscure faint emission, particularly when the interior is filled with knots and patches. A faint pulsar is always possible at some level. The apparent absence of synchrotron radiation does, however, allow us to set an upper limit on $\dot{E}$ of possible hidden pulsars.

Thermal radiation from neutron stars

Just after formation a neutron star is predicted to have a surface temperature in excess of 10^7 K. The time required to cool to 10^6 K is calculated to be several hundred years. A blackbody with radius 10 km and surface temperature 10^7 K will emit 10^{37} erg s^{-1} in the energy range 1–10 keV. The rate of radiation goes as the fourth power of the temperature so the power radiated by even so small an object is enormous. A neutron star at this temperature, located anywhere in the near half of our galaxy or in the Magellanic Clouds, will be easily detectable. However, no such hot neutron stars have been observed within the young supernova remnants. Either no neutron stars were formed there or they cool very rapidly.

It is possible that the point-like X-ray emission associated with the Vela Pulsar is from the surface of the neutron star. If so, the surface temperature is 1×10^6 K. The absence of soft X-rays between secondary and primary pulses of the Crab Pulsar (figure 3.18) sets an upper limit of 2×10^6 K on the surface temperature. Even the bright pulsed radiation shown in figure 3.24 cannot be thermal since the pulsed spectrum extends down to the optical band where the same wave form is observed. The faint source within RCW 103 is still a viable candidate for a neutron star radiating X-rays from a hot surface. It could, however, be a pulsed source; it is too faint for us to search for regular pulsations or to measure the X-ray spectrum using existing data.

Because, even in the interior of young supernova remnants, no high-temperature neutron stars have been observed, we think that young neutron stars

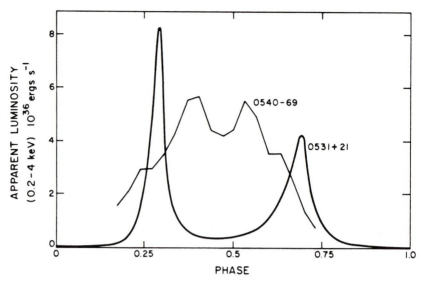

Fig. 3.23 Pulsed wave forms from young neutron stars. The pulsar in the LMC remnant 0540–69.3 has been detected at both X-ray and optical wavelengths and has an almost sinusoidal pulse. Although the integrated X-ray luminosity of this pulsar is apparently double that of the Crab Pulsar (0531+21), the maximum rate of radiation during the pulse is less.

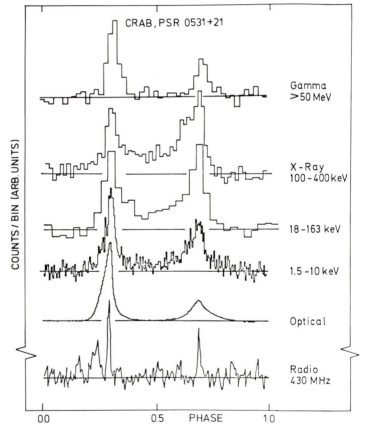

Fig. 3.24 Pulsed wave form from the Crab Pulsar at different frequencies. The Crab Pulsar emits two rather sharp pulses separated 140° in phase at all wavelengths. There is also emission between the two pulses at high energy X-ray wavelengths. (Courtesy G. Bignami, CNR, Milan.)

cool more rapidly than expected. It is possible to derive information about the interior composition of neutron stars from the cooling curve but existing data are not precise enough to draw definite conclusions. Future possibilities are interesting. The composition of nuclear matter at densities in excess of 10^{15} g cm^{-3} and the interior structure of a neutron star might be eventually derived from measurements of the surface temperature.

Young pulsars

The mechanism by which pulsars emit bursts of high energy radiation is an interesting piece of unsolved physics. The shape of the pulses we observe (figure 3.23) must depend on the pulse formation mechanism, the structure of the neutron star's magnetic field, and the relative orientation of spin axis, magnetic axis, and viewer. The Crab Pulsar has now been extensively observed at many wavelengths and data from other young pulsars are starting to be collected. The 'big picture', however, is not clear.

The young pulsars described are all spinning rapidly, are all bright at high energy, and exhibit a variety of pulse shapes. The pulses are probably incoherent synchrotron emission. High energy electrons are accelerated in the vicinity of the magnetic poles and radiate when they reach the speed-of-light cylinder where particles and fields can no longer corotate with the neutron star. Fast pulsars with strong magnetic fields (and necessarily high $\dot{E}$) are expected to be bright at high energies.

The Crab Pulsar at all frequencies radiates two pulses separated 140° in phase. It is natural to associate one pulse with each magnetic pole as the rotation carries them through our line of sight. Differences in pulse shape (figure 3.24) can be ascribed to difference in field strength. The uneven phase separation might be due to the position of the poles on the surface of the star or could be a geometrical effect dependent on our viewing direction. The primary pulse is narrow, rises quickly to maximum, and there is no phase change at different frequencies although the wave form changes somewhat as energy increases. The secondary pulse becomes relatively stronger at higher energies and there is some radiation observed between the two pulses.

The Vela Pulsar is quite different. Most of the energy appears as high energy gamma-rays where two pulses are observed with shape and spacing similar to that of the Crab Pulsar. The radio and optical pulses, however, are single and occur at different phases than the gamma pulse, and the X-ray pulse is weak. After several diligent searches, X-ray pulses were finally observed by Hakki Ogleman (University of Wisconsin) using ROSAT. The

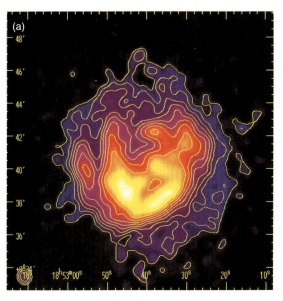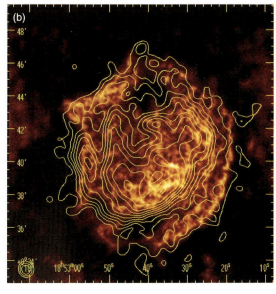

Fig. 3.25(a) A ROSAT PSPC X-ray picture of Kes 79, a remnant which has bright, interesting radio structure in the interior. This was observed in the hope of finding a neutron star nearby. (b) The contours of constant X-ray surface brightness are overlaid on a 20 cm radio map. The X-ray and radio emission define an outer shell and an internal region. There is no sign of an interior neutron star. Kes 79 may have been produced by a type I supernova. (Radio data courtesy T. Velusamy.)

pulsar appears as a steady source of soft X-rays with modulation of about 4%. This pulsar is not very luminous; it is only well observed because it is nearby.

At the other extreme, the pulsar in the Large Magellanic Cloud is the brightest isolated X-ray pulsar known. The pulse shape at both X-ray and optical frequencies is approximately sinusoidal and identical within the accuracy of measurement. Relative phase and shape of the pulse at other frequencies are at present unknown. The pulsar in MSH 15–52 also has an approximately sinusoidal pulse shape in the X-ray band.

Thus four known young isolated pulsars have been detected at high energies but with a variety of waveforms. We suspect that the magnetic field configuration on the surface of the individual neutron stars is different but the importance of viewing geometry is unknown. Our understanding will certainly improve as more data are collected on these objects. We need, however, to find more young neutron stars; hopefully one with characteristics that reveal something fundamental about pulse formation. Future X-ray observatories are eagerly awaited. With the sensitivity of AXAF we might find Crab-like pulsars in M31, a galaxy with more stars than ours and a potentially rich source of young supernova remnants and neutron stars.

4 Active stellar coronae

4.1 The Sun

Our Sun is, because of its proximity to us, the brightest source of X-rays in the sky. But why should there be detectable X-rays from the Sun at all? Certainly not on the basis of its everyday visible appearance. The optical spectrum of the Sun can be represented quite well by a simple blackbody at a temperature of about 6000 K. Such an object should produce no detectable X-ray flux at all whereas the amount actually seen implies the presence of material at a temperature of at least a million degrees!

In fact, the X-ray observations of the Sun represented the last pieces in a puzzle about the nature of the Sun's corona that began in the middle of the nineteenth century. The corona is, of course, only visible to us on Earth during a total eclipse by the Moon, surely one of the most spectacular natural sights afforded to man (see the photo montage of the Sun in figure 4.1). When the first spectroscopic observations were made of the corona during an eclipse over a hundred years ago, scientists were extremely puzzled with the results (see box 4.1). They indicated the presence of an unknown element in the corona since the spectral *signature* had never been seen before in the laboratory! The *new* element was dubbed *coronium*.

Further work by Lyot in the 1930s implied that the corona was even more bizarre. He measured the width of the emission lines in the coronal spectrum and found that they were very broad. If this is a velocity effect (see box 4.2) then it requires a temperature of more than a million degrees. Unfortunately, at that time no one was prepared to accept such an extraordinary result as no seemingly rational physical explanation could be found to support it.

Of course, the discovery of solar X-rays gave credence to this result, but by then the emission lines of *coronium* had been identified through laboratory work as due to nothing more exotic than *iron* and other heavy elements! However, the reason that they had not been identified earlier was because these coronal elements were in a rather unusual state. They had been stripped of a large number of their outer electrons leaving them with a net positive charge. To do this requires very high temperatures, of order 1 to 2 million degrees at least!

Structure of the Sun's atmosphere

The existence of a million degree solar corona is now accepted without question. But what supports it and provides it with its energy? The answer to this problem is currently the subject of intensive study and debate, but we have

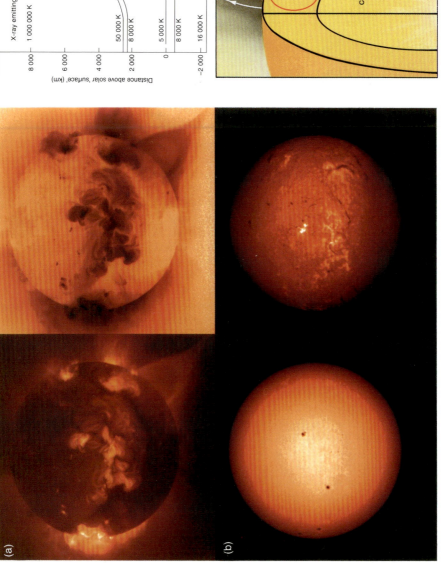

Fig. 4.1 X-rays from the Sun on January 25, 1992. (a) This stunning X-ray image was obtained with the Japanese Yohkoh soft X-ray (0.1–2 keV) telescope, and is printed in both positive (left) and negative (right) versions so as to enhance fine detail. It shows an extraordinarily rich variety of loops, holes and bright regions in the X-ray emitting corona. Large bright spots are associated with active regions, and there are large coronal holes (visible as dark patches). The Hα image (right) on the other hand shows maximum emission from the active regions and cool dark filaments. Alongside these photographs are schematics showing details of the Sun: its temperature structure as it varies with height, and an explanation of these X-ray features. Note the dramatic increase in temperature about 2000 km above the visual surface (photosphere) of the Sun. (The solar X-ray images are from the Yohkoh mission of ISAS, Japan. The X-ray telescope was prepared by the Lockheed Palo Alto Research Laboratory, the National Astronomical Observatory of Japan, and the University of Tokyo with the support of NASA and ISAS. Photographs courtesy of Bob Bentley, MSSL, and the ISAS Yohkoh team.)

Box 4.1 Optical spectroscopy of the corona

A spectrum of the solar corona is shown in figure 4.B1 which indicates the strong and initially unidentified emission lines. Such lines are produced when an excited atom makes a transition from a higher to a lower level.

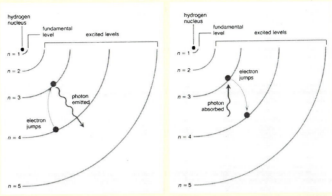

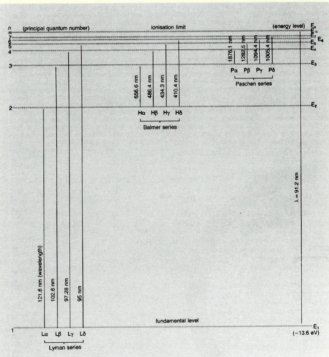

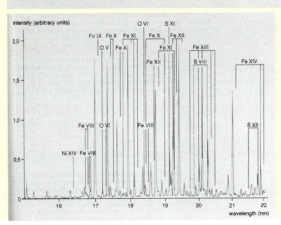

Fig. 4.B1 Coronal spectrum (bottom); and atomic energy levels (above) showing emission when an electron makes a transition (jump) from a higher to a lower energy level. The energy level diagram shown is for hydrogen.

These energy levels depend uniquely on the number of electrons that the atom possesses and the mass of the atom's nucleus. Hence they provide a *fingerprint* of the element that is present in the gas producing the spectrum. But, in the early days of atomic spectroscopy, these *fingerprints* were determined empirically by observation in the laboratory, which meant that they applied only to materials under laboratory conditions, not the extremes of temperature and pressure that we now know can occur in space.

Box 4.2 Doppler broadening of spectral lines

If all the atoms in a gas that is emitting light were stationary, then the emission line that we see would be almost infinitely narrow, as shown in figure 4.B2.

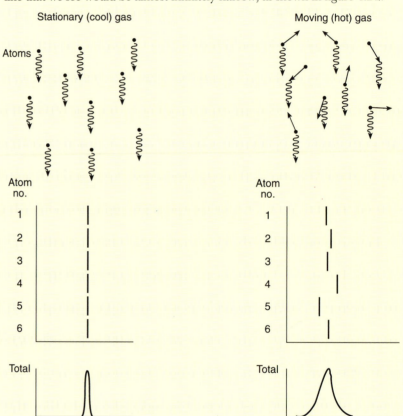

Fig. 4.B2 Stationary (cool) gas produces emission lines at essentially the same wavelength, whereas a hot gas has rapidly moving atoms that produce Doppler shifted lines and hence a broadened line profile.

However, it is very likely that such atoms are in fact hot (which is why they are emitting in the first place) and hence will be moving rapidly in all directions (as are the air molecules that you are breathing now). Light emitted by a moving atom will be Doppler shifted to a slightly longer or shorter wavelength. When the light from all the individual atoms is summed to produce the observed profile, it is seen to be broadened. The amount of broadening can be combined with a theory of how the gas atoms are moving to estimate the temperature of the gas. This kinetic theory produces the Maxwellian distribution of velocities in the gas, from which the well-known Gaussian line profile is derived.

progressed a long way in our understanding of the Sun's corona since the first X-ray detection and most of that progress has occurred in the last 10 years. This is because the first extensive and detailed X-ray pictures of the Sun were not taken until the Skylab missions in 1973–74. These pictures are dramatic in that they show a highly structured and variable X-ray corona with complex bright regions and dark *coronal holes* (figure 4.1). To try and understand what this is telling us we will need to refer to the schematics in figure 4.1.

It is obvious from the plot of the temperature of the Sun as a function of height above the surface why scientists 50 years ago were unwilling to accept the high temperature hypothesis. It does not make sense for the temperature to increase above the Sun's surface, one would expect it to decrease. There was no obvious mechanism by which the material in the corona could be heated. And without a heat source it would very rapidly cool (on a timescale of tens of minutes in fact). There is a clue to what is happening, however, which is visible and has been known for centuries.

Spots on the solar surface can actually be seen with the naked eye, but, since a telescope is required to demonstrate that they are really associated with the Sun, credit for the discovery of sunspots is given to Galileo. However, in the 1600s such solar *blemishes* were denounced as heresy against the church's doctrine of the Sun being a *perfect body* (even Galileo had to recant their existence in order to escape the Inquisition) and the study of sunspots did not really get started until the 1800s. It was soon recognised that the number of spots goes through an 11 year cycle, but with longer term trends that are little understood even today.

The spots appear dark on the Sun's disc because they are cooler than the surrounding area of the *photosphere*. However, it is now known that the spots are associated with intense magnetic fields (as strong as 2000 gauss, compared to a mere 1 or 2 gauss at the Sun's poles) and form the *footprints* of active region loops shown in the schematic figure 4.2. The Skylab pictures were very important in revealing the highly structured X-ray corona, rather than a large diffuse sphere of emission. This was in stark contrast to the Sun's optical appearance. The most pronounced regions of X-ray emission occurred where the magnetic field was strongest, namely near the active

Fig. 4.2 Schematic of two mechanisms proposed to explain how the solar corona is heated: acoustic and magnetic heating. Both rely on energy provided by turbulence in the Sun's convection zone just below the surface. Shock waves generated by this turbulence had been thought capable of propagation through the surface and into the corona. But the favoured model now is one in which the turbulence helps amplify magnetic fields in the surface, creating loops along which electric currents can pass, thereby heating the corona. In the coronal loop model a pair of spots of opposite magnetic polarity act as the active region footprints for a giant magnetic loop in which the magnetic field is strong enough to contain the hot gas. These loops are very large and reach high up into the corona.

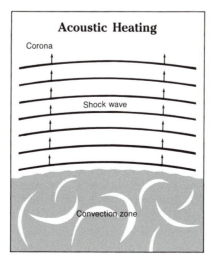

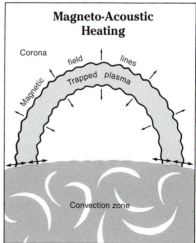

regions where the magnetic field lines formed closed loops. If the magnetic field lines were not closed but open (i.e. they extended off into space) then the X-ray emission was very weak. These are the so-called *coronal holes*.

Solar flares

In addition to these highly complex coronal structures the Sun undergoes, at irregular intervals and without warning, enormous eruptions that are called *solar flares*. A large amount of energy (optical, radio and X-ray) is released in a very small volume of the Sun's surface that leads to the ejection of a large amount of material out into the solar system (see figure 4.9). As we have already seen in chapter 2 this material has a significant effect on the Earth and its surroundings.

The heating mechanism

Clearly the means by which energy is transported into the corona is intimately associated with the spots and active regions, which are themselves areas of intense magnetic fields. How are these fields generated and how is the energy actually transferred?

A detailed photo of the Sun's surface (figure 4.1) shows a granular structure which we know to be due to cells of turbulent motion. These cells are the key to the current models that explain the heating of the solar corona. Schematics of the two basic ideas are shown in figure 4.2.

The *acoustic heating model* was proposed because of the surface oscillations actually observed on the Sun. These might create shock waves which would propagate up into the corona and heat it. *Magnetic heating*, on the other hand, would link up these turbulent cells with closed magnetic loops by means of huge electric currents driven into the loops by the turbulent motion. These currents would then heat the corona. But which of these models is correct? Curiously enough, it is very hard to tell from observing the Sun alone (in spite of its being so close!), but X-ray observations of a wide variety of other types of stars have given us the answer.

4.2 The Einstein stellar X-ray survey

The Einstein X-ray Observatory was the first X-ray instrument which enabled astronomers to search for stellar coronal X-ray emission from other stars at a level comparable to that from the Sun. The Sun's quiescent X-ray luminosity is only about 10^{27} erg s^{-1}, or 10^{10} times fainter than Sco X-1! Nevertheless, the huge gain in sensitivity afforded by Einstein's imaging telescopes enabled literally hundreds of *ordinary* stellar sources to be found. These stars detected in this survey covered a wide range of spectral types in the Hertzsprung–Russell (H–R) diagram (see box 4.3).

An H–R diagram is basically a plot of a star's brightness against its temperature, with brightness increasing towards the top and temperature increasing to the left. Not surprisingly, for those stars that reside on the *main sequence* (see box 4.3), as the stars get hotter they get brighter. The important point here is that stars of virtually all spectral types were found. But within any given type, large fluctuations in X-ray output occurred from star to star. Both these points run counter to the acoustic coronal heating model. The turbulence is caused by the convection zone just below the stellar surface. Since stars earlier (i.e. hotter) than about F5 have no convection zone they would not be expected to generate any acoustic flux for a corona. Also, stars of the same spectral type have identical convection zones and would be expected to have roughly the same coronal X-ray flux. Neither expectation is borne out by this survey.

The surprising results that have come out of the Einstein observations of stars are best illustrated with a single Einstein image (figure 4.3), the HRI picture of the 40 Eridani region. 40 Eri is the closest triple star system other than

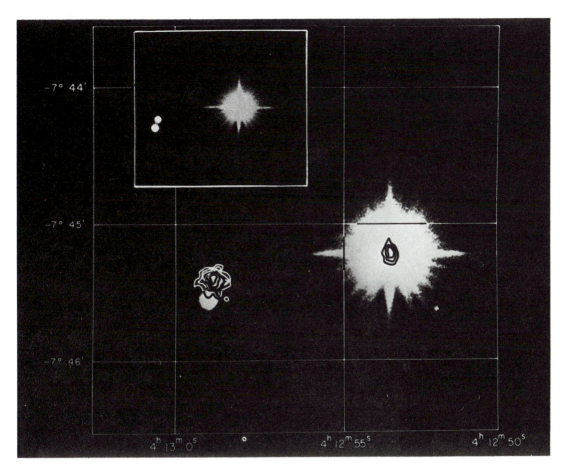

Fig. 4.3 Einstein HRI image (contours) superposed on a Lick Observatory photograph of 40 Eridani, one of our nearest stellar neighbours. Here the bright visual star (which is similar to our Sun) is very weak in X-rays, whereas the faint M star (northern one of the orbiting pair) is very bright.

α Cen (our nearest stellar neighbour), and is at a distance of 4.8 parsecs. The brightest component, A (at 5th magnitude), is a K1 main sequence star, but it is orbited by a dimmer pair consisting of B, a 9th magnitude hot white dwarf, and C, an 11th magnitude M star. The entire system stretches across only 90 arcseconds on the sky, and so fits well within the HRI field of view. The X-ray image (which is superposed in the form of contours onto an optical photograph) reveals a totally different picture. 40 Eri A is only a weak X-ray source (comparable in fact to our Sun's X-ray output) and the hot white dwarf is completely undetectable in X-rays (it is hot enough only to emit in the extreme ultraviolet and was seen by ROSAT). By far the brightest X-ray source is 40 Eri C, the faint M star. The ratio of the stellar brightnesses is completely reversed, with the M star somehow supporting an extremely luminous corona.

4.3 RS CVn systems

So far we have looked at the Sun and stars like the Sun, which were only detectable because of Einstein's vastly improved sensitivity. But are there any objects much brighter than the Sun but emitting by essentially the same mechanism? Such objects were discovered in 1977 by the soft X-ray all-sky survey conducted by the scanning HEAO-1 satellite. Although Uhuru had surveyed the X-ray sky in 1971–72 this had been done at energies of between 2 and 10 keV (approximately 1 to 5 Å in wavelength). There had been no satellite survey of the sky for lower energy X-rays which have a wavelength beyond 10 Å (traditionally referred to as soft X-rays).

This difference in wavelength or energy is crucial for studies of stellar coro-

Box 4.3 Spectral classification and the H–R diagram

Around the turn of this century it was noted by the two astronomers E. Hertzsprung and H. N. Russell that the colour of a star was closely related to its intrinsic optical brightness. This latter quantity is usually quoted as *absolute magnitude* (or M_v) which is simply the apparent magnitude the star would have at a distance of 10 pc. The colour is defined as the difference in magnitude of the star in blue and visual (yellow) light (or B-V). Most stars are found to lie on the *main sequence* as shown in figure 4.B3, as does the Sun which is marked with its usual symbol ($\odot$). Indeed, we now know that when stars are formed from collapsing clouds of interstellar gas they settle onto the main sequence, where they spend the greatest fraction of their lives. Their position on the main sequence is determined almost entirely by their mass; the heaviest stars are the hottest (upper left of the main sequence), the lightest stars are the coolest (lower right).

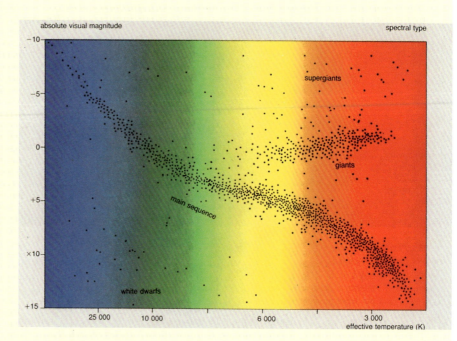

Fig. 4.B3 H–R diagram showing normal stars and the main sequence.

If we treat stars as *blackbodies*, their colour is a direct indicator of the star's temperature. If it is blue it is hot, whereas a red star is cool. This is also borne out by examining the spectra of these stars, and both spectral types and temperatures are also shown for the *x*-axis. Hence, stars are principally classified by their surface temperature as derived from optical spectra.

Energy leaving the star must pass through the outermost *surface* layer of gas. Atoms or ions in this gas absorb light at discrete energies which depend on the number of bound electrons (or ionisation state) of the various atoms. The higher the temperature, the fewer the number of electrons that can be bound to the nucleus of the atom. Thus the particular lines observed identify the ionisation state and hence the temperature can be calculated.

The ionisation of any atom is a well-known function of temperature. A cool gas, for example, can contain atomic hydrogen. Above a temperature of about 15 000 K, collisions between atoms in the gas are energetic enough to remove the bound electrons from hydrogen atoms. At higher temperatures hydrogen atoms are rare and there are therefore no strong hydrogen absorption lines in the spectra of hot stars.

(Astronomers refer to neutral hydrogen as HI. Ionised hydrogen is HII. Similarly neutral iron is FeI, iron with one electron removed is FeII. Iron with two electrons removed is FeIII, and so on. When the gas is very hot, the ionisation state is high. Emission from FeXVII and FeXXIV has been observed in X-ray spectra indicating temperatures of millions of degrees. Fig 14.9 gives more information on this topic.)

The hottest known stars are the O stars with spectra showing an absorption line of singly ionised helium (HeII). Spectra of A stars show lines of neutral hydrogen (HI), and the coolest stars, the M stars, have temperatures low enough that molecules such as TiO can exist giving spectra that are characterised by strong molecular absorption bands. The complete main sequence of spectral types is OBAFGKM, with temperature decreasing from O to M. (This apparently strange ordering of the letters is an accident based on the original ordering of stellar spectra according to the observed strength of the hydrogen lines. They are strongest in A-type stars, next strongest in B-type and so on. Unfortunately, the relationship with temperature is not straightforward.)

It is useful to subdivide each class into subclasses. Thus the hottest A stars are of class A0 and the coolest are A9. The Sun, with a temperature of 5800 K and strong lines of HI, CaII and FeII is of spectral type G2. Table 4.B1 lists characteristics of stars in the midrange of each spectral class.

Table 4.B1 Physical characteristics of normal stars

Spectral type	Mass $M_\odot$	Radius $R_\odot$	Surface temperature (K)	Prominent spectral absorption lines
O5	40	18	40 000	HeII
B5	6.5	3.8	15 500	HeI, H
A5	2.1	1.7	8 500	H, CaII
F5	1.3	1.2	6 500	CaII, H
G5	0.93	0.93	5 520	CaII, Fe, H
K5	0.74	0.74	4 130	CH, CN
M5	0.32	0.32	2 800	TiO

Stars are further divided into luminosity classes. The table defines the *main sequence* and refers to the most common stars, namely those of luminosity class V (sometimes called *dwarf* stars). The *giant* stars which have tenuous, greatly extended, outer layers, are luminosity class III. The largest stars, the *supergiants*, are luminosity class I. The intermediate classes, II and IV, are *bright giants* and *subgiants* respectively. All of the luminosity classes I to IV indicate that the star is *evolved* and has moved away from the main sequence; the process of stellar ageing. This occurs when the star exhausts its central source of nuclear fuel, causing the core to collapse and the outer atmosphere of the star to expand (see box 9.1).

A giant or supergiant star is not only larger but is more massive than main sequence stars of the same spectral type. The Sun is a G2V star. A giant G5III star has radius of 10 $R_\odot$ and mass 3 $M_\odot$. A B5I supergiant has radius 30 $R_\odot$ and mass 25 $M_\odot$.

nae. We have already seen that such coronae have temperatures of a million to a few million degrees in ordinary stars. At these temperatures most of the X-rays emitted will be soft (see the expected temperatures associated with different wavelengths in figure 3.B2). In order to be detected by Uhuru an object's effective temperature had to be greater than about 10 million degrees or the object had to be extremely luminous. It was therefore not surprising that Uhuru detected virtually no stellar coronae.

The HEAO-1 soft X-ray proportional counters used simple collimators to limit the detectors' field of view and the spacecraft spun once every 30 minutes thus providing a viewing *strip* on the sky of 3° by 360° (see figure 2.1). The detectors incorporated all the techniques of background and cosmic ray rejection because this is even more of a problem at these low X-ray energies. The *windows* of the counters have to be as thin as 1 micron in order for soft X-rays to penetrate into the counter gas volume! The satellite was launched in August 1977 and would take 6 months to generate a complete picture of the soft X-ray sky. However, within a couple of months, several strong soft X-ray sources had been discovered, most of which had not been detected by Uhuru. The spectra of these sources indicated that the X-rays originated in a hot gas at about 10 million degrees (see box 3.4). But what were their optical counterparts?

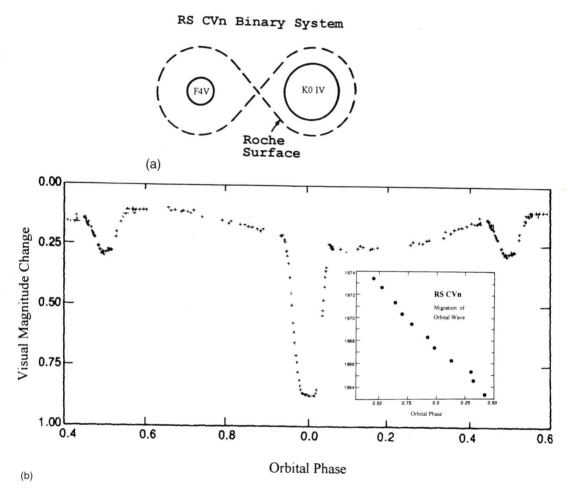

Fig. 4.4 (a) Details of the RS CVn binary system. Although a detached binary (i.e. there is no mass transfer taking place between them), they are close enough to be locked rotationally. This means that each star rotates once per binary cycle. (b) Optical light curve of RS CVn. Both primary and secondary eclipse are easily visible, but note the gentle undulation or wave in the light curve. At this time (1975) the wave peaks at around orbital phase 0.6, but over a long period of time (years) the wave drifts or migrates, taking about 10 years to return to the same binary phase. This is comparable to the Sun's own activity cycle. (Based on original diagrams by Fred Walter, SUNY)

In the scan direction the HEAO-1 soft X-ray experiment had an angular resolution of about 0.5 degrees. However, since the shape of the collimator was well calibrated and understood, it was possible on bright sources to determine their position to about 0.1 degrees. Of course, even this is still not very good when it comes to hunting about in the sky for an optical star to associate with the source (e.g. look back at figure 2.4 when the search was on for Sco X-1!). However, with several X-ray sources that seemed to have a common origin or mechanism, it was possible to search for stars in the error boxes that were themselves of a similar type. The HEAO-1 source H0324+28 was noted to be near the well-known and bright optical variable UX Ari. It was in this way that these strong soft X-ray sources were identified with the group of stars known as RS CVn systems, of which UX Ari is an important member.

This somewhat abstruse name comes from a common astronomical naming convention. If a number of objects are found to have certain similar features (such as characteristic colours or timescale of variability, for instance) then that group of objects is named after the first one that was found with that particular characteristic. In this case, the star RS CVn was just such a prototype. *CVn* is short for the constellation Canes Venatici and *RS* indicates that it is a variable star which is listed in the massive Russian General Catalogue of Variable Stars (or GCVS). The GCVS catalogues all known variable stars and uses the notation of capital letters before the constellation to distinguish them from the well-known Greek letter listing of stars simply in order of brightness.

The characteristics of the RS CVn systems were laid down in a now classic paper by Doug Hall (Dyer Observatory) in 1972. Figure 4.4 shows a schematic of RS CVn itself. The main properties of these systems are summarised in table 4.1 and a list of the bright X-ray emitting RS CVns is compiled in table 4.2.

Table 4.1 Characteristics of the RS CVn binary systems

Binary period	1 - 14 days
Spectral type	K0 IV/V + main sequence star
Spectral features	calcium, hydrogen emission lines
Masses	$1 \sim 1.5\ M_\odot$
Optical brightness	6th ~ 9th magnitude
Optical *wave* amplitude	0.1 ~ 0.3 magnitude
Wave migration period	years
Colour	ultraviolet and infrared excess
Variability	radio and X-ray flares

Table 4.2 Observed properties of X-ray bright RS CVn systems

Name	Distance (pc)	Visual magnitude	Rotation period (days)	Spectral type	L_x (0.1–3 keV) (10^{30} erg s^{-1})
UX Ari	50	6.5	6.4	K0IV + G5V	21
HR1099	36	5.7	2.8	K0V + G5V	26
RS CVn	150	8.4	4.8	K0IV + F4III	19
AR Lac	40	6.9	2.0	K0IV + G2IV	15
LX Per	145	8.1	8.0	K0IV + G0V	6
Capella	14	0.1	104.0	G0III + G5III	4
HK Lac	139	6.5	25.1	K0III + FIV	14
σ Gem	59	4.3	19.5	K1III + ?	20

(These latter three are members of a *long-period* RS CVn group.)

Giant starspots

At first sight these RS CVn systems seem to have little in common with our Sun and its coronal X-ray emission. RS CVns are close binary systems, one of whose components is ageing and slightly evolved. However, the clue to the nature of their activity comes from the variability that put them in the GCVS in the first place. Figure 4.4(b) shows the light curve of RS CVn in 1975.

Both stars pass in front of each other at opposite sides of the orbit, producing the primary and secondary eclipses that are easily visible (the primary eclipse is almost a full magnitude in depth). The eclipses are a straightforward way of indicating a binary star and represent a *selection effect* in the way these stars were found in the optical. The X-ray emission of RS CVns, however, suffers no such effect and will enable many more non-eclipsing systems to be found. Given the sizes of the stars in figure 4.4 we would expect only about 10% of such systems to be eclipsing as observed from one particular direction. This implies that there are about 10 times more objects of this class of comparable optical brightness yet to be found.

The slight undulation or *wave* in the light curve is very important and is also easily visible in figure 4.4(b). This wave, surprisingly, does not have exactly the same period as the binary period. If the position of the minimum in the wave is plotted as a function of binary phase, this changes smoothly with time. It takes about 10 years in RS CVn for this migration to march completely through one cycle and back to the phase it started at. Whilst this wave migration is taking place, the amplitude of the wave is varying slowly with time. That in RS CVn changes from an amplitude of 0.20 to 0.05 magnitudes on a timescale of 20–25 years. This timescale is considered to be analogous to the solar 11 year sunspot cycle, and the wave to be caused by the presence of giant starspots on the surface of one of the two stars in the system (figure 4.5).

These starspots have to be gigantic in order to explain the amplitude of the wave that is observed, i.e. they must cover a significant fraction of the stellar surface. For comparison, sunspots occupy only about 0.1% of the Sun's surface, even at solar maximum, whereas these starspots must cover typically 20% of the star's surface. More importantly, they have to cover 40% of one

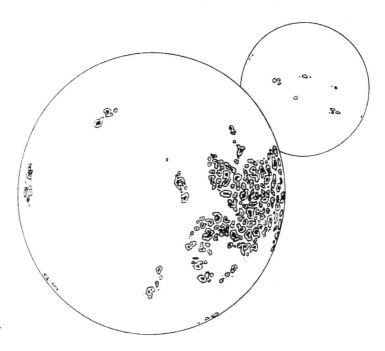

Fig. 4.5 Artist's conception of giant starspots and the RS CVn binary. These are all scaled up versions of actual features seen on the Sun (see figures 4.1 and 4.9).

hemisphere since the starspots cannot be uniformly distributed but must be concentrated on one half of the star in order to explain the light curve. And just like sunspots, which have cycles of activity that change every year, so do the starspots. The wave of the RS CVn system λ Andromedae changes its overall shape and structure on timescales of a year.

Observations of the wave and the binary motion show that these starspots are associated with the *subgiant* member of the RS CVn binaries (which is a K0 IV star in almost every system). This star has started to evolve off the main sequence because it is running out of its nuclear fuel (hydrogen) at the centre of the star. It starts to expand (on its way to becoming a red giant) and develops a much more substantial and deeper convective envelope which generates the turbulence. This star contains the giant starspots.

The starspots can explain much of the characteristics of the RS CVn systems. As *footprints* for even larger active regions they will account for the greatly enhanced chromospheric activity (the calcium and hydrogen emission lines) and, of course, for the bright X-ray coronae. To have been seen by HEAO-1 means that the RS CVn coronae are about 10 000 times brighter than the Sun. Given that it is not possible to resolve any of these stellar discs, is there any direct evidence for these giant starspots as sketched in figure 4.5? An ingenious new technique called *Doppler imaging* (see figure 4.6) has enabled astronomers to directly determine the distribution of starspots on a stellar surface.

The starspots in the diagram are derived directly from the line profiles that are shown under each map. The technique is based on the Doppler effect and is fully explained in box 4.4. The narrowness of the emission features shows that the spots are localised with respect to the disc and will give enhanced chromospheric emission lines immediately above the spot. These lines can then be resolved by high resolution spectroscopy as part of the much wider absorption line profiles that come from the entire stellar disc.

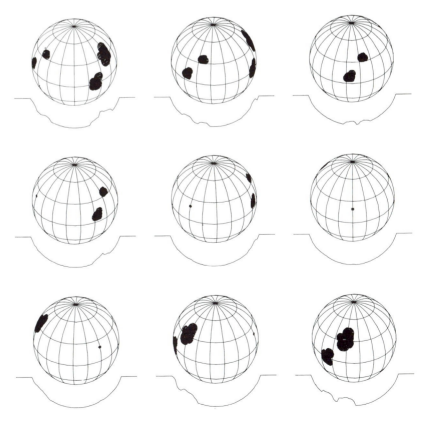

Fig. 4.6 Doppler imaging of the starspots on UX Ari. The rapidly rotating star produces a normally broadened absorption line (shown schematically here as a dish-like feature). However, the starspots are active regions confined to particular areas of the star's surface and hence giving rise to emission at a single velocity (Doppler shifted relative to us), rather than a range of velocities. The emission from each starspot starts to fill in part of the absorption line giving rise to the wiggles in the line profile. By following this structure as a function of time it is possible to map the movement of these spots across the star. (Based on a diagram by S. Vogt.)

Box 4.4 Doppler imaging of starspots

Figure 4.B4 shows a rotating star with a single large spot at three different times. The activity associated with the spot produces an emission line, whereas the rest of the stellar surface produces only absorption (broadened because it is coming from all parts of the surface which are travelling at different velocities relative to our line of sight). The emission line appears in the centre of the broad line when the spot is travelling directly *across* our line of sight, and at Doppler shifted positions when the spot is travelling towards or away from us, at the edges of the stellar disc. In other words, by studying the line profile at different times, the position of the spot on the stellar surface can be determined, hence the term *Doppler imaging*.

Fig. 4.B4 Schematic of a large spot on a rotating star and the resulting variable line profile it produces.

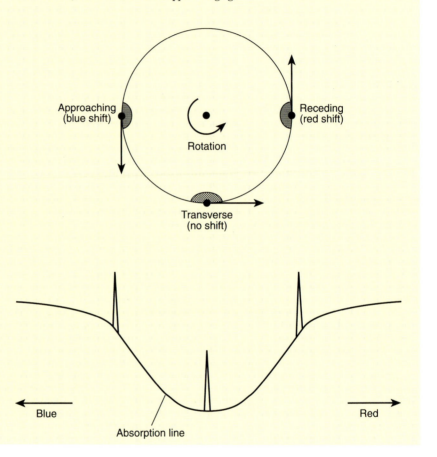

Coronal loops

These giant starspots will presumably support correspondingly giant coronal loops, scaled from those seen on the Sun, as depicted earlier in figure 4.2. This super active region has a hot plasma at about 10 million degrees contained by the powerful magnetic field. It is possible to scale very roughly the theoretical work done on the solar coronal loops up to the sizes required for the RS CVn systems. The temperature of the RS CVns is known quite accurately from the improved X-ray spectra obtained by the Einstein SSS and by the EXOSAT transmission grating. These are shown in figure 4.7.

The emission lines are only explainable by a very hot gas, as in the case of the solar X-ray spectrum (see box 3.4). For Capella there is emission by a range of Fe ions, indicating the presence of at least two spectral components, with temperatures of a few million and 10–20 million degrees or greater.

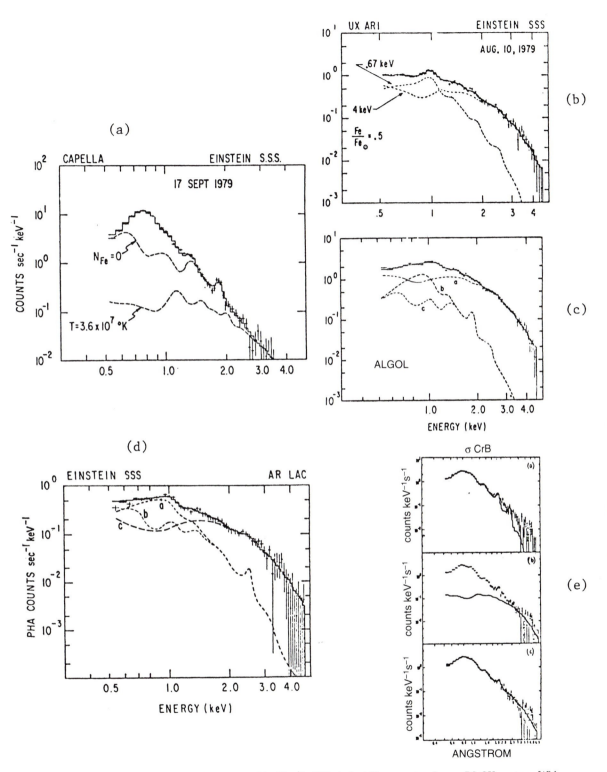

Fig. 4.7(a)-(e) X-ray spectra of active stellar coronae. The Einstein SSS obtained X-ray spectra of many RS CVn systems. With coronae at temperatures of 5–40 million degrees, the X-ray spectrum is ideal for study with the SSS which can resolve the highly ionised lines of silicon, sulphur and argon. In all cases the SSS spectra also reveal the presence of material at two different temperatures, almost certainly from different physical regions. Each temperature component is plotted separately under the spectrum and then combined as a best-fit model.

Fig 4.7(f) EXOSAT high
resolution X-ray spectrum of
Capella. Using the transmission
grating on EXOSAT it is
possible to resolve far more of the
highly ionised species in the
Capella corona which contains
gas at temperatures of 3 and
18 million degrees. (a-e from
GSFC SSS group led by Steve
Holt, f is courtesy of EXOSAT
Observatory, ESA)

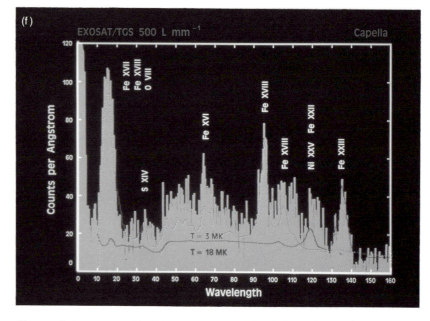

Coronal structure of AR Lac observed by EXOSAT

If the *wave* in the light curve of RS CVn systems is attributable to giant starspots on one hemisphere of the active star, then we would expect the X-ray emitting regions to be non-uniformly distributed around the stars also. How could this be tested? The answer is to examine a nearby, X-ray bright, *eclipsing* RS CVn system, in which the presumably less active component eclipses the active star. By observing the X-ray output during the eclipse it would then be possible to determine the structure of the X-ray emitting regions.

The ideal system for such a study is AR Lac, although its two components are both subgiants, suggesting that both may be X-ray active to some extent. This RS CVn system, first observed by HEAO-1, has a 2 day period and the properties summarised in table 4.3. In July 1984 EXOSAT observed AR Lac for a continuous period of 2 days, essentially covering one binary cycle completely, producing the X-ray light curve shown in figure 4.8.

The times of primary and secondary eclipse correspond to phases 0 and 180 degrees, and both are approximately 8 hours in duration. This indicates why this study was so difficult to execute before EXOSAT. A similar attempt by Einstein some years earlier was plagued by the many large gaps in the data due to occultation by the Earth itself. This led to a phase coverage of only 17% of that covered by EXOSAT.

No structure was seen at higher energies at the times of the eclipses, implying that the hotter (15 million degrees) component is extended compared to the size of each individual star, and probably surrounds both stars uniformly. On

Table 4.3 Properties of the RS CVn system, AR Lac

	Primary	Secondary
Spectral type	G2IV	K0IV
Radius ($R_{\odot}$)	1.54	2.81
Separation ($R_{\odot}$)	9.2	
Distance (pc)	50	
Inclination	90°	
X-ray luminosity (erg s^{-1})	10^{31}	

Fig. 4.8 (a) A 2 day EXOSAT observation of the AR Lac binary system with the low energy (LE) telescope. These data are most sensitive to gas at about 6 million degrees, and show both a primary and secondary eclipse and hence must arise in relatively compact regions or loops on both stars. They would then be comparable in size to those on the Sun, but much hotter and more luminous. The high energy (ME) light curve reveals no evidence for eclipses, and so the 15 million degree component of the corona must be larger in extent than the stars themselves. (b) Maximum entropy maps showing the surface X-ray emission from each component in AR Lac. This mathematical technique allows one to calculate the smoothest distribution of X-ray emission on the surface of each star which can reproduce the observed LE light curve. The model shown assumes an extended K star corona (height 3.9 $R_\odot$) and a compact G star corona (height 0.04 $R_\odot$). The light curve generated by this model is shown together with the actual data. A more conventional, but two-dimensional, view of AR Lac shows the location of these various regions.

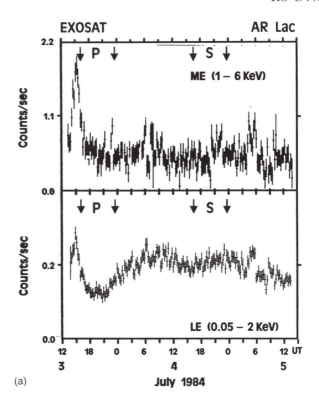

(a)

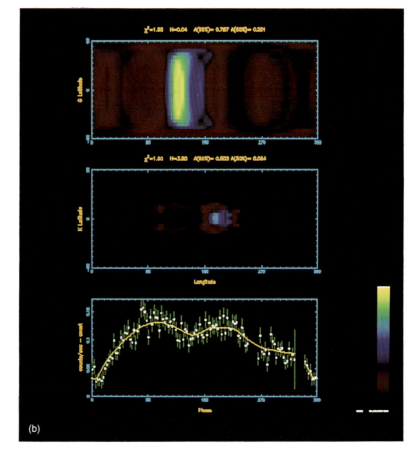

(b)

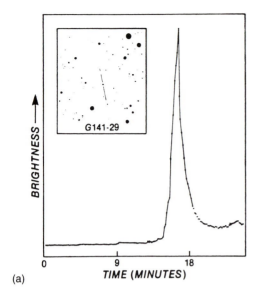

(a)

Fig. 4.9 (a) Optical flare from G141–29 and (right) an image of a solar flare in the principal visible line of hydrogen, Hα. G141–29 is a 13th magnitude red star which was observed in April 1980 by B. Pettersen (University of Tromso) to outburst by almost 5 magnitudes (a factor 100) in less than 2 minutes! At other times during these observations with the McDonald Observatory 2.1 m telescope, more than half of the light output from this star was found to be in the form of flares, making it the most active star of this class known. (b) Montage of X-ray flares as seen from Einstein and EXOSAT. Of the Einstein events shown, by far the most luminous is that from the Hyades star HD27130, whereas the closest star is, of course, Proxima Centauri. The EXOSAT flare light curve (taken in 1983) and spectrum of Algol, shows that it had a peak luminosity of 10^{31} erg s^{-1} and a temperature of 60 million degrees (the inset shows an X-ray spectrum of a later flare observed by the Japanese satellite, Ginga, in which the high temperature iron emission at 6.7 keV is clearly detected). The K subgiant in this system was responsible for the flare which reached a height of about 0.2 stellar radii. Gl644AB is a nearby visual binary in which both stars are of dMe spectral type, and this flare was observed simultaneously by EXOSAT and from the ground in Hα. The similar light curves further support the view that these are analogous to solar flares. (Based on figures by Bob Stern, Lockheed.)

the other hand, the low energy (LE) light curve, which records the behaviour of the cooler, 6 million degree component, shows evidence for structure associated with both primary and secondary eclipses, with a particularly deep primary eclipse. This implies that there must be a bright X-ray emitting region on the surface of the G star, which is compact compared to the size of the star. To account for the broad minimum offset from secondary eclipse, however, requires an extended emitting region above the K star. These are illustrated in figure 4.8(b) by maps of the stellar surface regions produced from these data by a group led by Nick White of the EXOSAT Observatory. The technique used (*maximum entropy method*) derives the smoothest distribution of X-ray brightness across the stars which is consistent with the observed light curves.

The bright, compact region on the G star is likely to be coronal loops similar to those on the Sun, but at a higher temperature and over 1000 times larger than those on the Sun. It is now becoming clear that the magnetic fields play a crucial role in the activity of the RS CVn systems and this strongly supports the magnetic heating model of figure 4.2. However, we have not yet addressed the question of why these loops are so much larger and more powerful than in the Sun, and what energy source is driving them.

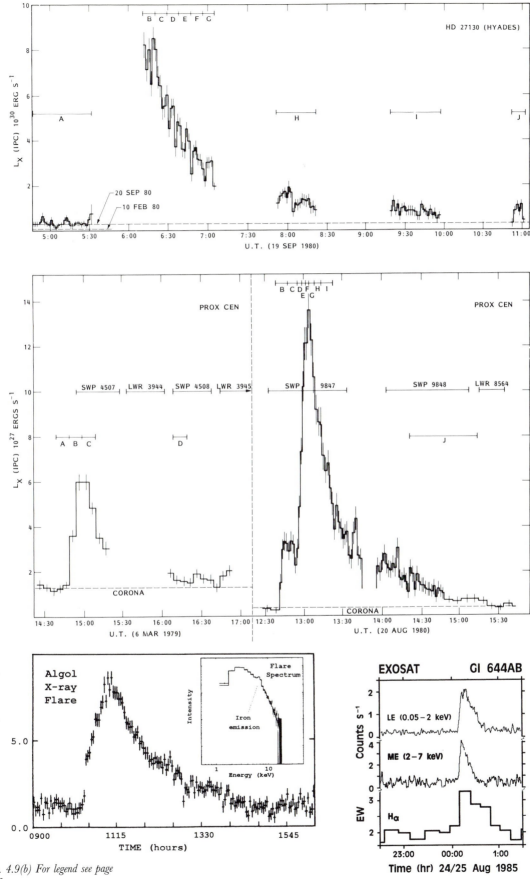

(b)

Fig. 4.9(b) For legend see page 112.

4.4 Flare stars

Even before the discovery of X-radiation from the RS CVn systems, X-rays had been found by the Dutch/US satellite ANS from two optical flare stars. These stars produce truly dramatic visible light flares as shown in figure 4.9, and are believed to be much scaled-up versions of solar flares.

It is remarkable that these stars are otherwise some of the smallest and dimmest objects known! They are all *dMe* stars. Spectral type M corresponds to an intrinsic optical brightness at least a hundred times fainter than the Sun and masses of only a fraction of that of the Sun (hence the *d* for dwarf). M type stars are however the most common object in our galaxy, probably outnumbering all other stars put together. Indeed, 40 of the nearest 63 stars are M type. The *e* suffix is added because emission lines have been seen in their spectra, the first indication of some kind of activity.

Observing optical flares requires patience and, usually, short integration times since they are often over within a matter of minutes. One of the most well-known flare stars, UV Ceti, was not seen flaring until 1948, although archival plates at Harvard were searched to show that similar flares had in fact been occurring since 1900. There are now well over a hundred flare stars known.

Before the Einstein Observatory none of the known flare stars were detected as continuous X-ray sources. But given the similarity of stellar flares to solar flares (but scaled up!) several satellites took part in campaigns to search for X-rays during flares, the ANS being successful. Perhaps the best X-ray flares though have been seen by Einstein, and more recently EXOSAT, and a collection of them is also shown in figure 4.9.

A link between flare stars and RS CVn systems is provided by yet another group of stars known as BY Dra (constellation Draco) stars. These are also flare stars but are short period binaries, one component of which is K or M and has a modulation or wave in its light curve (remarkably similar to the RS CVns). A flare in such a system might look like that depicted in figure 4.5 for

Fig. 4.10 The Orion Nebula in X-rays as observed with the Einstein HRI. The bright central source is θ¹C Orionis, an O star. The other sources are nebular variables, stars just recently formed. The field shown is 17 arcminutes square.

Fig. 4.B5 Principal components
of the Orion Nebula region.

Box 4.5 Star formation

Our understanding of how stars form has undergone radical changes with
the explosion of new observations in the last 20 years, particularly in the
infrared. The Orion Nebula is now known to be an extremely complex
region of collapsing gas clouds (cool molecular clouds) and outflowing material, the main components of which are shown in figure 4.B5.

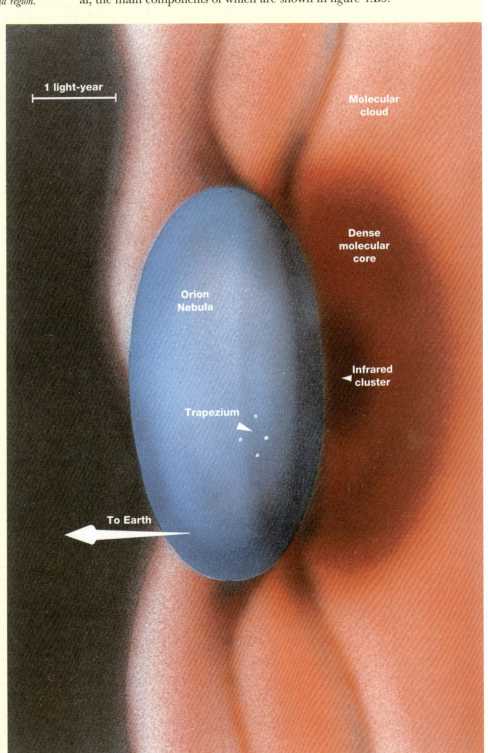

The conservation of angular momentum as a slowly rotating cloud collapses defines the rotation axis of the star, along which jets of outflowing material are ejected. The passage of a star from these early beginnings onto the main sequence can be long and arduous with frequent outbursts and erratic variability. Its track in the H–R diagram is shown in figure 4.B6 along with theoretical evolutionary tracks.

Fig. 4.B6 Evolutionary tracks of a star onto the main sequence.

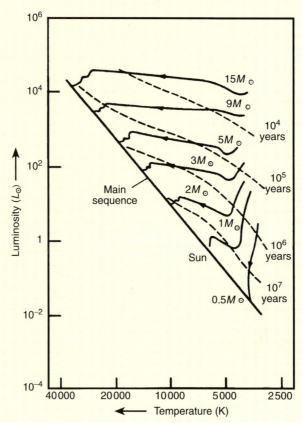

RS CVn since it must employ the giant coronal loop structures that have been invoked to account for the RS CVn X-ray coronae. It was once thought that binarity was an essential element of BY Dra stars (as it is in RS CVns), but recently single BY Dra stars have been found. It now appears that the key feature is *rapid rotation*.

In close binary systems this is easy to produce because each star rotates synchronously with the orbital period (in exactly the same way that the Moon rotates once in each lunar month so that it always keeps the same face towards the Earth). The orbital periods are so short (of order of days) that the rotational speed is correspondingly fast when compared, say, with the rotation period of the Sun which is about 27 days. If a BY Dra star is not a member of a binary then it has to be concluded that the star was formed rapidly rotating and has not yet slowed down. What evidence is there that rotation plays a significant role in stellar activity and what is the mechanism by which it works?

4.5 Young stars

Stars are believed to form from large interstellar gas clouds that collapse under their own gravitational forces. Theoretical calculations of such a collapse are very difficult, but observationally we are seeing earlier and earlier in

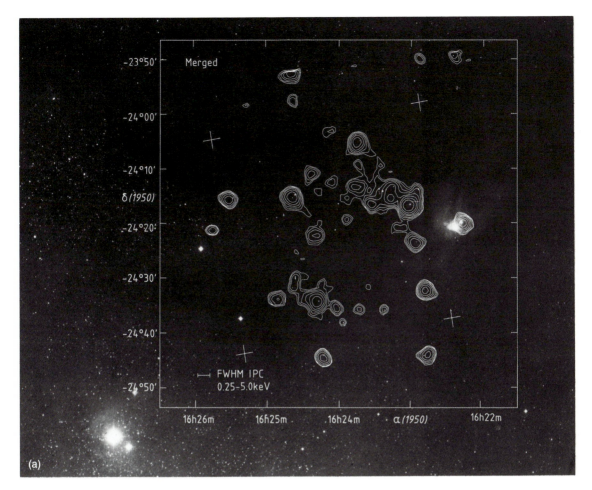

(a)

Fig. 4.11 Montages of X-ray images of ρ Ophiuchi (with Einstein) and the Hyades (with ROSAT), the former superposed on optical photographs. The ρ Oph box outlines one square degree and contains contours of constant surface brightness from the IPC. About 30 X-ray emitting nebular variables are detected. The bright star on the right is HD147889, that on the left is SAO184429, both B stars. The optical picture is from the POSS red print. (Pictures courtesy of Thierry Montmerle, Section d'Astrophysique, Centre d'Etudes Nuclaires de Saclay, and John Pye, University of Leicester.)

the life cycle of stars. This has particularly come about with the advent of sensitive infrared telescopes and instrumentation which can detect the very cool clouds and the protostars embedded within them. For example, one of the most well-known and most spectacular regions of star formation is in the Orion Nebula (see box 4.5).

The Einstein X-ray picture (figure 4.10) of this region shows many active stars which, although still forming, are undergoing nuclear burning but are not yet on the main sequence. Hence the astronomical term for these objects: *pre-main-sequence stars*. The H–R diagram for a star cluster such as the one in Orion shows that it contains a large number of young stars. H–R diagrams for other clusters show whether they are older or younger by the fraction of stars they possess that are still approaching the main sequence (see box 4.5).

An older star cluster, where most of the star-forming material has been used or dispersed is the Hyades, which has been the focus of a major Einstein observing programme. Because of the large angular extent of the Hyades on the sky (5 degrees) and the only 1 degree field of view of Einstein, several Einstein X-ray pictures were taken and then superimposed in a montage. A similar technique was used on the ρ Ophiuchi Dark Cloud star-forming complex. Both are shown in figure 4.11.

These two surveys were some of the most exciting results to come out of the Einstein X-ray Observatory. Almost 50% of the Hyades member stars observed were detected and as many as 70 X-ray emitting objects in the ρ

Fig. 4.11(b) For legend see page 117.

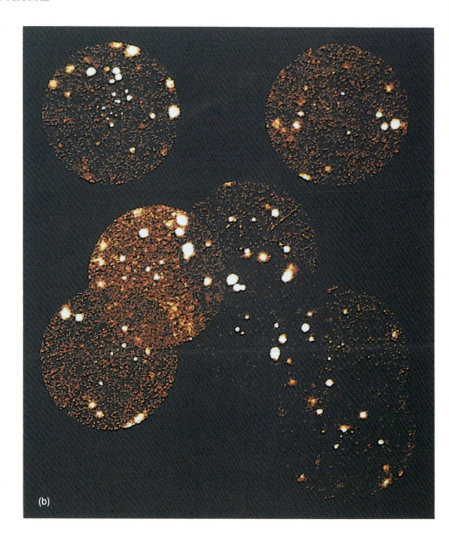

(b)

Oph Dark Cloud. To detect the Hyades stars at all required them to be emitting an X-ray flux at least 10 times that of the active Sun! In addition, both surveys showed activity in the form of stellar X-ray flares and large long-term variations in X-ray intensity, as is evident in figure 4.12.

Since the ages of these clusters are fairly well known from numerical models of stellar evolution we can see how the stellar activity (in the form of X-ray emission) varies as the stars age in a diagram (figure 4.13) which shows the evolution of the X-ray output.

4.6 Rotation – the evolutionary link?

We have now presented samples of the X-ray observations of what appears to be a very diverse range of stellar systems and types. They are linked, however, by the single most important parameter that determines the level of activity, the speed of rotation. The correlation of X-ray luminosity with a star's rotational velocity is quite graphically demonstrated in figure 4.14 for all the stars mentioned in this chapter, whether they are single or binary.

But why, as shown by the young cluster surveys, do stars' activity decrease with age? Probably because a star's speed of rotation slows down as it gets older. It does this by losing angular momentum through its stellar wind, a constant stream of material flowing out from the star just as in the case of the

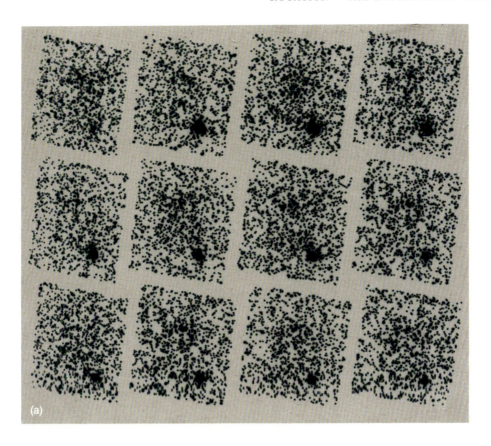

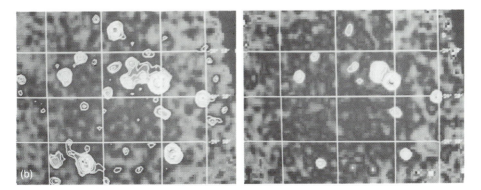

Fig. 4.12 The sequential Einstein IPC images of (a) the X-ray flare in the Hyades binary HD27130 and (b) 2 day variations in the ρ Oph Dark Cloud. The ρ Oph images were taken in February 1981 by Thierry Montmerle.

solar wind. This implies, probably correctly, that the Sun was spinning faster when it was younger and, at that time, had a much more active and more powerful corona. A much stronger X-ray flux from the Sun (possibly a thousand times stronger) may well have influenced the evolution of the Earth's atmosphere and the time at which life began on Earth.

The fact that the RS CVn systems are much older (indeed they are evolving off the main sequence, not onto it) does not matter here. They acquired their high rotation through being in a binary system and being forced into synchronous rotation. But why exactly does rotation have such an important effect?

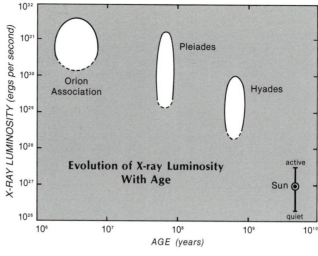

Fig. 4.13 Evolution of X-ray luminosity with age. The youngest objects are the most active, with luminosities as much as 10 000 times that of even the active Sun. This clearly declines as they age. (The dashed lines indicate uncertain data.)

X-ray Activity

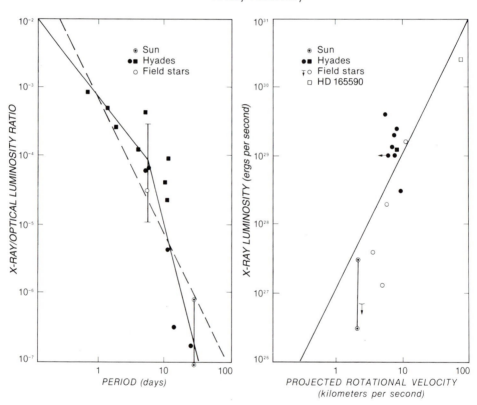

Fig. 4.14 Correlation of X-ray luminosity with rotation speed. The left-hand box shows how the ratio of X-ray to optical luminosity falls with increasing binary period, whereas the right-hand box shows the X-ray luminosity as a function of actual rotational velocity of the star. (Both these figures from Bob Stern, Lockheed.)

4.7 The dynamo model

The clue again comes from watching the surface of the Sun. It is well known that the Sun's rotation period cannot be stated exactly because it is rotating differentially. This is obvious by looking at the rotation periods of sunspots as a function of latitude on the Sun's surface. At the equator the angular velocity is almost half again as much as what it is near the poles. The net result of such a shear is dramatic on any magnetic fields that exist in the solar surface. The magnetic fields are created, it is believed, by a dynamo mechanism that is similar to that used to account for the Earth's magnetic field. The convective zone below the Sun's surface creates the turbulent cells which draw up

Fig. 4.15 A theoretical model of RS CVn magnetic field configurations developed by Uchida. Such complex fields could develop given the rotational locking of the two stars that occurs because of gravity.

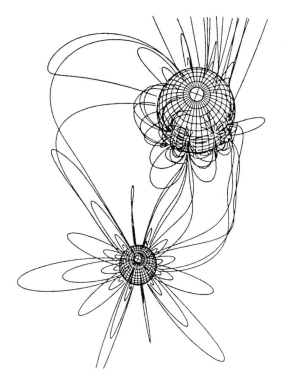

magnetic flux and produce the coronal loops shown in the magnetic heating model of figure 4.2.

Now we can see why the RS CVn systems have the most powerful stellar coronae of all. Not only are they rotating rapidly, which will generate a large magnetic flux by this dynamo mechanism, but, since they are evolving and expanding, they also have an extra deep convection zone which will generate yet more powerful magnetic fields. Models for spots on flare stars require magnetic fields of at least 20 000 gauss, 10 times that in sunspots. This translates into the giant starspots that we see in the stellar waves, and these support the huge coronal loops, sometimes comparable to the size of the star itself, which contain the hot X-ray emitting plasma.

This presents a nice all-embracing picture and we must emphasise that there are lots of details yet to be filled in, such as what keeps the RS CVns' giant starspots located on just one hemisphere. There are also major problems that have to be explained by any model, such as the fact that, being the brightest X-ray coronae, the RS CVn systems also have an extremely hot X-ray component corresponding to temperatures of typically 50 million degrees. This is hard to explain in the above scenario. It would require coronal loops not just the size of the subgiant but comparable to the size of the binary system itself! Perhaps a theoretical picture (figure 4.15) of interacting magnetic fields between the two stars is on the right track.

5 Early-type stars and superbubbles

5.1 O stars

The most luminous, most massive stars are the O stars. Starting with more than 25 $M_\odot$ of material, they burn their nuclear fuel at a prodigious rate. They live only a short time and end in a brilliant supernova explosion. The surrounding space is left full of stellar debris enriched in heavy elements. We all contain in our bodies elements made in these massive stars.

These are not common stars and none are nearby. The brightest ones visible to the naked eye are δ and ζ Orionis at the two ends of Orion's Belt. Both are 1600 pc distant and spectral type O9.5. Zeta Puppis is 2400 pc distant and type O5. Because the nuclear fuel is consumed rapidly, the lifetime is, astronomically speaking, short. In a few million years an O star changes character, becoming perhaps a red giant or a Wolf–Rayet star. We see, with naked eye or telescope, only the outer layer which gives little information about events in the core. Hidden from view, the central region evolves rapidly until the nuclear fuel is exhausted. As explained in chapter 3 on supernova remnants, the core collapses and the gravitational energy released powers the resulting supernova. That is the end of the O star.

Astronomers originally believed all stars evolved along the main sequence. In this scheme a star would start life as a hot O star and, as it aged, would change into progressively cooler spectral types. The O stars and B stars were thus called 'early' stars. Although this idea has long been abandoned, the nomenclature lives on! Hot stars are 'early-type' and cool stars are 'late-type'.

The energy output of O stars is enormous. A star of spectral type O5I has a luminosity 5×10^5 that of the Sun. If such a star were as close to us as Arcturus, a distance of 10 pc, it would have a visual magnitude of -7, and would appear 20 times brighter than Venus at its brightest! Since the surface temperature is 40 000 K, most of the energy is radiated in the ultraviolet. Only 4% of the energy is found in the visible band. Energy is also ejected in the form of a hot stellar wind. A mass loss of $10^{-6} M_\odot \, \mathrm{yr}^{-1}$ at a terminal velocity of 3000 km s^{-1} is not unusual. The mass loss of the sun, $10^{-14} M_\odot \, \mathrm{yr}^{-1}$, is puny by comparison. This solar wind, which forms the tails of comets, has a velocity of 300 km s^{-1}. It is only a gentle breeze compared with the hurricane in space surrounding an early-type star.

All O stars are apparently X-ray sources and consistently radiate a fraction (1 to 2×10^{-7}) of their bolometric luminosity as soft X-rays. These are thought to be coronal X-rays produced in the strong outflowing wind but the mechanism has yet to be explained. The most luminous O star known in our galaxy is HD93129A, located in the Carina nebula. The bolometric luminosity of this star is 2×10^{40} erg s^{-1}. The energy radiated as X-rays, although an insignificant part of the output of this star, is almost equal to the total output of the sun at all wavelengths!

O stars have a profound effect on their surroundings. After the star is formed, the energy generated by the star soon ionises and sweeps away any material remaining in the vicinity. It is the strong stellar wind which pushes matter outward. The physics of the region around the star was first worked out by John Castor, Richard McCray and Thomas Weaver (University of Colorado). They predicted a central region with wind flowing freely, a surrounding volume of hot (perhaps X-ray emitting) gas, and a relatively cool and dense outer shell. This shell contains most of the mass and should expand slowly into the interstellar medium. A 'bubble' is thus formed around the star and is predicted to have a diameter of 10–30 pc. An example of this is illustrated in figure 5.1 which shows a shell around a Wolf–Rayet Star – an evolved O star with a particularly massive stellar wind.

Fig. 5.1 A bubble around a Wolf–Rayet star. The diffuse nebula NGC 2359 in Canis Major. Blue-white filaments outline the shell of the bubble. The red material outside is ionised by radiation from the star and has not yet been swept up by the stellar wind. (Photo by David Malin. © Anglo-Australian Telescope Board.)

O stars are not scattered randomly throughout the galaxy. Most are found in rather loose groups or clusters called OB Associations which delineate the spiral arms of our galaxy. They are usually close to clouds of dust, nebular variables, and other indications that star-formation is in progress. This is expected. Since O stars do not have long lives, they do not usually drift far from their birthplace.

The first detection of X-rays from O stars occurred on December 15, 1978. On this occasion, three weeks after launch, the Einstein telescope was pointed at the bright X-ray binary Cyg X-3. This source was to be used to calibrate the IPC detector. The appearance of O stars was an unexpected result, a thrill for all the scientists examining these first pictures, but a shock for Rick Harnden (CfA), the person responsible for the calibration of the imaging proportional counter.

This was the first operation of this detector in space, at the focus of the telescope and pointed at a bright X-ray source. There had been two years of laboratory calibrations. All the variable instrument parameters had been set hopefully to give optimum and reliable operation. Here was the first opportunity to observe the detector response to a bright cosmic source. The source, Cyg X-3, was clearly visible in the field and the spatial spread caused by instrumental effects was about as expected, but it was with sinking heart that Harnden first contemplated the confusion in the centre of the field (figure 5.2).

Fig. 5.2 The first O association discovered to emit X-rays. The brightest source (heavily overexposed to the lower right) in this Einstein IPC field is Cyg X-3. The other, weaker, X-ray sources are bright O stars in the Cyg OB2 association.

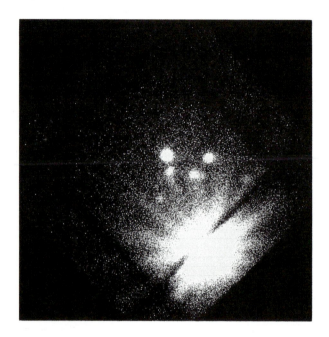

Several irregular bright spots (looking particularly abnormal when viewed with the quick-look TV monitor at the control centre) appeared in a region of the sky thought to be free of sources. An anxious three days passed, filled with thoughts of 'hot spots' caused by dirt on the wires within the proportional counter (and impossible to clean at this point). All ended happily with the realisation that the detector was operating properly and that this emission was real. When the X-ray picture was overlaid on the Palomar Sky Survey (figure 5.3), there was a bright star at the position of each X-ray bright spot in the center of the field. The pattern of stars was the same as the pattern of X-ray spots, an unmistakable identification. Thus O stars were found to be strong X-ray sources. These particular stars are all members of the Cyg OB2 association which is highly reddened. It is 2 kpc distant and X-rays are attenuated by a factor of 5–10 in their journey to the vicinity of Earth. Only because the stars are so very luminous were the X-rays detected in this observation.

5.2 The Carina Nebula

Located in the far southern sky, the Carina Nebula is one of the brightest optical emission nebulae in the Milky Way. The nebula subtends about one square degree and can easily be seen with the naked eye. A photo through a

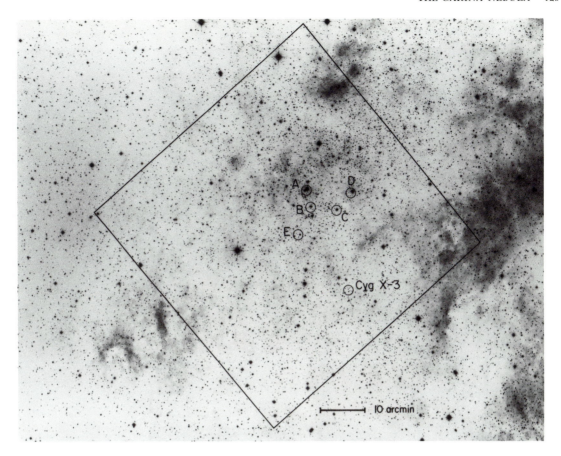

Fig. 5.3 The Cyg OB2 association as it appears in the red Palomar Sky Survey. Circles show X-ray sources prominent in figure 5.2 and the box indicates the boundary of the IPC field of view. (Courtesy of F. R. Harnden Jr., CfA.)

telescope (figure 5.4) shows detailed structure within the nebula and many bright O stars (of magnitude 6–9). This Nebula contains a collection of the hottest and most massive stars known, all of spectral type O3. There are also three Wolf–Rayet stars and the peculiar object η Carinae, which radiates most of its energy in the infrared. Eta Carinae probably contains the most luminous single star known in our galaxy. Two prominent dust lanes give the Carina Nebula an easily recognisable dark 'V' just below and on either side of the brightest region. There is general agreement that the Carina Nebula contains shells of rapidly moving gas and dust. The great clouds of gas are illuminated by UV radiation from early stars and reradiate much of this energy as optical emission lines. Here is an area in which massive stars have just been, and are still being, created.

The Einstein X-ray pictures of the Carina Nebula were surprising (figures 5.5 and 5.6). Not only was η Carinae itself detected but also strong diffuse X-ray emission extending over most of the nebula. Many of the massive stars, including all of the most luminous O stars, show as unresolved X-ray sources.

What do these data imply? The mass of the diffuse optical-emitting gas was already known. It had been calculated from the strength of the observed emission lines. This material is mostly hydrogen and fills a region about 50 pc in diameter. This is the stuff from which stars are being made. There are $3 \times 10^5 \ M_\odot$ of visible matter. The dark clouds are presumably much denser and might contain at least this much material. Therefore an estimate of $10^6 \ M_\odot$ for the mass of the nebula is not unreasonable.

The diffuse hot gas which emits X-rays is interspersed with this cooler material. The hot gas and the cool gas are not mixed uniformly but co-exist in irregular filaments and patches. The pressure of a cloud of hot gas is

Fig. 5.4 A colour photograph of the Carina Nebula taken with the CTIO 61 cm Schmidt telescope. North is up and East to the left. The emission region shown is about one degree in extent and the nebula is one of the brightest emission regions in the southern Milky Way. (Photo © AURA.)

roughly equal to that within the denser cooler material. The brightest regions of diffuse X-ray emission have a size of about 1 pc and a density of about 1 atom cm^{-3}. The size and density of the hot regions, however, vary greatly throughout the nebula. In any case, only about 30 $M_\odot$ of hot gas are required to account for the diffuse X-ray emission. Only a small fraction of the material in the nebula is in the form of hot gas.

The energy and mass needed for this diffuse hot gas are easily supplied by the winds of the early-type stars. The kinetic energy of outflowing stellar winds is thermalised upon collision with surrounding cooler material. The hot stars soon create high pressure regions which change the morphology of the cool clouds from which they formed. Thus in the Carina Nebula the expected 'bubbles' around O stars exist but, because the star-forming cloud is dense and irregular, the appearance is not simple. Only if the star were created in a uniform medium, with no dense clouds, would a large spherical bubble be formed by the wind. The importance of the X-ray picture shown in

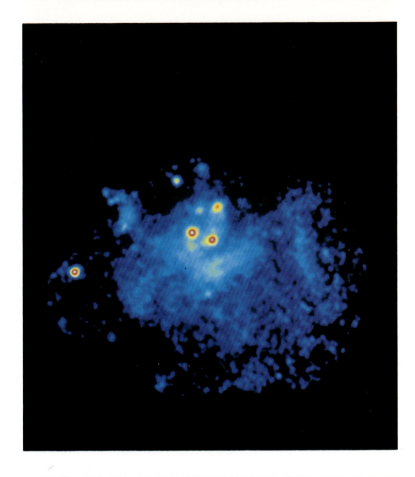

Fig. 5.5 An X-ray picture of the Carina Nebula made from four two-hour observations with the Einstein IPC. The easternmost source (at the far left) is an X-ray pulsar with period 6.4 s and is not associated with the Carina Nebula. The pair of bright sources at the center are η Carinae (on the left) and a Wolf–Rayet star, HD93162 (on the right). All other sources are O stars. The strong diffuse emission is probably powered by stellar winds.

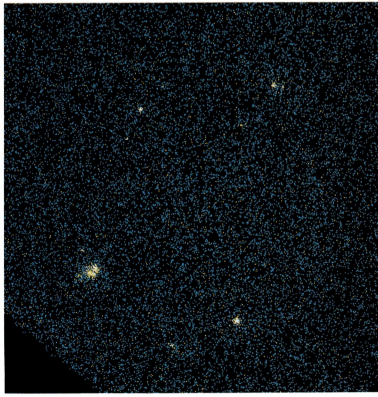

Fig. 5.6 A high resolution X-ray observation of the brightest part of the Carina Nebula from the Einstein HRI. The field shown is 17 arcminutes on a side. Point sources are all O stars, the cluster Tr 14 (upper right) is resolved into its two brightest members, and two fainter pairs of stars can also be seen. Eta Carinae appears as a diffuse source about 1 arcminute in extent. This field illustrates the power of an X-ray telescope with few-arcsecond resolution.

Fig. 5.5 is that it is an observation of the hot gas, formed far from the stars by the energy carried in stellar winds. This will eventually displace the star-forming cooler material from a large volume surrounding the cluster of stars.

As a digression of historical interest, this field is a nice illustration of the power of the X-ray telescope. X-rays from this entire complex were first detected in the rocket data shown in figure 1.4. The small maximum labelled 'η' in the middle energy band was identified as a barely detectable source. Thus one of the weakest sources seen in 1970 has now been resolved into about 20 point-like stellar sources and a large region of diffuse emission.

5.3 Eta Carinae – an O star in hiding?

In the middle of the brightest part of the nebula lies the variable star η Carinae. It was first noticed as a bright variable star, and we now know it to be an extremely luminous object obscured by dust. It is probably unique in our galaxy. To illustrate its unusual behaviour a light curve spanning 150 years is shown in Fig. 5.7. Initially of 4th to 2nd magnitude, the star brightened to 1st magnitude in the 1830s and reached a maximum brightness of −1 in 1843. It remained above 1st magnitude until 1857, declined steadily from 1st to 7th magnitude between 1857 and 1869, and has varied irregularly over the last 100 years between 6th and 8th magnitude. No other star has been discovered which behaves in this fashion.

Eta Carinae is not an unresolved source of light. It now forms the bright core of a small optical nebula, the 'homunculus' (meaning 'little man' because such is its shape) (figure 5.8). This nebula is small with dimension only 12 × 17 arcseconds and contains several bright knots. It is expanding with a velocity of about 500 km s^{-1}, consistent with this material being ejected during the 1843 or later outbursts. The homunculus is in turn surrounded by a faint outer shell of dimension 22 x 45 arcseconds. This also contains discrete optical features expanding with velocities of 300–1300 km s^{-1}. There was apparently an earlier event or phase during which material was thrown off at moderate velocity.

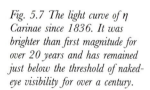

Fig. 5.7 The light curve of η Carinae since 1836. It was brighter than first magnitude for over 20 years and has remained just below the threshold of naked-eye visibility for over a century.

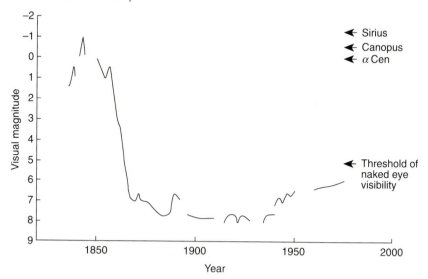

Viewed in the infrared, η Carinae becomes spectacular. At a wavelength of 20 microns, the homunculus is the brightest extrasolar infrared source in the sky. At a distance of 2600 pc, the bolometric luminosity is $6 \times 10^6 \, L_\odot$. The source of this emission is a shell of dust surrounding a central massive star (or perhaps stars). The dust absorbs optical and ultraviolet energy radiated by the star and reradiates it in the infrared band. Furthermore, optical emission

Fig. 5.8 Eta Carinae: (a) This HST picture shows a bright optical centre (overexposed in this reproduction) surrounded by yellow-white dusty material ejected in a 'bipolar' pattern. This is in turn surrounded by a more ragged, red, shell. This outer shell is composed of nitrogen and other material that has been ejected from the star. The highest velocity material appears as irregular clumps at the edge of the field. (b) The Einstein HRI image shows that the high-velocity outer shell is a strong source of X-rays. The central star, or material in its near vicinity, also emits X-rays. To match the scale of the HST picture, the X-ray picture should be expanded by a factor of two in size. (HST picture courtesy of STScI and J. Hester, University of Arizona.)

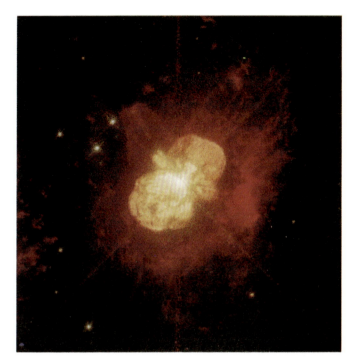

(a)

(b)

from the homunculus is strongly polarised, as expected if the light is being scattered from dust. There has been considerable speculation about the nature of the luminous central object, but since the shell is thick enough to prevent direct observation, the source of the energy remains a mystery. Proposed candidates include a very young star approaching the main sequence, a binary system, and a single massive star with surface temperature about 30 000 K – an O or B star. Nolan Walborn (STScI), Ted Gull (GSFC), and Kris Davidson (University of Minnesota) have suggested that this massive star may well become the next galactic supernova. The great luminosity implies a very short lifetime.

The 1843 outburst was indeed bright enough to be a supernova but lasted two years rather than the usual two weeks. It was thus called a 'slow supernova' by A. D. Thackeray, who made an extensive study of this object in 1956. The behavior of η Carinae has recently been better explained in terms of an expanding shell by Davidson. According to this model, a single massive star radiating most of its energy in the UV, ejected a thick shell of material. This soon expanded to the point where UV radiation from the star absorbed by the surrounding shell was downshifted in energy to the visible band. Since this material stopped all radiation from the underlying star, the maximum visible luminosity was close to the bolometric luminosity of the star. The shell has since expanded further. Dust grains have formed, and the radiation from the shell is now downshifted to the infrared. James Westphal and Gerry Neugebauer (CIT), for example, have pointed out that the infrared luminosity is now the same as the maximum optical luminosity observed in 1843.

It is important to remember that η Carinae is not an isolated star. It is embedded in a bright emission nebula which contains many luminous, early stars. Some of these, the O3 stars, have bolometric luminosities greater than 10^{40} erg s^{-1}, approaching that of η Carinae itself. In this neighbourhood, η Carinae is not out of place. Only the presence of circumstellar material and the 1843 outburst have given it a unique appearance and an air of mystery.

Eta Carinae itself is an X-ray source but not a strong one. The X-ray luminosity is less than 10^{-6} of its total luminosity. The Einstein HRI picture shows an extended, elongated X-ray source with several bright features. The size and shape are about the same as the 'outer shell' observed at optical wavelengths and the X-ray bright spots are not far from bright optical knots. In analogy with supernova remnants, the soft X-ray emission can be explained as from a shock generated by an expanding shell, consistent with ideas derived from optical data.

There is also a maximum of X-ray emission, which looks like an unresolved source, at the exact centre of the homunculus. This might be a sighting of the central object and, if so, argues for a single massive star as the source of energy. Thus there is a model for η Carinae which is consistent with observations at all wavelengths.

X-ray observations have the potential for both 'seeing' inside the shell and determining characteristics of the shell itself. For example, a good spectrum of X-rays from the central source would measure the thickness of the shell. We must await the AXAF mission, with its imaging spectrometer, to make this observation.

5.4 The Cygnus superbubble

The study of our galaxy's interstellar material has been radically affected by space-based observations. Interstellar space is now believed to be filled with a dilute, hot gas. This tenuous material has a density of 10^{-3} particles cm^{-3} and temperature near 10^6 K. Cool, dense clouds with density 10 atoms cm^{-3} and temperature 100 K are embedded in this sea. These cool clouds are probably surrounded by a warm gas with temperature of 10^4 K. All this is in approximate pressure equilibrium.

Fig. 5.9 (a) The observed X-ray superbubble. The X-ray brightest areas are yellow and the darkest, blue; black denotes regions for which no data were received. (Picture by F. Walter, UCB.) (b) A schematic of the features, drawn to the same scale. The area covered is $27° \times 34°$ – a large area of sky! The contours show X-ray emission, while red indicates filaments of nebulosity. A cross shows the position of the Cyg OB2 association, which is encircled (dashed line) by the Cyg X radio source. (Figure by W. Cash, Boulder, and PAC.)

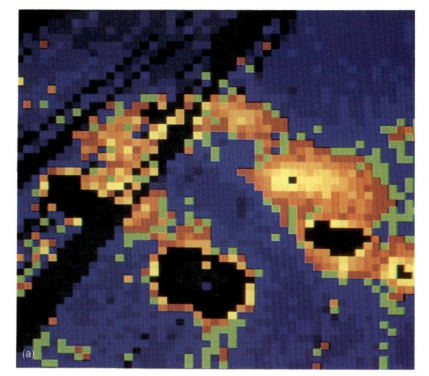

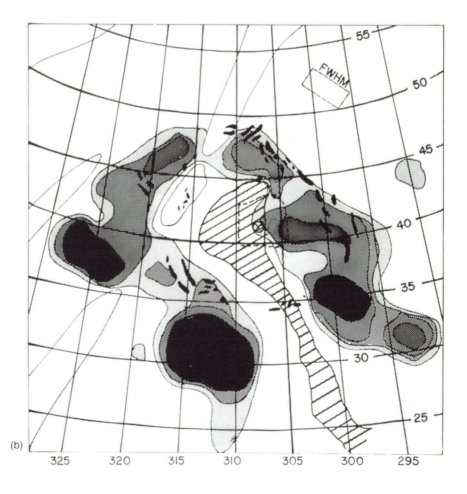

The largest known region of hot gas, a 'superbubble' more than 400 pc in diameter, was discovered in 1980 in the constellation Cygnus. It is perhaps the largest known object in our galaxy. (In astronomy the term 'object' embraces many phenomena, usually referring to a spatially isolated source of radiation at a particular wavelength.) Its properties reveal much, not only about the interstellar medium, but about the birth and evolution of stars. The superbubble was discovered through its X-ray emission.

It was found using one of the HEAO-1 instruments; a set of large-area soft X-ray detectors designed to map the soft X-ray background. During the analysis of the HEAO-1 scans of the sky, a previously uncatalogued source in Cygnus was noted by Webster Cash (Colorado) and one of us (PAC). It was of low temperature, obviously extended, and appeared to be a supernova remnant. But when adjacent 3°-wide scans were examined a surprising result was discovered. The source was not isolated like other supernova remnants, but appeared to be part of a much larger structure. A colour-coded map of this region is reproduced in figure 5.9.

This map encompasses most of Cygnus, and therefore shows several other well-known X-ray sources, including Cyg X-1 (a black hole candidate), Cyg X-2, the Cygnus Loop, and a lesser-known supernova remnant, G 65.2+5.7. The most striking feature, however, is a nearly complete ring of X-ray emission 13° by 18° in extent! The spectrum of this huge object indicates a temperature of 2 million degrees. Moreover, there is a substantial amount of cold interstellar material between the ring and us. Most of the X-rays are absorbed before they reach Earth, which, considering the location, is not surprising.

Dominating the northern Milky Way in summer is a celestial feature familiar to amateur observers – the long dark lane known as the Great Rift of Cygnus (figure 5.10). On a clear night this rift appears to split the Milky Way

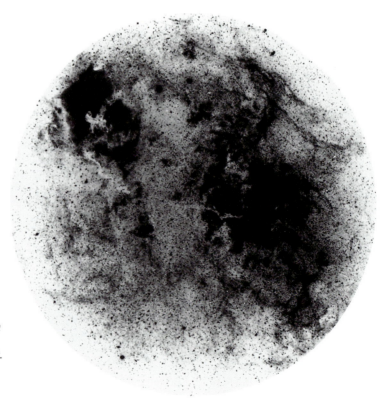

Fig. 5.10 The constellation Cygnus photographed in the light of the hydrogen alpha emission line, and showing the Great Rift. (Figure by W. Cash, Boulder, and PAC.)

down the middle, from Aquila in the south to as far north as the star Deneb in Cygnus. The Rift is caused by a massive cloud of gas and dust, 200 by 400 pc across, containing enough material to make millions of stars and thick enough to obscure the Milky Way beyond. The Rift cuts in front of the giant X-ray ring just where it is broken and hides the centre from our view. The ring shape is presumably caused by the shadow of this absorbing gas and dust on a giant bubble of approximately uniform emission.

This region of the sky also contains numerous wreaths of nebulosity. These show most clearly when photographed through an Hα filter (6563 Å) and are thus separated from the background starlight. Of special interest are long string-like filaments, normally associated with the passage of a shock wave through the interstellar medium (as in the old supernova remnant shown in figure 3.15). When these filaments are superimposed on the X-ray brightness contours of the giant ring the alignment is excellent. For 20 years astronomers have been searching for the energy source which created and which lights up these filaments, so the link with the X-ray ring is particularly gratifying.

Another phenomenon spatially linked with the ring is a shell of neutral hydrogen. Using a radio telescope to map the distribution of 21 cm line emission, Carl Heiles (UCB) has studied the large-scale distribution of cold hydrogen. He has found several huge expanding shells within our galaxy. One of these 'supershells' aligns with the outer edge of the Cygnus X-ray ring, showing the presence of a large mass of cool material along that boundary. Apparently the pressure of the hot X-ray emitting gas drives the expansion of the radio supershell.

5.5 Formation of a superbubble

How was the Cygnus superbubble formed? The Cygnus dust lane is about 1500 pc distant, so the distance of the X-ray bubble which lies behind it is about 2000 pc. At that distance, the ring is some 400 pc in diameter – 10 times the size of the Carina Nebula and 14 times that of the Cygnus Loop. The bubble's dimensions are so large when compared to other galactic phenomena that it is called a 'superbubble'. With the large size comes a large energy requirement for formation. The gas inside is as hot as that in the Cygnus Loop and yet occupies over 2000 times the volume. Even though the density is less, over 10^{52} ergs of thermal energy (20 times that of an old supernova remnant) are locked up in the hot gas. The key to understanding the superbubble lies in finding the source of this energy.

Could this energy have come from stars inside the superbubble? There are an exceptional number of young massive stars in the region including the Cyg OB2 association near the centre of the ring. This luminous, compact group, shown in figures 5.2 and 5.3 and mentioned earlier as an X-ray source in its own right, is heavily reddened by the dust in the rift. If this dust were not there, these stars would appear as bright to us as the Pleiades, which are 15 times closer. Despite the high luminosities of these O stars, there are not enough to supply the energy requirements of the superbubble.

There are three ways in which O and B stars can deposit large amounts of energy into the interstellar medium. First they are powerful sources of ultraviolet radiation which is absorbed by interstellar gas through ionisation of hydrogen. Although an O5 star might radiate 10^{53} ergs of energy during its lifetime, most of this is reradiated in the optical band by the absorbing material. In any case, this ultraviolet energy cannot heat dilute gas above 10^5 K. The second way is through the stellar wind. Through collisions the high velocity particles in the wind can deposit 10^{50} ergs in surrounding material over the lifetime of the star. This energy can heat the interstellar material to a temperature of 10^6 K. Finally, some, perhaps all early stars will eventually explode as type II supernovae, and each will deposit an additional 10^{50} ergs in the interstellar medium. It appears then, that 100 O stars could supply the energy for the superbubble through their winds and through supernova explosions.

Might it be possible to make this superbubble with a single giant explosion, a 'super-supernova'? This can in fact be calculated using the concepts explained in chapter 3. A remnant 500 pc in diameter could be created by an energy release of 10^{54} ergs but the remnant would be expected to have an X-ray bright shell, 100 times brighter than observed. A super-supernova is therefore an unlikely origin. The bubble was probably built by a chain of 'normal' supernova explosions. As explained above, it would require 100 separate supernovae, all occurring in a localised region of space, but exploding at intervals throughout the life of the bubble. To a good approximation, the effect of these would resemble a stiff, steady wind coming off the OB association at about 10 000 km s^{-1}. A simple analogy is the process of digging a hole. A dynamite blast can create a hole very quickly, but a man with a shovel can, over a longer period of time, accomplish the same thing with less expenditure of energy.

The observed size and velocity of expansion show the Cygnus superbubble to be at least 3 million years old. It could not have been created by stars in the presently observed Cyg OB2 association which is only about 1 million years old. It is more reasonable that both the superbubble and the Cyg OB2 association were created by an earlier generation of O stars which have since exploded and vanished. A possible sequence of events leading to the superbubble as we see it today is described below and is shown in figure 5.11.

5.6 Supernova → star formation → supernovae → superbubble

First the giant gas cloud known as the Great Rift formed in space. It contained enough material to make millions of stars. Stars formed in the densest part and the most massive of these became a supernova. This explosion created a shock wave which compressed material within the Rift. New stars formed through gravitational collapse of this material. Some of these were massive and constituted an O association. Over the next million years, the O stars one by one became supernovae and the expanding mingling remnants, aided by winds, gave birth to the superbubble. Shock waves from the young superbubble continued to propagate into the dense dark Rift and caused a new generation of stars to coalesce. These stars form the Cygnus OB2 association which we see today. When these stars reach the end of their lifetimes they too will explode and add their energies to the superbubble. Thus, in the Rift, new stars form steadily, while outside a giant bubble grows.

The superbubble is a catalyst to star formation as well as a side effect. Known star-forming sites may well contain evidence for other superbubbles. However, the sheer size of superbubbles and ease of obscuration (total and partial) by dust make it difficult to identify them by their X-ray emission alone. Perhaps the most promising other candidate is the Orion-Eridanus X-ray hot spot which extends 30° across the sky. The gas cloud has an X-ray temperature of 2 million degrees and is associated with a large arc of nebulosity known as Barnard's Loop, which envelops most of Orion. At the eastern edge of this region is a large system of gas and dust which includes the very well-known Orion and Horsehead Nebulae (see chapter 4). Nearby are the bright O and B stars of Orion's Belt and Sword. Although 21 cm radio emission is notably lacking, there is little doubt that we are seeing a smaller version of a superbubble.

The frequency of such phenomena in our local area suggests that superbubbles could contain more than half the energy to be found in interstellar space. This simple fact implies that the superbubble might be the basic dynamical unit of the interstellar medium. Finally, the process by which a

Fig. 5.11 The formation of a superbubble. In (a) and (b) a massive star evolves rapidly and eventually explodes as a supernova, driving a shock wave into a nearby massive interstellar cloud and precipitating the formation of new stars (c). As these stars in turn evolve and explode (d), their combined blast waves blow a large hot bubble in space and also eat farther into the cloud, causing more star formation. In (e) the process continues as the sputtering supernova wind from the next generation of stars inflates the bubble into a superbubble.

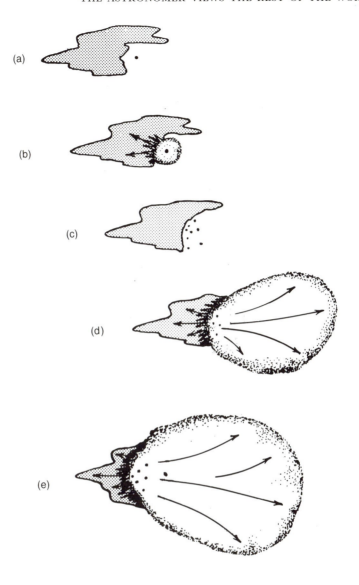

(a)

(b)

(c)

(d)

(e)

superbubble causes (and is caused by) a chain reaction of star formation is very effective. It may produce most of the new stars in the galaxy. Indeed, the odds are that the solar system was created at the edge of a superbubble 4.5 billion years ago.

5.7 The astronomer views the rest of the world

One of the authors recently sat down to watch what he thought was a television programme about astronomy. He was surprised to find O stars listed in the schedule and pleased that it would cover a topic of particular interest to him. He thought the show was starting rather obliquely with a long section about sailboats in the middle of the Atlantic Ocean. He soon learned that to a yachtsman, and to the producers of that program, the letters 'OSTAR' mean '*Observer* Single-handed TransAtlantic Race', a heroic test of sailor and ship sponsored by the British newspaper, The *Observer*. The programme was fascinating, but when an astronomer sees these five letters, a quite different phenomenon springs to mind. We hope, dear reader, that you will now share this enthusiasm for early stars.

6 *Normal galaxies*

6.1 X-ray sources in the Milky Way

Our galaxy, the Milky Way, is a 'normal' spiral galaxy. As far as we can tell most other spirals have similar properties. It does not have an active nucleus (see chapter 13), nor is it unusually luminous at any wavelength. Since we live in it, we find this a pleasing situation.

The first X-ray sources discovered were naturally the brightest, the accretion-powered binaries, and were within our galaxy. As time progressed, other Milky Way objects were also found to emit X-rays. The emission from our galaxy seemed to be from discrete objects within the Milky Way, not from some galactic-sized diffuse region.

Some of the accretion-powered sources have X-ray luminosities of 10^{38} erg s^{-1}. They are probably operating close to the Eddington limit where the pressure of infalling material is balanced by the pressure of outflowing radiation (see box 7.5). This limit depends on the mass of the accreting star and, if all neutron stars have about the same mass, all these X-ray binaries might have about the same luminosity. Some are within globular clusters where the high stellar density is apparently favourable for the formation of these binary systems.

Cataclysmic variables (CVs) and supernova remnants (SNRs) are also strong X-ray sources with X-ray luminosities in the range 10^{35} to 10^{37} erg s^{-1}. CVs are accreting binary systems consisting of low-mass normal stars and white dwarf companions. Many SNRs also contain isolated neutron stars which generate X-rays. X-ray luminosities range from 10^{37} erg s^{-1} from PSR 0531+21 within the Crab Nebula to 10^{32} erg s^{-1} from PSR 1929+10, a nearby radio pulsar.

The largest class of galactic sources are stars, some perhaps binary, but without compact companions. The coronal X-ray luminosities of stars so far detected range from 10^{33} down to 10^{26} erg s^{-1}, a level of emission similar to that of the Sun.

One of the largest uncertainties in the study of galactic sources is the distance to the more luminous sources. Although estimates of distance are always made, they are usually uncertain by a factor of 2 or more. The result is sometimes an order of magnitude uncertainty in the X-ray luminosity. Thus the luminosity function of the various types of sources is unknown and some interesting questions remain unanswered. (The luminosity function shows the number of sources having a given luminosity as a function of luminosity.) For example: are all accretion-powered binaries 1.4 $M_\odot$ neutron stars operating at the Eddington limit, or are some of these black holes with larger masses? Or: how does the luminosity of supernova remnants depend on diameter and can this relationship be used to derive the energy release of the supernovae?

A nearby galaxy provides a large sample of X-ray sources, all at the same distance. Here is an excellent means of overcoming the distance uncertainty. All that is needed is a detector sensitive enough to find the sources and to isolate them from their neighbours in the always crowded fields. The imaging X-ray telescope can do this.

6.2 The Milky Way and the Local Group

The 'Local Group' consists of five large star aggregations plus a number of dwarf galaxies (figure 6.1). Most of the mass is in the Milky Way and in the nearby spiral galaxy, M31. M31 is larger than our galaxy and contains 1.5 times as many stars. Another small galaxy, M33, is a face-on spiral and half again as distant as M31. Our nearest neighbour, the Large Magellanic Cloud (LMC), contains about one tenth the number of stars in the Milky Way. Not far away another small galaxy, the Small Magellanic Cloud (SMC), is about half this size. The distances from Earth to the Magellanic Clouds are well determined as 55 kpc to the LMC and 63 kpc to the SMC. M31 is 670 kpc distant, ten times farther away. These systems are all close enough so that Cepheid variables can be distinguished as individual stars and the Cepheid period-luminosity relation used to derive the distance. The next nearest neighbours lie in groups at farther, more controversial, distances.

The Local Group galaxies were extensively mapped and a number of nearby bright galaxies were observed by Einstein. Spatial extent ranges from 10° for the LMC to about 35 arcmin for M33 and structure was easily resolved. The X-ray appearance of these galaxies is quite different from the optical. Rather than a cloud of countless unresolved faint stars, each galaxy looks like a loose collection of individual bright sources.

The Magellanic Clouds are near enough so that Einstein saw not only the very luminous accretion-powered binaries but supernova remnants and miscellaneous weaker sources as well. The bright sources clearly stand out from any fainter emission which might be present and the supernova remnants can easily be identified by their spatial extent. In M31 only the brightest sources, having luminosities above 10^{37} erg s^{-1}, were visible but over 100 of these have been detected and located. In more distant galaxies only the brightest of the bright sources were seen as individuals.

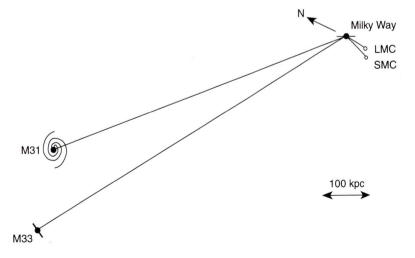

Fig. 6.1 A scale drawing showing shape and separation of the principal members of the Local Group. These are projected on a plane containing the Earth's axis of rotation and oriented perpendicular to the plane of the galaxy (the Milky Way). All galaxies shown are approximately in this plane.

A non-trivial problem arising is that of 'interlopers'. These are sources in the field, which are not members of the galaxy being observed. They are usually much closer than or much more distant than the galaxy. With a little extra work most can be recognised and excluded from further consideration. Foreground stars are usually identified as prominent objects on optical plates although M stars, which have high ratios of X-ray to optical luminosities, can be difficult. More distant sources are usually nuclei of active galaxies and quasars which can be identified through spectra of the optical counterparts. However, if viewed through the central region of a distant galaxy, obscuration by dust or confusion with the high density of stars can make an accurate identification impossible. Studies of the number of X-ray sources found in nearby galaxies must be corrected for the presence of interlopers.

6.3 X-ray observations of the Magellanic Clouds – our nearest neighbours

The Large Magellanic Cloud was first seen at X-ray wavelengths in October 1968 by Hans Mark and colleagues (LLL). They used a rocket launched from Johnston Island, a small military base at latitude 17°. It is far enough south so that both clouds can clearly be seen above the rocket horizon. Although the X-ray luminosity and extent of the LMC were close to expectations, the result was accepted with reservation. The signal was small and appeared above a background which varied with the pointing-direction of the detector. The signal was significant only if the shape of the background was well known.

In September 1970, Richard Price and colleagues (LLL) used a very large detector to scan slowly across both clouds and clearly detected X-rays above background. To their surprise, emission from the LMC seemed to come not from many weak sources spread throughout the volume of the galaxy, but from a few bright unresolved objects. The SMC appeared as a single bright source with a very hard (many high energy X-rays) spectrum. At about the same time, in December 1970 and January 1971, the newly-launched Uhuru satellite mapped the X-ray emission from both clouds and found three bright sources in the LMC and again the single SMC source.

These strong sources are the equivalent of the bright sources clustered around the centre of our own galaxy, the 'galactic bulge' sources. The discovery of such objects in the Magellanic Clouds led to the realisation that this class of source was characterised by the amazing X-ray luminosity of about 10^{38} erg s^{-1}. These are now known to be accretion-powered binaries at about the Eddington limit (see box 7.5). These are of great interest, both in themselves and because such objects might serve as 'standard candles' to determine accurate distances of other galaxies. Subsequent observations with rockets and the satellites SAS-3, Ariel V, and HEAO-1 found another bright source in the LMC, several bright transient sources in both clouds, and have given accurate locations leading to identifications of optical counterparts.

One of these bright sources, LMC X-3, has been identified as a binary comprised of a B3 main-sequence primary and a possible black hole. A. Cowley and colleagues have measured a velocity of 235 km s^{-1} for the B3 star and a period of only 1.7 days! The mass of the compact companion is estimated as 9 $M_\odot$ (but see chapter 11).

To those who struggled to untangle source locations from the count rate data of early rocket observations, a glimpse at the 1980 Einstein maps of the Magellanic Clouds was like a look into a treasure chest, with precious gems and bright coins sparkling in the light (figures 6.2 and 6.3). It was difficult to decide which item to examine first. Over 120 sources were found within the

Fig. 6.2 (a) An Einstein IPC map of a 5.7° square field containing the Small Magellanic Cloud. Eleven 2–6 hour exposures were used in this mosaic. SMC X-1, normally the brightest source, was in a low state at this time. The brightest source in this figure is the supernova remnant shown in the upper right corner of figure 6.5. (b) Key to figure 6.2 showing identifications of some of the X-ray sources.

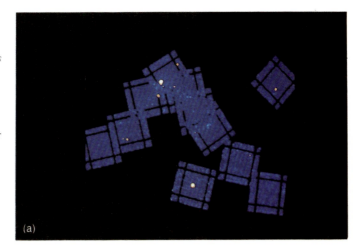

(a)

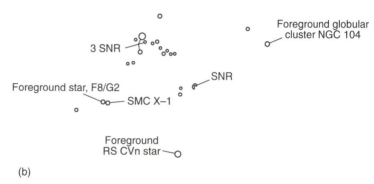

3 SNR

Foreground globular cluster NGC 104

Foreground star, F8/G2

SNR

SMC X–1

Foreground RS CVn star

(b)

LMC and over 40 within the SMC. The five very luminous sources were easily identified. The survey extends in places to objects 10 000 times fainter. A most pleasing result was the large number of supernova remnants. These were easily identified by their spatial extent. Many were even brighter X-ray emitters than expected. Usually, pre-observation predictions are optimistic and fewer counts are collected than anticipated. In this case the reverse was true and detailed X-ray maps and spectra were obtained where none were expected.

The Einstein HRI was pointed at most of the brighter sources to determine the morphology of the SNR and to obtain precise locations of the unresolved sources. Spectra were obtained of a few objects with the SSS before the end of its life. The life time of this spectrometer was limited by the cryogen supply to 10 months. After the last of the cryogen had evaporated, the spectrometer warmed up and was no longer useful (see chapter 2).

To observe a bright SNR with the SSS it was first necessary that it be discovered in the IPC survey. SSS time was then scheduled for the next interval when both the source was an acceptable target and the spectrometer was at the telescope focus. Such new targets competed with observing schedules of other investigators and the expanded schedule was greeted with less than universal enthusiasm. While viewing the spectra shown in figure 6.4, reflect on the operational difficulty of getting these spectral data. Since the HRI lasted as long as the spacecraft, there was no great difficulty in getting high spatial resolution observations of most of the bright sources.

Three Crab-like remnants were identified in the LMC. One, 0540–69.3, shows both the hard spectral continuum characteristic of synchrotron radiation and a bright unresolved X-ray source within the remnant. Here, in fact,

Fig. 6.3 (a) A 2.3° square field within the Large Magellanic Cloud as mapped with the Einstein IPC. The brightest source is the accretion-powered binary LMC X-1. There are at least seven supernova remnants visible but the resolution of the IPC was not high enough to resolve structure. The remnant N132D is the second brightest source in the field. The background is grainy in regions of low exposure, due to photon statistics. (b) Key to figure 6.3(a) showing identifications of most of the X-ray sources in this LMC field. Note that four interlopers have been identified.

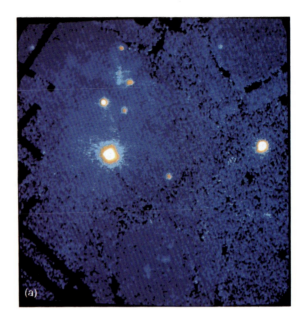

(a)

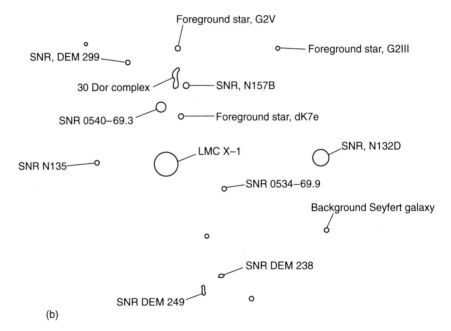

(b)

is an object very like the Crab Nebula. It contains a bright 50 ms pulsar easily detected at both X-ray and optical wavelengths. This pulsar is also surrounded by diffuse optical emission (chapter 3 contains more detail). Classification of the other two remnants is based on the spectrum for N157B and on the spatial appearance of N103B, but neither was observed with both HRI and SSS.

The other remnants are shell-like with diameters ranging from 0.5 to 3 arcminutes. Some of these are shown in figure 6.5. Undoubtedly larger structures are also present in the data and will be uncovered by diligent analysis. The remnant with the smallest diameter, probably the youngest, is the only bright remnant in the SMC. This is rather curious but is probably an accidental situation. The complete sample, about 25 remnants all at the same

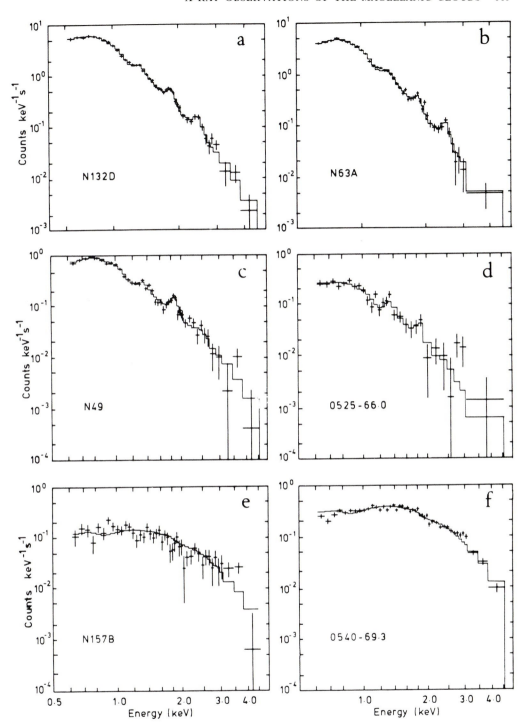

*Fig. 6.4 Einstein SSS spectra of six supernova remnants in the LMC. Radiation from hot gas shows characteristic emission lines. Remnants a–d are old shell-like remnants and the spectra are thermal. Remnants e and f emit a continuum without lines. This nonthermal radiation is typical of nebulae surrounding young pulsars. [Adapted from D. Clark, et al., Astrophysical J. **255**, 440 (1982).]*

Fig. 6.5 Nine bright supernova remnants in the Magellanic Clouds as imaged by the Einstein HRI. Each field is 2.3 arcmin or 34 pc on a side. Three of the remnants are Crab-like and bright at the centre. The remainder are all shells and generally the smaller, younger shells are more luminous. Starting at upper left and reading like a book, the remnants are: N49, N23, 0102–72.3 (in SMC), 0540–69.3, N132D, N103B, 0519–69.0, N63A, and DEM71.

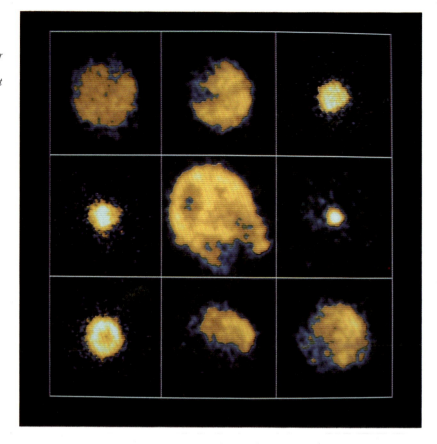

distance, has allowed the investigation of phenomena depending on the diameter of the shell. For example, if the rate of supernovae is constant in time, the number of remnants having a given diameter depends on how the shell expands into the surrounding material. A realistic analysis must take into account a range of interstellar densities and a range of initial energy release and is not easy. The data sample is already excellent, the best yet available. Furthermore, ROSAT will add data on large faint remnants which Einstein could not distinguish from the background. These data are deserving of a long careful study which should resolve some of the present uncertainties concerning the evolution of supernova remnants.

6.4 X-ray observations of M31

Our nearest massive neighbour is M31, the Great Galaxy in Andromeda. It is four times closer than NGC 55 and NGC 253, the next nearest galaxies with a size comparable to our own. M31 holds a certain fascination. It is a spiral galaxy similar to our own. Since we view it from outside, the structure and distribution of stellar populations is clear. In contrast, our understanding of the arrangement of stars in the Milky Way is based on observations severely limited by obscuring clouds of dust. We hope that observations of M31 will clarify the form of our galaxy and of our place within it.

On a clear moonless autumn night, M31 can be seen with the naked eye as a patch of diffuse light just below the curved chain of stars which form the constellation Andromeda. A time exposure through a large telescope gives the result seen in figure 6.6. The eye sees only integrated light from stars in the central part of the 'galactic bulge', an ellipsoidal arrangement of old red stars. Photographs reveal a larger extent. Faint spiral arms extend to a diameter of 1.5° and contain bright patches and dark lanes. These are regions of recent star formation and clouds of absorbing dust. The spiral arms are blue, the colour of young, bright stars. Doppler shifts of spectral features can be measured accurately and show the SW arms to be moving towards us and the NE arms to be moving away. The rotational velocity is about 250 km s⁻¹ implying a period of revolution of 10^9 years, very similar to that of the Milky Way. At the other extreme, a short exposure through the telescope shows only a bright, almost star-like nucleus.

Maps at other wavelengths exhibit some, but not all, of these features. In 21 cm radio maps neutral hydrogen appears arranged in a ring centred on the nucleus and surrounding the 'bulge' stars. Results from the Infrared Astronomy Satellite, IRAS, show the same ring structure seen in the radio

Fig. 6.6 An optical picture of M31 with X-ray source positions superposed. The scale is the same as that of figure 6.7. (Courtesy L. Van Speybroeck, CfA.)

band but also a bright infrared emitting nucleus. The nucleus is of particular interest. We strongly suspect it is similar in character to the nucleus of our own galaxy which is hidden (or at least very highly obscured) from our view at most wavelengths.

X-rays from M31 were first observed by Stuart Bowyer and colleagues (UCB). Data from a rocket launched in February 1973 showed a source at the position of M31. This was subsequently confirmed by the Uhuru and Ariel V all-sky surveys. The calculated X-ray luminosity was 2×10^{39} erg s^{-1}, about the same as that calculated for the Milky Way.

The Einstein IPC survey in 1979 revealed 30 bright X-ray sources in the outer spiral arms and an unresolved cluster of sources in the nuclear region (figure 6.7). The HRI in turn resolved this cluster into bright individual sources. A careful analysis of these images by Leon Van Speybroeck (CfA) has resulted so far in the detection and location of 117 individual sources (figure 6.8).

Fig. 6.7 The Einstein IPC survey of the nearby spiral galaxy M31. North is up and east to the left. Data from three observations have been added to make this picture. Several bright sources can be seen in the spiral arms and there is an unresolved group of sources clustered close to the nucleus. (Courtesy L. Van Speybroeck, CfA.)

Although the total luminosity is about as expected, the bright source distribution is not. The dominant central cluster is naturally associated with the 'bulge'. The bright sources are thought to be population II (associated with old stars) accretion-powered binaries, the same as the bright bulge sources in the Milky Way. There are, however, more than expected and they are grouped closer to the nucleus.

There is, within 5 arcminutes of, and perhaps coincident with, the optical nucleus, a source with L_x about 10^{38} erg s^{-1}. This was observed to vary by a factor of 10 over 6 months. This type of activity is characteristic of active galactic nuclei but, here in M31, the luminosity is at an uncharacteristically low level (about a million times less). Although the chance of accidental association is high ($\sim$10% because of the high density of bright sources clustered about the nucleus) it is still an intriguing result. Perhaps massive black holes exist at the centre of all spiral galaxies. If the rate of accretion of matter is low they would not be luminous enough to be normally observable.

Some of the other M31 sources have been identified. About 20 are associat-

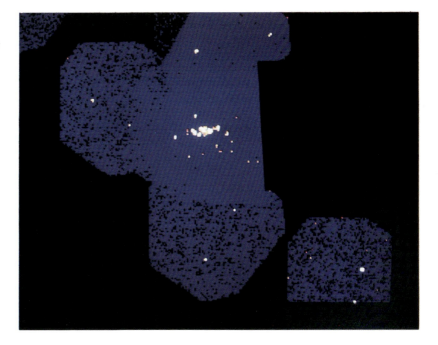

Fig. 6.8 An Einstein HRI mosaic of the inner part of M31. The diffuse source in figure 6.7 is now resolved into several individual sources, believed to be accretion-powered binaries. (Courtesy L. Van Speybroeck, CfA.)

ed with globular clusters, about 4% of the 550 known globular clusters in M31. This is about the same fraction of globular clusters known to be X-ray sources in our galaxy. The proportion of high luminosity sources in M31, however, seems larger. Two sources are associated with bright optical supernova remnants, but no X-ray emission was found at the location of the known historical supernova of 1885 (S Andromedae) to a limit of $< 10^{37}$ erg s^{-1}.

About half of the remaining 95 sources are within the galactic bulge and within 2.5 kpc of the nucleus. These have an average luminosity of about 3×10^{37} erg s^{-1}. Since most sources were only resolved with the HRI, spectra were not recorded. Considerable gas and dust are expected in the paths because of the 20° inclination of the disc of M31. The true luminosity distribution will not be known until reasonable spectra are available, so the question of the suitability of bulge sources as 'standard candles' remains an open one. The sources in the spiral arms correlate well with the distribution of neutral hydrogen so they are likely to be associated with young stars (population I stars) – probably accretion-powered binaries.

Thus many individual bright X-ray sources have been detected in M31. Some have been identified. The nature of most is conjecture. The distribution is not quite as expected, particularly in the region close to the nucleus, which is itself interesting. Only about five interlopers are expected in the Einstein M31 sample. None has yet been identified. The next generation of instruments hopefully will be capable of mapping M31 at high resolution while simultaneously measuring spectra of the brightest sources. Long term monitoring might also identify binary systems. These observations are needed to understand the source population in M31.

6.5 X-ray sources in other normal galaxies

Table 6.1 lists the larger galaxies in the Local Group and other nearby galaxies, all observed by Einstein. Obviously, the farther the galaxy, the greater the threshold for detection of individual sources. In this sample, there is also a dependence on observing time. After the Local Group only the brightest of the bulge sources were seen. At distances greater than a few Mpc only unusually bright sources were detectable.

Table 6.1 Some nearby galaxies observed by Einstein

Galaxy	Type	Distance	Mass $(10^9 M_\odot)$	X-ray sources detected	Limit of survey (erg s^{-1})	Luminosity of nucleus (erg s^{-1})
Milky Way	Sbc	we are inside	20	2000?	varies	2×10^{35}
LMC	Ir	55 kpc	1.0	103	4×10^{35}	no nucleus
SMC	Ir	63 kpc	0.2	45	1×10^{36}	no nucleus
M31	Sb	670 kpc	32	117	4×10^{36}	1×10^{38}
M33	Sc	730 kpc	1.3	13	1×10^{37}	1×10^{39}
Maffei I	E?	2 Mpc	20	3	1×10^{38}	–
NGC 253	Sc	3 Mpc	15	8	1×10^{38}	1×10^{39} (starburst)
M82	Ir	3 Mpc	3	8	2×10^{38}	7×10^{39} (starburst)
M81	Sb	4 Mpc	25	9	2×10^{38}	2×10^{40}
M51	Sbc	9 Mpc	40	3	8×10^{38}	8×10^{39} (extended)
M100	Sc	16 Mpc		4	3×10^{38}	2×10^{40}

M33 is a face-on spiral in the Local Group (figure 6.9). It presents an opportunity to see exactly where X-ray sources are located within clearly visible spiral structure (figure 6.10). There is little complication from absorption in gas and dust. The galaxy, however, is not very massive so not many bright sources are expected. Only 13 sources were detected above a threshold of 1×10^{37} erg s^{-1}. Three are foreground stars and one is a background galaxy. One of the remaining sources is coincident with a bright optical supernova remnant. None of the approximately 20 globular clusters was detected. The most remarkable, and the brightest, source is coincident with the nucleus (within the 3 arcsecond accuracy of the X-ray position). It has a luminosity of 10^{39} erg s^{-1}, 10 times that of the M31 nuclear source and 1000 times the X-ray luminosity of the nucleus of the Milky Way. Here is another candidate for a low-level active nucleus in an otherwise normal spiral galaxy, and in this case, one with little total mass.

M51, the Whirlpool Galaxy, is another ideally oriented face-on spiral. It is massive enough to contain many luminous sources but far enough away so only the very brightest can be seen as individuals. The Einstein observation, shown in Fig. 6.11, found only the nucleus and three bright point-like sources. There is also extended emission from the central region, probably due to unresolved sources in the galactic plane. The luminosity of each of the bright sources is about 10^{39} erg s^{-1}.

Thus, when data are collected from galaxies beyond the Local Group, most of the bright accretion-powered binaries fade below the detection threshold and only unusually bright sources can be seen. (SMC X-1, at 10^{39} erg s^{-1}, is the most luminous steady source known with definite observed binary characteristics.) The bright nuclei of the massive spirals M81 and M100 are worthy of note. There is also a source in, or projected on, one of the spiral arms of M100 which, if not an interloper, has an X-ray luminosity of 1.5×10^{40} erg s^{-1}. It might be the most luminous non-nuclear galactic source yet detected.

6.6 Starburst galaxies

Some galaxies appear to have 'starburst' nuclei, so named because the nuclear regions are strong diffuse sources of optical emission lines characteristic of gas at a temperature of 10 000 K. Such optical spectra are usually associated with regions of recent star formation. The galactic centres are thought to be the sites of unusually high rates of supernova activity, perhaps one explosion every 10 years. These nuclei appear also as diffuse X-ray sources

Fig. 6.9 The nearby spiral galaxy M33, the 'Pinwheel Galaxy' in Triangulum. This galaxy is a member of the Local Group but contains less than 1/10 the mass of our own galaxy. This field is 0.6° × 0.9° in extent. (Photograph from Hale Observatories.)

rather than the single bright source which may indicate a black hole. Both M82 and NGC 253 are nearby starburst galaxies.

M82 has long been recognised as a peculiar system. It was first proposed to be an 'exploding' galaxy because of the high luminosity and large extent of the nuclear region. Nuclear radio, infrared, and X-ray emission is quite clumpy and has extent of about 30 arcseconds. The Einstein HRI shows seven unresolved sources and strong diffuse emission concentrated at the centre of the galaxy. Furthermore, this diffuse emission extends perpendicular to the plane of the galaxy to form a faint halo on both sides of the nuclear region.

The supernova explosions which are thought to energise the nuclear region will produce hot gas at high pressure. This gas, confined by material in the galactic plane, flows out along the minor axis. It becomes a galactic

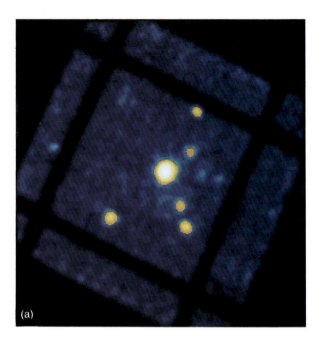

(a)

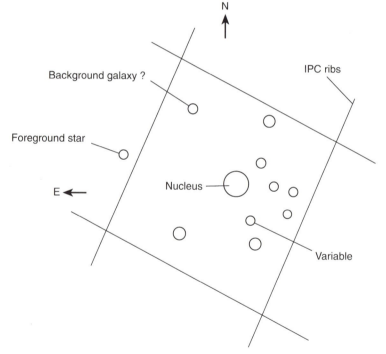

N

Background galaxy ?

IPC ribs

Foreground star

E

Nucleus

Variable

(b)

Fig. 6.10 (a) An Einstein X-ray picture of M33, which fills the inner part of this 1° square IPC field of view. Those parts of the detector shadowed by the window-support ribs have been excluded. (b) Key to figure 6.10 showing identifications of X-ray sources in M33.

wind which, after leaving the galaxy, forms the X-ray halo. There are also faint optical filaments observed close to M82 which coincide with the X-ray halo and apparently move outward with the wind. All evidence points to a flow of material from the nuclear region.

NGC 253 (figure 6.12), a larger galaxy, shows similar but less intense nuclear activity. There is strong diffuse emission from the centre and this emission extends to the south along the minor axis. Several un-resolved sources are embedded in the diffuse emission. These are probably

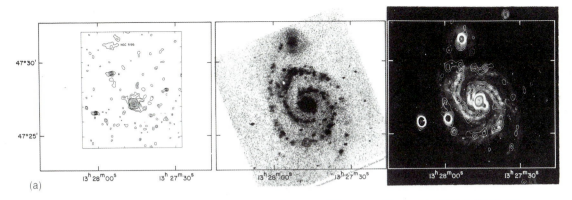

(a)

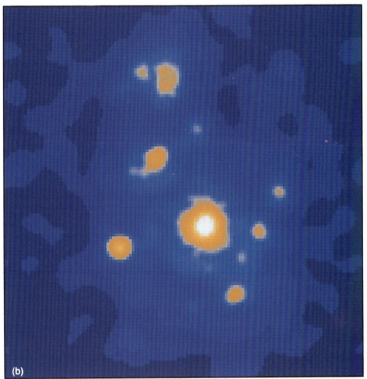

Fig. 6.11 (a) M51, the
'Whirlpool Galaxy', viewed at
four different wavelengths. At the
right, radio surface brightness
contours are overlaid on a blue
picture. The centre photograph
shows only the red emission line
from hydrogen atoms, and spiral
arms are clearly indicated by
bright HII regions. The Einstein
HRI image is on the left. The
strongest X-ray source is the
nucleus. Three weaker sources are
associated with spiral arms.
(Courtesy of G. Fabbiano, CfA.)
(b) ROSAT PSPC X-ray picture
of M51. Compare with the
Einstein picture (a). Four more
sources in the spiral arms are
visible as well as diffuse emission
from the interior. (Courtesy of W.
Forman, CfA.)

(b)

accretion-powered binaries like those seen in M31. Here is a large spiral
galaxy with starburst characteristics but they are not as pronounced as those
in M82.

6.7 The nucleus of the Milky Way

What about the nucleus of our own galaxy? Does it contain a black hole, currently inactive because of a lack of accreting material, or might it be a starburst nucleus? We cannot see the centre at optical wavelengths because of
the intervening clouds of dust. The knowledge we do have has been gained
through radio and infrared observations.

The motion of neutral hydrogen at the centre has been studied at 21 cm.
The density of matter is found to increase as one approaches the centre,
peaking at $4 \times 10^5 \, M_\odot \, \text{pc}^{-3}$. Total mass within a radius of 100 pc is estimated
to be $10^9 \, M_\odot$. A weak radio source, Sgr A West, lies at the dynamical centre
and is thought to be coincident with the nucleus itself. Sgr A West, although

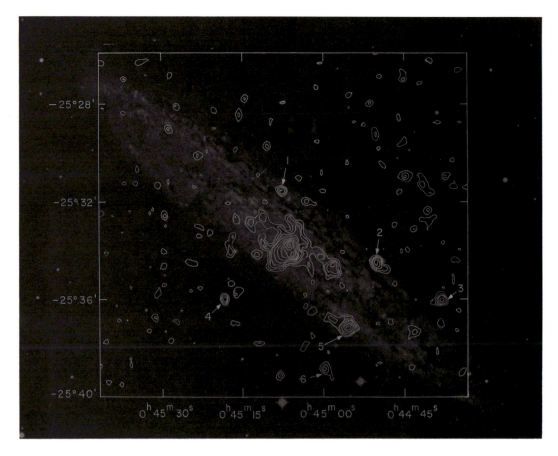

Fig. 6.12 Optical picture of NGC 253, an edge-on spiral, with contours of constant X-ray surface brightness superimposed. This galaxy has a 'starburst' nucleus which shows as a diffuse central X-ray source. (Courtesy of G. Fabbiano, CfA.)

quite compact, has not yet shown large variations in luminosity or radio flares such as are associated with nuclei of active galaxies.

X-rays above about 2 keV are also capable of penetrating the material in the path to the centre. Several attempts have been made to study this region with X-ray observations. The excellent EXOSAT map in Fig. 1.6 shows that the centre is not a bright source. A detector with high sensitivity and arcminute resolution is needed to distinguish the centre from nearby bright sources. Accordingly, observations were made with the Einstein IPC with the result shown in figure 6.13

This observation reveals a weak point-like source at the centre with surrounding diffuse emission. The diffuse emission may be truly diffuse or may be due to faint unresolved sources in the central region. The very bright source in the field is not associated with the nucleus and is an accidental superposition. The source at the centre of the field is probably the nucleus itself since it is within an arcminute of Sgr A West. Although not conclusive, this high energy emission from the nucleus is evidence for a very weak 'active' source at the centre of the Milky Way. A black hole is still a possibility for the actual nucleus! A location good to an arcsecond, a good spectral measurement, and an observation of variability are needed to remove any doubt of this identification.

The galactic centre was also observed by an interesting British detector carried on the American Spacelab 2 mission. This was a coded mask telescope designed and built by Gerry Skinner, Peter Willmore, and colleagues (University of Birmingham). It was built to detect X-rays from 2 to 30 keV in energy. This is more penetrating radiation than that focused by the Einstein telescope which would not reflect X-rays above 4 keV.

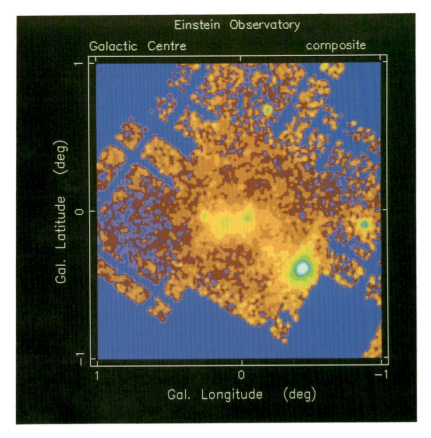

Fig. 6.13 Einstein IPC pictures of the nucleus of the Milky Way. The central field shows a region of diffuse emission containing several sources. Within this region the source on the right coincides with the radio source Sgr A West, thought to be the actual nucleus. The brightest source is A1742–294, detected in other surveys and visible close to the centre in figure 1.6. (Courtesy G. Skinner, University of Birmingham.)

The coded mask telescope used a plate opaque to the X-rays except for a large number of holes with a random pattern. This is in front of a large-area position-sensitive detector. Radiation from sources is stopped by the plate except where there are holes. Each source thus casts a shadow of the plate on the detector. The position of the shadow depends on the location of the source. A computer is used to untangle the 'images' of many overlapping shadows to determine the number of sources and the direction to each one.

This instrument was pointed at the centre of the galaxy during August, 1985. A $6° × 6°$ region was observed for a total of 7 hours. Nine point-like sources and diffuse emission were detected. Figure 6.14 (a) shows the resulting map of the central 4 square degrees. As in the Einstein picture, the field is dominated by the strong source A1742–294. The nucleus is the weak source at the centre of the field and is well resolved from other stronger sources. This field is also almost filled with diffuse emission which, at these energies, appears more extensive than that seen in figure 6.13.

The ART-P instrument on Granat is also a high energy coded mask detector and larger and more sensitive than the Spacelab instrument. Figure 6.14 (b) shows three square degrees around the galactic centre scanned by ART-P. The same sources are detected by both instruments, which is encouraging, and ART-P has found a new source, GRS1741.9–2853. It is close to Sgr A and variable. These data are a convincing demonstration of the usefulness of coded mask detectors.

6.8 Diffuse galactic emission

Thus far, a galaxy has been considered as a collection of stars and X-ray sources, all at the same distance. Observations are valuable because they reveal the morphology of different populations of X-ray sources as well as

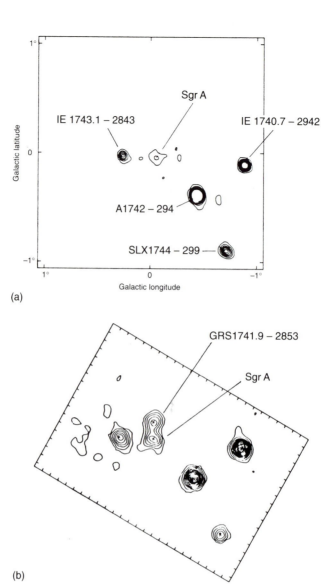

Fig. 6.14 The galactic centre as seen with coded mask telescopes operating in the energy range approximately 3 to 35 keV. Sgr A is the galactic centre. A1742–294 is again the brightest source in the field. 1E1740.7–2942 has an unusually hard spectrum. (a) is from Spacelab. The field is 2° × 2°. (Courtesy G. Skinner, University of Birmingham.) (b) is from the ART-P instrument on Granat. The field is 1.5° × 2.1°. (Courtesy R. Sunyaev and the Granat team, IKI.)

properties of the individual sources. What about the nature of the galaxies themselves? What do X-ray data reveal concerning the physical properties of a galaxy – a gravitationally bound aggregate of matter consisting in part of 10^9–10^{12} stars?

Optical results are already familiar. Galaxies are classified according to morphological type. The broadest classes are irregular (Ir), elliptical (E), and spiral (S). The optical luminosity indicates the number of stars or the mass in the form of stars. Colours and spectra are used to distinguish stellar populations. Elliptical galaxies and the central regions of spiral galaxies consist of old red stars. The young blue stars reside in arms of spiral galaxies.

The shape of these features is determined by the configuration of material at the time of star formation. A galaxy forms from a primordial amorphous cloud of gas. As the cloud collapses, angular momentum is conserved but energy is dissipated through viscosity. The cloud becomes a rotating disc if the original angular momentum is high. When stars are formed they retain the ballistic trajectory of the star-forming region at the time of condensation. Thus the oldest stars have a spherical distribution (e.g. the distribution of the

Fig. 6.15 The X-ray surface brightness of NGC 4472 showing a halo of X-ray emitting gas extending beyond the optical boundaries of this elliptical galaxy. (Courtesy of C. Jones, CfA.)

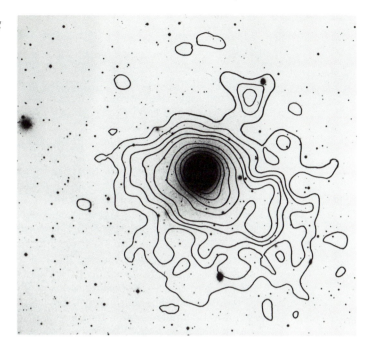

globular clusters) and the youngest form a narrow disc (e.g. O and B stars). Stars forming a thick disc are thought to be of intermediate age. The name given to a stellar population reflects the spatial distribution. There are 'halo', 'bulge', 'old disc', and 'young disc' stars. X-ray sources can be also so classified.

The mass of a galaxy is determined by measuring the rotational velocity of stars in the outermost regions. Observed Doppler shifts of spectral features give the rotational velocity as a function of distance from the centre. The reasonable assumption of circular orbits in a gravitational field allows the calculation of the distribution of mass and of the total mass in the galaxy. Galactic masses obtained in this manner are consistently a factor of 2–10 times higher than masses estimated from the optical luminosities. Galaxies apparently contain much unseen or 'dark' matter!

The rotation curves of spirals, in particular, are flatter than expected in the outer regions. There seems to be a halo of dark material extending to, and probably beyond, the outermost optical features. The distribution of matter at large radii in ellipticals cannot be determined with this technique because there are no bright spiral arms extending far from the nuclear regions. Recent X-ray data, however, show dark matter is also present in elliptical galaxies.

Einstein images of individual galaxies in the Virgo cluster by William Forman, Christine Jones, and Wallace Tucker (CfA) show X-ray emission clearly extending beyond the optical limits. The X-ray images are larger than the optical. This is particularly clear around some of the Virgo elliptical galaxies as is shown in Fig. 6.15. The X-rays are interpreted as thermal radiation from a diffuse hot gas forming a halo around the galaxy. The gas has a temperature of about 1 keV and density of only 0.01 electrons cm^{-3}. Although the total mass of hot gas in the halo is only about $10^{10}\ M_\odot$ (compared with $10^{12}\ M_\odot$ in luminous stars), the halo requires about $5 \times 10^{12}\ M_\odot$ to gravitationally bind it to the galaxy.

The existence of X-ray halos therefore not only indicates the existence of diffuse hot gas but implies the existence of a halo of dark matter extending beyond the optical border of the galaxies. Furthermore, the radial distribution of X-ray emission traces the radial distribution of matter. Thus elliptical galaxies, like the large spirals, appear to have large dark halos containing as much as 90% of the mass.

Such X-ray halos around nearer galaxies have not been observed above background (with the exception of M82). Diffuse emission from the inner regions, however, is sometimes clearly seen. This is thought to be from many unresolved individual sources. By comparing the average radial falloff of X-ray surface brightness with that at other wavelengths, the nature of the X-ray emitting population can be inferred. In M51, for example, Giorgio Palumbo (Bologna), Giuseppina Fabbiano (CfA), and colleagues have shown that the X-rays correlate well with the blue light distribution indicative of old disc population I stars. This leads to the belief that the (unresolved) sources belong to this population and have the same age as the stellar members. There is also a good spatial correlation with the radio continuum caused by cosmic ray electrons radiating in interstellar magnetic fields. Apparently both cosmic rays and X-ray sources originate in the same population. X-ray sources might, as many suspect, play a key role in the generation of cosmic rays.

The X-ray halos also play a key role in the study of emission from clusters of galaxies, the subject of chapter 14. The X-ray observations are a new means of studying the distribution of dark matter in the universe. Even though the total mass of dark material is greater than that which can be directly observed, the nature of the dark matter has so far been elusive. It is thought to be fundamental in the formation and structure of galaxies and in determining the large-scale structure of the universe.

6.9 Summary and future prospects

Bright X-ray sources in nearby galaxies have been clearly detected and resolved. Supernova remnants in the Magellanic Clouds have been easily identified. The Einstein observations have found several previously unrecognised remnants and identified a 'twin' of the Crab Nebula. A large sample of probably accretion-powered binaries has been extracted from M31 and a few of the brightest members of this class have been seen in a half dozen nearby galaxies. Unresolved sources at the centre of normal galaxies, including our own, have raised the possibility that the nuclei are the same as those of active galaxies but are currently in a state of low emission.

Future missions (AXAF, XMM, see chapter 17) are planned with telescope and X-ray detector combinations more sensitive than Einstein. With these, long term monitoring of nearby galaxies should yield pulsation and/or binary periods for some of the bright sources. Supernova remnants in M31 will be seen as extended sources. The high resolution of AXAF will make possible mapping of source populations in galaxies beyond M31 with results similar to those now in hand for M31. With several systems well mapped, the X-ray source population(s) and morphology can be meaningfully compared with the optical data. This should increase our understanding of both evolution of the individual X-ray sources and that of the host galaxies.

One of the most interesting results will be a measure of the luminosity function of the brightest X-ray sources. These objects might all contain a $1.4\,M_\odot$ neutron star which is accreting matter at the Eddington limit (explained in box 7.5). There will then be a feature in the plot of source number vs. luminosity showing that these sources can be used as standard candles. Distances of galaxies between M31 and the Virgo cluster, which are now uncertain by a factor of 2, might then be measured to 10–20%. These distances are vital to the determination of the Hubble constant (box 13.1) which is currently a factor-of-2 uncertain.

The ability to map faint diffuse emission will aid the study of dark matter in galaxies and will in turn lead to a better understanding of material in clusters of galaxies.

7 Massive X-ray binary stars

7.1 The discovery of binary behaviour

The very existence of the bright cosmic X-ray sources discovered by rocket flights in the 1960s represented one of the most exciting and challenging problems in all of astrophysics. No physical process was known at the time which was capable of generating the enormous X-ray luminosities that had been observed. The subsequent optical identification of Sco X-1 and Cyg X-2 in the mid-1960s using the modulation collimator techniques described in chapter 2 stimulated theorists and observers alike to learn more about these new *X-ray stars*. However, the first remarkable feature of the optical stars associated with these extremely powerful X-ray sources was just how unremarkable they appeared visually (see figure 1.2). They were rather faint (13th to 15th magnitude) and would not obviously be picked out on optical photographs or surveys. The optical spectrum of Sco X-1 was somewhat similar to the *cataclysmic variables* that were being intensively monitored by amateur groups and had been shown, a few years earlier, to be interacting binary systems (and these will be described in detail in chapter 10).

As shown in figure 7.1, Sco X-1 displayed a smooth blue continuum with superposed weak emission lines of hydrogen and ionised helium. The absence of absorption features, as in the normal (solar type) spectrum, indicated that little or none of the light observed was coming from a main-sequence star. The presence of ionised helium indicated that the source of excitation of the lines was very hot and very likely to be connected with the X-rays. However, despite much observation in which it was shown that the source flickered and varied, no indication of binary behaviour was found. The same was true for Cyg X-2.

Nevertheless, there was a great deal of theoretical speculation about the nature of the energy source in Sco X-1 and this centred around what had previously been considered to be very exotic objects: white dwarfs, neutron stars and black holes. Indeed, at this time (1966), pulsars had still not been discovered and so neutron stars existed only in the minds of theoreticians.

But major strides in understanding the nature of the bright X-ray stars had to await the Uhuru satellite. Within a few months of launch it had discovered two sources which were pulsating regularly with extremely precise periods, as shown in figure 7.2. When these same sources were observed over much longer time intervals (figure 7.3), both showed X-ray eclipses. It was recognised immediately that both X-ray sources were close binary systems in which the X-ray emitting object was a rapidly spinning neutron star (it could not really be considered to be a white dwarf because the period was so short

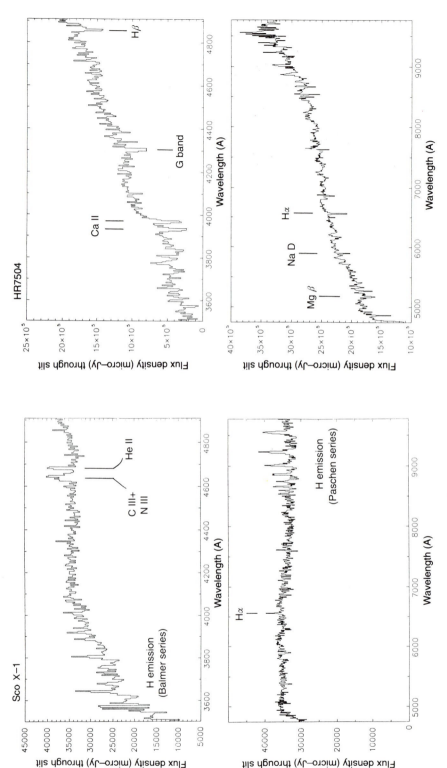

Fig. 7.1 Optical spectra of Sco X-1 and a 'normal' star (HR 7504) which has a spectral type very close to that of the Sun. Note that Sco X-1 appears very blue compared to HR 7504, and that its continuum shape is very different. HR 7504 displays easily recognisable absorption features due to calcium, sodium and magnesium. Sco X-1 has no absorption features, but instead emission lines due to hydrogen (especially Hα and the Paschen lines beyond 8500 Å) and ionised carbon, nitrogen and helium (between 4650 and 4700 Å). HeII λ4686 especially indicates the presence of very hot material. Both spectra were obtained with the 4.2 m William Herschel Telescope on La Palma by PAC in 1988.

(a) Her X-1

SOURCE IN HERCULES (2U1705+34)
November 6, 1971

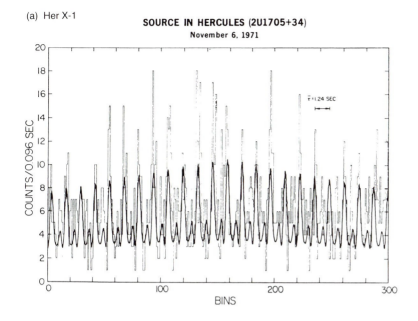

(b) Cen X-3

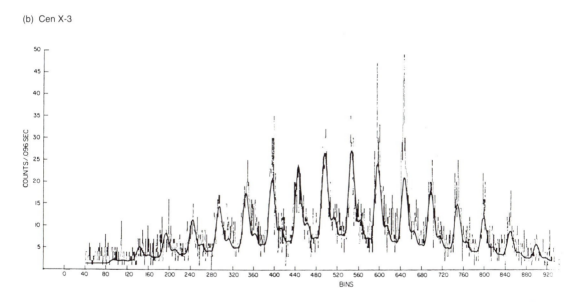

Fig. 7.2 Uhuru satellite observations that revealed the presence of X-ray pulsations in Her X-1 (1.2 s) and Cen X-3 (4.8 s), thus confirming that they are rapidly spinning neutron stars. (From figures by E. Schreier, STScI.)

Fig. 7.3 By observing the regular changes in the spin periods of Her X-1 and Cen X-3 with Uhuru, the orbital periods of both objects were discovered to be 1.7 days and 2.09 days respectively. What is more, both are clearly eclipsing binaries as the X-ray source regularly disappears behind its companion. The eclipses can be seen in the long-term variations of Her X-1 (a) and the light curve of Cen X-3 (c). The spin period changes (due to the orbital motion of the pulsar) are plotted in (b) and (c).

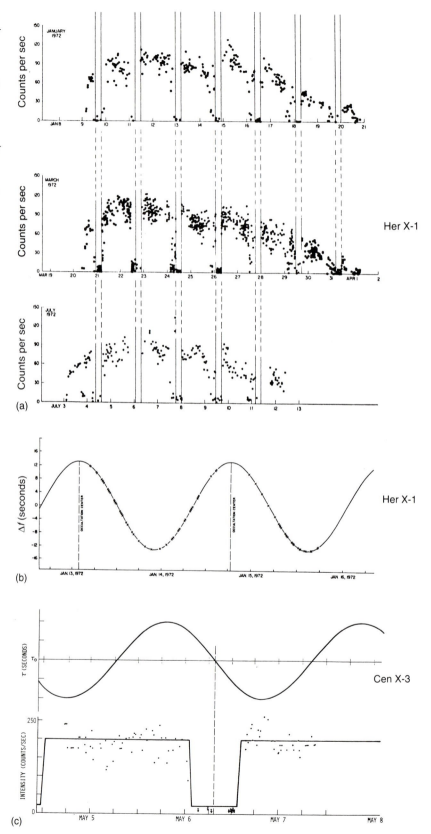

that a white dwarf would break up due to centrifugal forces). The short binary period indicates that the two components are *interacting*, i.e. likely to be exchanging mass (see box 7.1). The X-ray energy source is then *gravity*. Matter is transferred from one component onto the compact object, thereby releasing a large amount of gravitational energy in the form of X-radiation (that matter heats up in falling in this way can be nicely demonstrated by a simple experiment where the temperature of water is determined at the top and bottom of a waterfall).

Box 7.1 Interacting binaries

Once the orbital period of a binary system is known, Kepler's Third Law can be applied to estimate the separation, a, of the two masses :

$$a = 5 \, (M_1 + M_2)^{1/3} \, P^{2/3}$$

where a is measured in solar radii, the masses of the two stars are in $M_\odot$ and the period is in days. Since most X-ray binaries have short periods (< 20 days) and $M_1 + M_2$ is in the range 1–10 $M_\odot$ then the separations of the two stars will be not much larger than the size of the mass-losing star. In the case of HZ Her/Her X-1 which is eclipsing, we have a direct measure of the radius of the companion star and the binary separation is only about twice this value.

The observed X-ray pulsations, which are due to the spin of the compact object (nothing else could produce such an accurate clock), show that the compact object cannot be a white dwarf and hence must be a neutron star. Obviously the star is not spinning so fast that it will break up, so the centrifugal force felt by a mass m at the surface of the spinning star must be less than the force of gravity (see figure 7.B1).

Fig. 7.B1 A particle on the surface of a rapidly rotating star will experience a centrifugal force which would cause it to fly off. When this is balanced exactly by the gravitational force due to the star itself, then the star is rotating at its critical break-up speed.

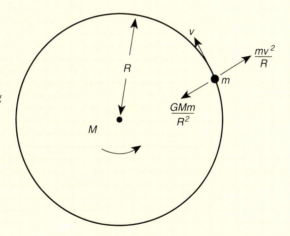

This gives $mv^2/R = GMm/R^2$ and since $v = 2\pi R/P$, R may be expressed in terms of M and P:

$$R < (GMP^2/4\pi^2)^{1/3} \text{ for stability.}$$

If the spin period is 1s and M is one solar mass, then $R = 1500$ km is the maximum radius of the compact object, for at this size gravity is only just balancing the centrifugal force. In any case, this is already much too small to accommodate a white dwarf, which has a typical radius of 10 000 km. The object must therefore be a neutron star, which as we shall see shortly has a typical radius of 10~15 km, well within the stability criterion evaluated here.

It was quickly realised that there were two basic forms that such binary systems could take and in which mass could be transferred from one star onto the other (see figure 7.4).

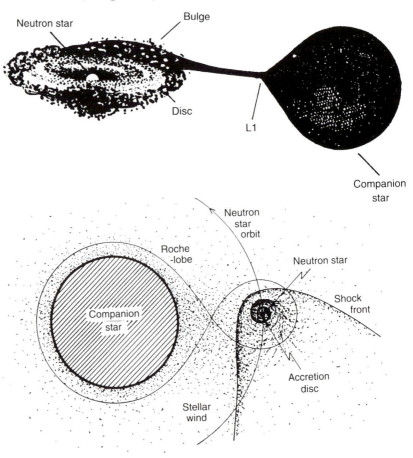

Fig. 7.4 Artist's conception of the two principal mechanisms by which matter is transferred onto a compact object in a binary system. In the upper drawing, a low-mass star (which has a very weak stellar wind) has evolved and is losing mass to its degenerate companion because it fills its Roche lobe *and mass flows through the* L_1 *point. It is the presence of the compact object that distorts the mass-losing star into this shape (rather like a pear), the point of which is the* inner Lagrangian point, *or* L_1. *Matter flowing out of the star forms a stream that impacts the accretion disc (creating a bulge or thickened region) and, by viscous forces, is gradually accreted onto the compact object where the X-rays are generated. The view is for binary phase 0.25. In the lower drawing, the compact object is orbiting a massive star which has a very powerful stellar wind – so powerful that there is sufficient material being lost in all directions for the compact object to accrete and produce copious X-rays as it ploughs through this wind, thereby creating a* comet shaped shock front. *The disc is small and so fluctuations in the wind density are immediately evident in the X-ray flux and pulsar frequency. (Courtesy EXOSAT observatory, ESA.)*

(a) Class I: massive X-ray binaries (MXRBs).
In these binaries the material for transfer onto the compact object is provided by a powerful stellar wind from an early-type, massive star similar to those in the OB associations discussed in chapter 5. Optically we see only the early-type star because it is so luminous, even compared to the X-rays emitted from the compact object. Cen X-3 and Cyg X-1 are both in this class.

(b) Class II: low-mass X-ray binaries (LMXBs).
If the mass-losing star is of low mass then it will not have a strong stellar wind and hence cannot power the X-ray source by the same mechanism as in class I sources. In this case the companion star has evolved to fill its *Roche lobe* (surface of equal gravity) and is transferring material through the *inner Lagrangian point*, L_1, onto the compact object (see figure 7.4). Sco X-1 is in fact a class II source, as is Her X-1.

Both of these pulsating X-ray sources (Cen X-3 and Her X-1) became cornerstones in the study of X-ray binaries because, as we shall see, they provided detailed information about the material near the compact object itself. In addition, even though the locations on the sky of X-ray sources as determined by Uhuru were relatively poor (especially compared to those produced subsequently by SAS-3 and Einstein) and there were consequently many stars in the *error boxes* (sky charts of the locations), the X-ray pulsations and binary periods gave an extra *key* with which to hunt for the optical counterpart. In fact, this proved invaluable in locating Cen X-3 as the visible star was finally confirmed by the discovery of an optical modulation at the binary period.

These discoveries transformed and stimulated the field of X-ray astronomy into a period of exciting growth that has been maintained until the present day, turning it into a major branch of observational astronomy.

7.2 Field guide to X-ray double stars

This brief and historical introduction to X-ray binaries hides what is now a bewildering variety of objects and behaviour patterns; at times it seems there are as many *classes* as there are objects to put in them! Indeed, we shudder to think how many times different objects' behaviour has been described as *unique*! However, we shall try to describe the most important features of the X-ray binaries to indicate where they fit in with each other, and to show how they reached their present state. It is important to remind the reader, though, that this is a very active field and the detailed evolution of binary stars as they transfer material between them is still a subject of debate and sometimes great uncertainty.

Basically, many of the properties of a binary depend on the nature of the compact object – is it a white dwarf, neutron star or black hole ? Tables 7.1 and 7.2 provide an overview of the range of X-ray binaries that we are now dealing with and which we shall attempt to put into perspective. This only includes *interacting* binaries in which matter is being transferred from one component to another, thereby making accretion energy available. (Non-interact-

Table 7.1.   Types of X-ray emitting double stars

Compact object Size	White dwarf 10 000 km			Neutron star 15 km		Black hole 5 km
Magnetic field Class	WEAK Dwarf nova	INTERMEDIATE Intermediate polar	STRONG AM Her system	WEAK Bulge source (QPOs)	STRONG X-ray pulsar	–
Accretion disc	Yes	Partial	No	Yes	Yes	Yes
Companion star	LOW MASS Cataclysmic variables	RED GIANT Symbiotic stars		LOW MASS LMXB Class II Sco X-1 Bursters	HIGH MASS MXRB Class I Vela X-1 *Be* systems	A0620–00, Cyg X-1
Typical periods	2–10 hours	days		11 mins – 10 days	3 days – months	hours – days

Table 7.2. X-ray binaries

Type	Donor star	Compact object[a]	Accretion disc	Examples
MXRB	OB I-III	n.s., b.h.	small	Cen X-3; Cyg X-1
Be system	*Be*	n.s.	small	A0535+26
LMXB	K-M V	n.s., b.h.	yes	Sco X-1
LMXB	A-F V	n.s., b.h.	yes	Her X-1; Cyg X-2
LMXB	w.d.	n.s.	yes	4U1820–30
CV (Dwarf nova)	K-M V	w.d.	yes	U Gem; SS Cyg
CV (Polar)	K-M V	magnetic w.d.	no	AM Her
CV (I.P.)	K-M V	magnetic w.d.	ring	DQ Her

[a] n.s. = neutron star; b.h. = black hole; w.d. = white dwarf; CV = cataclysmic variable

ing systems in which binarity enhances the normal X-ray output of one member, as in the RS CVn systems, have already been described in chapter 4).

In fact, only one white dwarf source, EX Hya, was identified by Uhuru, but, because of the limited spatial resolution and other candidate objects, it was not confirmed until much later. Hence, all the observational and theoretical work in the Uhuru era concentrated on the neutron star/black hole systems. They represented the *standard X-ray star* and accounted for many of the remarkable properties which attracted attention after the discovery of Sco X-1. The important point is that the enormous gravitational potential of neutron stars and black holes can yield very high luminosities from relatively small amounts of infalling material (see box 7.2). Remember it was the high X-ray luminosities that were completely unexpected.

Box 7.2 The power of accretion

Consider a compact object of mass M accreting material at a rate dM/dt, then the luminosity resulting from this accretion will simply be the rate at which gravitational energy is released:

$$L = \frac{GM(dM/dt)}{R}$$

where it is assumed that most of this energy is liberated near the object's surface (of radius R). For an M of one solar mass, dM/dt need only be about $10^{-8} M_\odot$ per year to release a luminosity of 10^{38} erg s^{-1}, which is close to the *maximum* seen from any galactic X-ray source.

It is interesting to compare the efficiency of this energy generating process for the three types of compact object with that for nuclear energy. The luminosity obtained from the given mass accretion rate may be expressed as a fraction of the total energy the matter possesses (according to Einstein's well-known formula, $E = mc^2$):

$$L = \eta \, dM/dt \, c^2$$

where $\eta = G M / R c^2$

then we find that $\eta \sim 0.1$ for neutron stars;
              0.06–0.42 for black holes;
              0.001   for white dwarfs;
              0.01–0.001 for nuclear reactions.

A typical neutron star system therefore has an object the size of a small city (Oxford for example) with a mass equivalent to that of the Sun but radiating in X-rays alone 10 000 times that of the Sun at all wavelengths! The amount of matter that needs to fall into the *potential well* to produce this luminosity is only about $10^{-8} M_\odot$ per year, but that is still a hefty amount at a million million tonnes per second!

In comparison, the white dwarf sources are less luminous simply because the compact object is less dense. Even though there are many, many more white dwarf systems than classical X-ray binaries, they are only seen at all because of their proximity to us (almost all those known are within a few hundred parsecs of us, whereas the neutron star/black hole systems can even be seen in our neighbouring galaxies at distances of millions of light-years). They are correspondingly much rarer.

For the remainder of this chapter we shall consider X-ray binaries with massive companion stars (the MXRBs), whereas the low-mass systems (LMXBs) will be dealt with in chapters 8 and 9, and their white dwarf analogues (the cataclysmic variables) are in chapter 10. It is fairly straightforward to distinguish between binaries that contain a white dwarf and those that contain a neutron star (the latter usually being *much* brighter in X-rays), however the distinction between a neutron star and black-hole compact object can be much more subtle. This we shall return to in chapter 11.

7.3 Discovery of X-ray pulsars

It was the discovery by Uhuru of the eclipsing X-ray source Cen X-3 (figure 7.2) which demonstrated beyond any doubt that, for many galactic objects, we are dealing with a binary star phenomenon. That they are also X-ray pulsars with periods of order of seconds provided valuable clues as to the nature of the compact object. Uhuru found several more eclipsing pulsars, including Vela X-1 and 4U1700–37, and there are now about 20 X-ray pulsars known. Almost all of these (except for two which will be discussed in the next section) were found to be associated with massive, early-type mass-losing stars, usually on the basis of the known orbital period. Details of the observed parameters of some of these X-ray pulsars are contained in table 7.3.

Table 7.3 X-ray pulsar orbital parameters

Source	P_{orb} (days)	$a \sin i$ (light-s.)	Mass function $(M_\odot)$	Eccentricity
LMC X-4	1.408	26.0	9.4	–
Her X-1	1.700	13.1831	0.85	<0.0003
Cen X-3	2.087	39.664	15.386	0.0008
4U1538–52	3.728	52.8	11.4	–
SMC X-1	3.892	53.46	10.84	<0.0007
4U1907+09	8.376	83	8.8	0.22
Vela X-1	8.964	113.0	19.29	0.092
4U0115+63	24.309	140.13	5.007	0.3402
2S1553-542	30.6	164	5.0	–
V0332+53	34.25	48	0.101	0.31
GX301–2	41.508	371.2	31.9	0.47

These systems are of particular importance in X-ray astronomy because they enable the mass of the neutron star to be determined directly by observation. This is possible through the measurement of the velocity of *both* members of the binary. As is clear in figure 7.5 the spin period of the pulsar exhibits *two* kinds of modulations to its basic clock or spin rate. The first is a

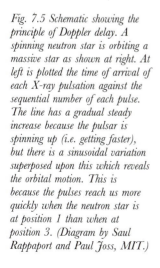

Fig. 7.5 Schematic showing the principle of Doppler delay. A spinning neutron star is orbiting a massive star as shown at right. At left is plotted the time of arrival of each X-ray pulsation against the sequential number of each pulse. The line has a gradual steady increase because the pulsar is spinning up (i.e. getting faster), but there is a sinusoidal variation superposed upon this which reveals the orbital motion. This is because the pulses reach us more quickly when the neutron star is at position 1 than when at position 3. (Diagram by Saul Rappaport and Paul Joss, MIT.)

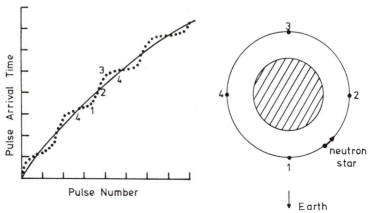

gradual and continuous decrease in the spin period (i.e. the neutron star is being 'spun-up') which we shall explain later. The second is a regular variation which is due entirely to the orbital motion of the neutron star about its companion (as shown in figure 7.5).

The mass-losing stars in these systems are all bright and of early spectral type (usually earlier than B2), and hence are easy to observe from the ground at high spectral resolution. It is therefore straightforward to measure the Doppler shift of the absorption lines of the primary through the binary orbit. Such a system is termed a classical *double-lined spectroscopic binary*. The one difference with classical astronomy is that one of the components is observed entirely in X-rays and not in the optical. Since they are eclipsing, the inclination of the system is known to a high degree of accuracy and we can therefore solve directly for the masses of each component, the separation of the two stars and the physical size of the early-type star (see box 7.3).

Box 7.3 Measuring neutron star masses

Determining stellar masses in binaries by measuring their relative velocities and the size of the orbit has been the fundamental method by which we have determined the masses of stars for almost a hundred years. (Indeed, masses for stars *not* in binary systems can *only* be estimated by indirect means and are notoriously inaccurate.) Radial velocity studies can easily be applied to X-ray binaries in which the mass-losing star is observable in the optical and the compact object is an X-ray pulsar. The velocity of each can be determined by the Doppler effect, one in the optical, the other in X-rays (since the

Fig. 7.B2 Definition of parameters in a binary system (top), and the geometry of the eclipse of the X-ray source by its large companion star.

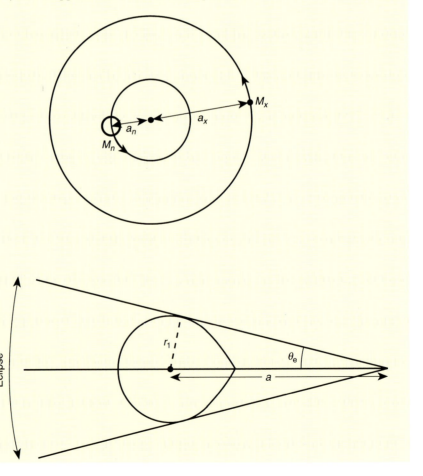

observed pulsar period will change smoothly through the orbit by the same process, as demonstrated in figure 7.3).

Referring to figure 7.B2 and assuming a circular orbit for simplicity, application of Kepler's Third Law gives

$$(a_x + a_n)^3/P^2 = G(M_x + M_n) / 4\pi^2$$

and the centre of mass is calculated from $M_x a_x = M_n a_n$ where x, n refer to the X-ray and normal stars respectively. But we actually observe $a_x \sin i$ (i is the inclination angle between the orbital plane and the plane of the sky) and $a_n \sin i$ through measuring the so-called K velocities of the optical and X-ray stars:

$$K_x = 2\pi a_x \sin i / P \text{ (since } v = 2\pi r/P\text{)}$$

$$\text{and } K_n = 2\pi a_n \sin i / P.$$

It is immediately possible to calculate the mass ratio of the two stars, q ($= M_x/M_n$), but without some indication of the value of i we cannot find the individual masses. This is clear by introducing the *mass function*, defined from the above equations as

$$f_x(M) = \frac{M_n^3 \sin^3 i}{(M_n + M_x)^2} = \frac{4\pi^2 a_x^3 \sin^3 i}{GP^2} = \frac{PK_x^3}{2\pi G}$$

which is obtained entirely from the observable quantities P and K_x, but which needs i in order to solve for the masses themselves. An estimate for i is forthcoming in the eclipsing systems from the value of the eclipse half-angle, θ_e. Using the geometry above we obtain

$$r_1 / a = (\cos^2 i + \sin^2 i \sin^2\theta_e)^{1/2}$$

where a is $a_x + a_n$, the total separation of the stars. The average size of the optical star, r_1, can be estimated from the spectral type, and r_1/a has been computed as a function of q. Hence for a handful of systems it is possible to obtain M_x and M_n directly, and these are the values given in table 7.4.

There are six X-ray binary systems for which such a solution is possible, and the results are given in table 7.4 with figure 7.6 showing the quality of the data available for SMC X-1 and 4U0115+63.

The sizes of the orbits and companion stars are shown to scale in figure 7.7, together with the measured masses. The range of uncertainty for these mass calculations is indicated by the error bars in which all the measurement errors have been taken into account. For comparison, the figure includes the masses of the components of PSR1913+16, the most extensively studied

Table 7.4 The neutron star masses

Source	Companion mass ($M_\odot$)	Companion radius ($R_\odot$)	i	Neutron star mass ($M_\odot$)
Her X-1	1.99	3.86	80	0.98
SMC X-1	16.8	16.3	65	1.06
Cen X-3	19.8	12.2	75	1.06
LMC X-4	14.7	7.57	68	1.38
Vela X-1	23.0	34.0	83	1.77
4U1538–52	16.9	15.2	71	1.8

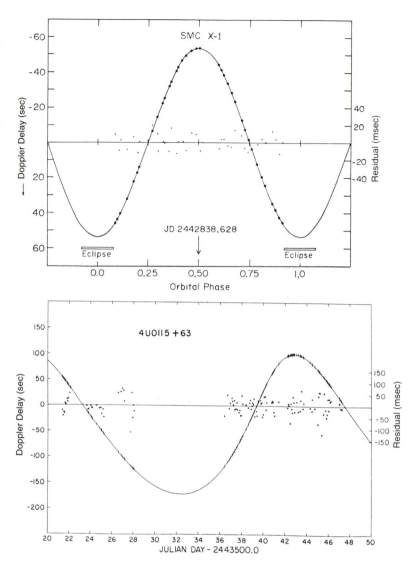

Fig. 7.6 Orbits of two massive X-ray binaries, SMC X-1 and 4U0115+63, as determined from observations of the Doppler delay. SMC X-1 is essentially circular, but 4U0115+63 is notably elliptical. The residuals (i.e. difference between the observations and orbital fits) are displayed at a scale 1000 times greater than the observations themselves. (Diagram by Saul Rappaport and Paul Joss, MIT.)

binary radio pulsar found in the 1970s. These are the most accurate measurements of neutron star masses and define the *canonical* neutron star mass of $1.4\,M_\odot$.

Such accurate mass determinations have an impact way beyond the study of X-ray pulsars. The maximum allowed mass of a neutron star is an important parameter that tells us the *equation of state* of the material inside the neutron star (i.e. how its pressure is related to the density and temperature), and because of the very high densities involved brings us directly into contact with fundamental particle physics. One of the permissible equations of state predicts a maximum neutron star mass of just over $1.5\,M_\odot$, and may already be excluded by the estimated mass of Vela X-1. Clearly even more accurate determinations of these and other X-ray pulsar masses must be made.

7.4 Spinning X-ray pulsars up and down

Accurate masses, based on purely dynamical considerations, have now been measured for these systems, but what is the actual accretion process at work? The basic model was developed by Kris Davidson and Jerry Ostriker

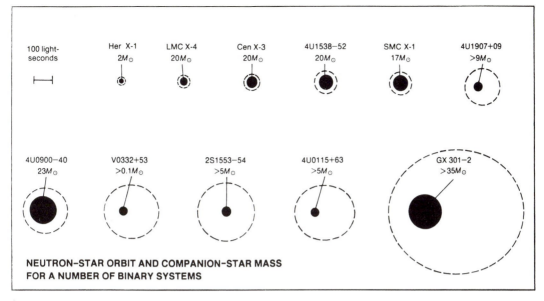

(a)

Fig. 7.7 (a) Orbits to scale for a selection of the massive X-ray binaries, and (b) the masses of the neutron stars that have been measured from optical and X-ray and radio observations. Note that the eccentric binaries are both larger and have the longest periods (10–40 days). The bar denotes the size of 100 light-seconds. It is remarkable that all measured neutron star masses (for both X-ray and radio pulsars) are consistent with the canonical value of 1.4 solar masses. (Diagrams based on originals by Saul Rappaport and Paul Joss, and modified by LANL and F. Nagase.)

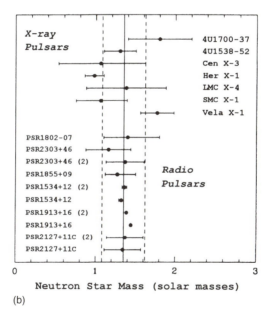

(b)

(Princeton) shortly after the first optical and binary identification of a massive X-ray system. It is shown schematically in figure 7.B3 (see also figure 7.4).

Ultraviolet observations of early-type stars by OAO-Copernicus and IUE show that such stars have prodigious stellar winds. Values as high as 10^{-5} to 10^{-4} $M_\odot$ per year are known, but 10^{-6} is typical. The wind is driven by the extremely high luminosity of these early-type stars (they can be 10 000 times brighter optically than the Sun; for comparison, the Sun's wind carries with it about 3×10^{-14} $M_\odot$ per year). Because of the high temperatures of early-type stars they are powerful emitters of UV radiation, and it is this that is absorbed and scattered readily by surrounding material, thereby imparting momentum and creating an expanding wind. This high mass loss occurs in all directions from the surface of the star, and the orbiting neutron star must pass through it. The calculation of how much of this material will be captured by the compact object was first performed by Bondi and Hoyle and has been applied in this case (see box 7.4). Even though only about 0.1% of the material is captured by the orbiting star, this is sufficient in many cases to power the observed X-ray luminosities. In some cases, including Cen X-3, the observed wind mass-loss rate is insufficient to power the source entirely and an additional mechanism is needed to give the neutron star the fuel it needs. This probably arises from the evolution of the early-type star towards filling its Roche lobe, whereby matter spills over the inner Lagrangian point directly onto the compact object. This is currently a matter of debate.

Box 7.4 Bondi–Hoyle accretion

The calculation of the amount of material accreted from a moving medium (e.g. a stellar wind) by a compact object was first performed 40 years ago by Hermann Bondi and Fred Hoyle (for the case of accretion by a body travelling through a uniform density medium, such as interstellar space). It can be applied quite straightforwardly to the case of the massive X-ray binaries. The basics of the geometry of the problem are shown in figure 7.B3.

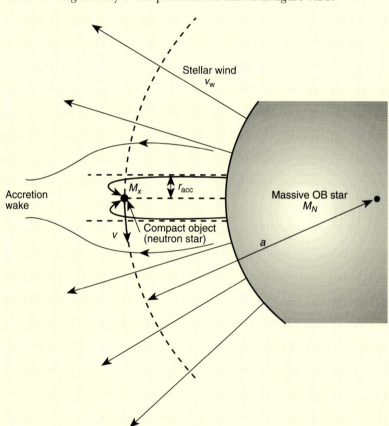

Fig. 7.B3 Geometry of accretion onto a compact object orbiting its massive companion. The wind of material leaving uniformly in all directions from this star's surface will accrete onto the neutron star only if it is within a critical distance (r_{acc}) of the compact object, thereby forming an accretion cylinder. Material that is perturbed in its flow but does not actually accrete forms an accretion wake beyond the compact object.

It is assumed that the massive star possesses a wind that flows out uniformly in all directions. Material within a radius r_{acc} will be accreted by the compact object, whereas material outside this *cylinder* will escape. This radius is calculated by noting that material will only be accreted if it has a kinetic energy less than the potential energy in the vicinity of the compact object, which has mass M_x; i.e. it is set by

$$\frac{1}{2}mv_{rel}^2 = \frac{GM_x m}{r_{acc}}$$

for a particle of mass m, which gives

$$r_{acc} = \frac{2GM_x}{v_{rel}^2}$$

where v_{rel} is the *relative* velocity of the compact object and the stellar wind. This is computed from

$$v_{rel}^2 = v^2 + v_w^2$$

where $v^2 = G M_n / a$. The normal (early-type) star has mass M_n and a is the radius of the orbit. The amount of material accreted by the compact object is then given by the amount inside the accretion cylinder, which is

$$\dot{M} = \pi r_{acc}^2 v_{rel} \rho$$

where the density ρ can be calculated from the assumption that the wind is uniform,

$$\rho = \frac{M_w}{4\pi a^2 v_w}.$$

These equations can be solved to give the fraction of the wind that is accreted:

$$\frac{\dot{M}}{\dot{M}_w} = \left(\frac{M_x}{M_n} \right)^2 \frac{(v / v_w)^4}{[1 + (v / v_w)^2]^{3/2}}$$

which yields values of order 10^{-3} to 10^{-5} for typical massive X-ray binaries.

The next step is to ask how the accreted gas finds its way onto the neutron star and what is the physical origin of the pulsations? Within the capture radius of figure 7.B3, the material will probably form an accretion disc as a result of its (small) angular momentum. However, as the gas moves through the disc towards the neutron star, it will reach a point at which its motion is controlled by the powerful magnetic field of the collapsed star. This point is called the *magnetospheric radius* and is shown in figure 7.8.

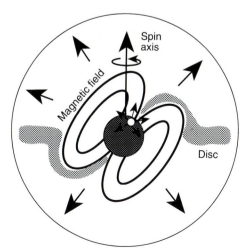

Fig. 7.8 Schematic of an X-ray pulsar and its magnetosphere. At the magnetospheric boundary, the infalling material from the accretion disc is forced to travel along the magnetic field lines to the polar caps of the neutron star. It is in these regions that the X-rays are generated. An off-axis viewing angle from Earth then causes the periodic modulation that we see as an X-ray pulsar.

Spin axis

Magnetic field

Disc

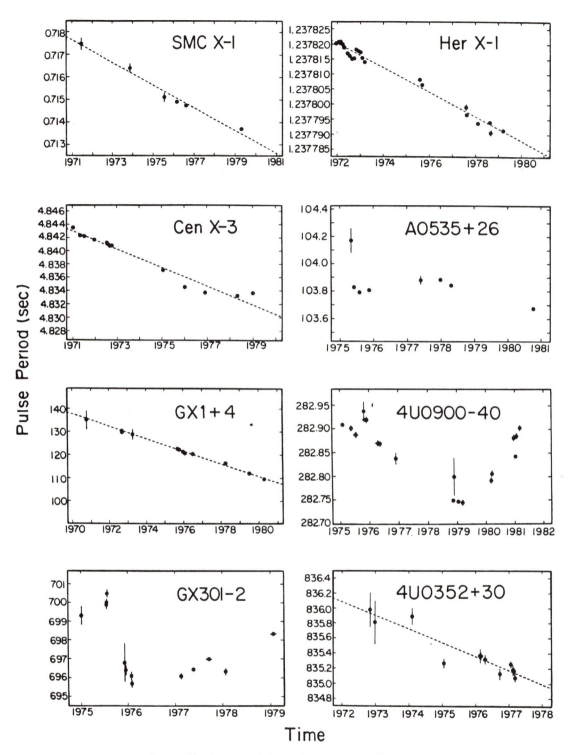

Fig. 7.9 The change in period of a selection of massive X-ray binaries is here shown over timescales of years. Most of them display a general trend towards decreasing period with time. In other words they are spinning faster. However, the reader should also note the deviations (sometimes quite wildly) from this trend, as in the cases of GX301–2 and 4U0900–40. (Diagram by Saul Rappaport and Paul Joss, MIT.)

The gas will now move along the field lines towards the magnetic polar caps of the neutron star. As material hits the neutron star surface a very hot shock is formed in which X-rays are produced. Because of the existence of the column of material above it, the X-ray emission is not uniform in all directions but shadowed into a kind of fan beam. In order to produce a modulation of the X-rays at the neutron star spin period, it is necessary for these magnetic pole caps to be displaced from the rotation axis. In this way a kind of lighthouse effect occurs and regularly changes the angle of view of the X-ray region.

These X-ray pulsars are all (on average) spinning up (figure 7.9). The accreted material from the companion star has angular momentum which is ultimately transferred to the neutron star. A simple extension of the Bondi–Hoyle accretion model can be used to calculate the magnitude of the expected spin-up rate and is compared with the observed rates in figure 7.10. The main effect is that the rate increases with the luminosity of the source. This is fairly obvious since the luminosity is proportional to the accretion rate, and more material will carry more angular momentum.

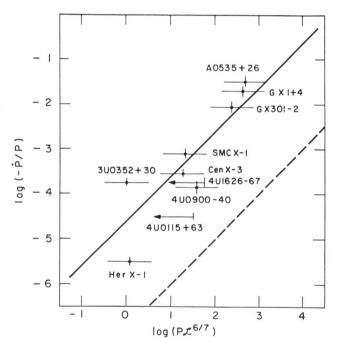

Fig. 7.10 The rate of change of the periods (the period derivative) of figure 7.9 is related to the X-ray luminosity of the pulsar. The largest rates of change are to the top of the y-axis, the highest luminosities to the right of the x-axis. Crudely put, the higher the X-ray luminosity, the more material is accreted by the pulsar and hence the faster it is spun up. The harder you hit a spinning top, the faster it will spin! (Diagram by Saul Rappaport and Paul Joss, MIT.)

Many of these sources, however, have periods of inactivity when they are weak or undetectable in X-rays. We presume that the accretion rate is much lower during these periods and so spin-up will not occur. Indeed, we know from observations of single radio pulsars that they *spin down* with time as a result of the energy loss in accelerating relativistic particles (see chapter 3). It is therefore to be expected that both spin-up and spin-down will be seen to occur, although many of the individual observations are peculiar and not well understood.

7.5 The X-ray source as a probe

The presence of the X-ray source so close to the mass-losing star might have a significant effect on the behaviour of the star's atmosphere and wind. The source itself can be used to probe the structure of the wind. Eclipsing systems are used to observe how the X-ray intensity and spectrum changes as the source is eclipsed by its supergiant companion. The best examples of this are 4U1700–37 and Cen X-3 which also cover a large range of luminosity,

Cen X-3 being almost 100 times brighter than 4U1700–37. The X-ray source is seen through different parts of the primary's wind and atmosphere as the eclipse progresses (figure 7.11).

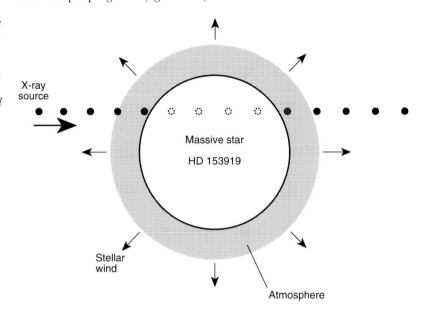

Fig. 7.11 Schematic of the eclipse of an X-ray source by its massive companion. The eclipse is not sharp because the star contains an atmosphere and is generating a powerful stellar wind of outflowing material. This material becomes more opaque to X-rays the closer it is to the star. It is possible then to estimate the density and structure of the material by observing the change in X-ray intensity and spectrum as the source goes into and then comes out of eclipse. Such observations (of 4U1700–37) were undertaken by EXOSAT and the light curve is shown in figure 7.12.

If the wind is dense and not highly ionised then it can obscure low energy X-rays thereby changing the observed spectrum of the source. We characterise this as a change in the column density of material between us and the X-ray source. On the other hand if the X-ray source is extremely bright it is capable of completely ionising all the material in the wind as soon as it leaves the stellar surface. Such ionised gas cannot absorb X-rays and so the spectrum will be little affected as eclipse approaches, with the source brightness dropping abruptly as it is finally occulted by the primary. The long (days) binary periods of MXRBs mean that long observing runs are required to study these effects, preferably with no break in coverage near the eclipse itself. EXOSAT was an ideal mission to undertake such observations and figure 7.12 shows an example, the light curve of 4U1700–37.

Outside eclipse, 4U1700–37 shows flaring-like behaviour, even on a timescale of minutes. The average luminosity of this object is completely consistent with that expected for accretion from a wind (see box 7.4) and so the variability indicates the structure or inhomogeneity that is present in the wind. The X-ray spectrum inferred from the data of figure 7.12 enable the column density to be calculated and compared with that expected for a normal stellar wind outflow (figure 7.13). The fit is extremely good, indicating that the X-ray source has not dramatically influenced its environment, and the wind is much as would be expected for a single star. However, there is one curious point concerning the variation of column density around the orbit. This is the increase that occurs just after phase 0.6, shortly after the neutron star has passed *in front* of the primary (remember that eclipse of the X-rays occurs from phase 0.9 to 0.1). It is best explained as being part of an additional gas stream from the primary that misses the neutron star (see figure 7.14). The reason for this is that the primary is *not* corotating with the neutron star. Normal tidal lobe overflow (transferring material directly from one star to the other) will *not* occur.

The X-ray spectrum of the much more powerful Cen X-3 reveals a very different system. Away from eclipse there is no such modulation as in

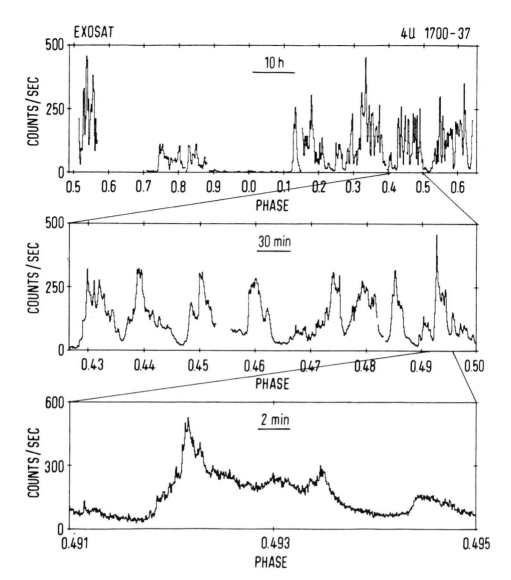

Fig. 7.12 An unbroken 80 hour observation of 4U1700–37 by EXOSAT. The X-ray emitting neutron star is eclipsed by its supergiant companion from phases 0.9–1.1, but shows extraordinary chaotic variability outside eclipse. This is almost certainly due to substantial structure in the wind emanating from the supergiant. Note the asymmetry in the absorption prior to eclipse. (Courtesy of EXOSAT Observatory, ESA.)

4U1700-37. We therefore infer that the high X-ray luminosity has completely ionised the stellar wind material making it incapable of absorbing the X-rays. However, very close to eclipse it is possible to see structure (the pre- and post-eclipse dips) and these are shown in detail in figure 7.15.

These clearly do not fit the same variation expected from a stellar wind. They have been modelled successfully by Charles Day (at the University of Cambridge) as an exponential atmosphere of the supergiant itself. The scale height of this atmosphere is typically 1/10 that of the supergiant radius and implies that the star will have a hot (10 million degrees) corona. The variations in the eclipse behaviour are due to the fact that, in Cen X-3, the supergiant is at least close to corotation and is rotating almost at break-up speed. Centrifugal effects will be important near the surface of the star, where gravity can barely keep material from flying off.

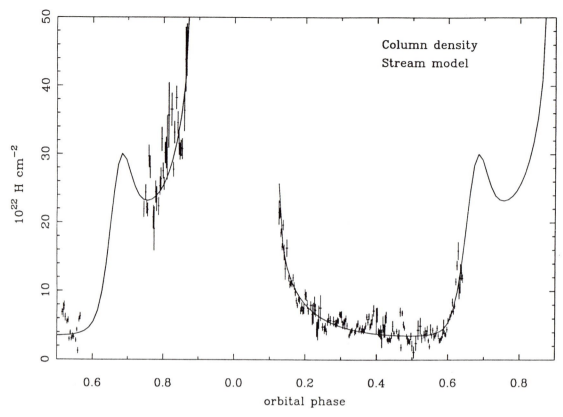

Fig. 7.13 *The variation in amount of material through which the X-rays are passing in 4U1700–37 (fig 7.12) is plotted against orbital phase. The eclipse is at phase 0 when the X-rays drop to zero (or amount of material becomes infinite!). The increase in this column of material leading into eclipse (the ingress) is clearly evident, as it is also in the egress. The solid curve represents the model calculation. The main structure is as expected for a normal stellar wind outflow. The unusual feature is the increase in column just after phase 0.6 (when the X-ray source has just passed in front of the companion star). (Courtesy of EXOSAT Observatory, ESA.)*

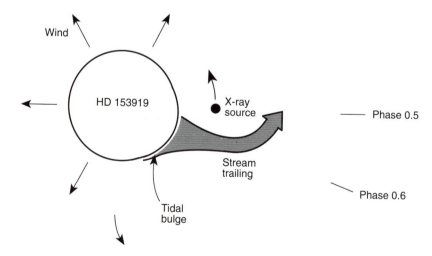

Fig. 7.14 *Schematic model showing a possible explanation for the enhanced absorption at phase 0.6 in 4U1700–37 (fig. 7.13). The massive star and X-ray source are* not *corotating (i.e. it does not rotate once for every orbit) and so a tidal bulge could give rise to a stream that misses the neutron star and trails behind it, thereby giving rise to the extra absorption.*

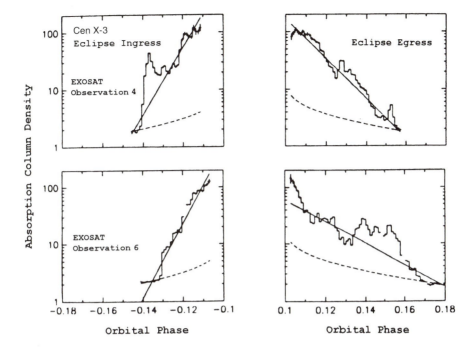

Fig. 7.15 The change in amount of absorption for eclipse ingress and egress in Cen X-3 as observed by EXOSAT. A normal stellar wind (dashed curves) gives a very poor fit to the data, whereas an exponential stellar atmosphere (solid curves) is very good. (Based on original diagrams by Charles Day, GSFC.)

7.6 X-ray transients

Of all the different types of X-ray sources discovered in the last decade, the most difficult to identify and study have been the X-ray transients. These flare up suddenly and then fade into invisibility after only a few weeks. One of us (FDS), after a rocket observation on May 18, 1967 that found a new bright X-ray source (Cen X-2) where none had previously been seen, thought for a few days that his group had discovered the first-ever X-ray transient. However, when the team returned from the field, there was a report on FDS's desk announcing the discovery of this source by J. Harries and K. McCraken (University of Adelaide) and R. Francey and A. Fenton (University of Tasmania) with a rocket launched from Australia on April 4, just 44 days before our observation. This source was greeted with great enthusiasm by the Australian group because the first estimate of its location placed the source in the constellation Crux, the Southern Cross, which is not only often visible in the southern sky, but is also reproduced on the Australian flag. Alas, the more accurate American information placed it in Centaurus, a much larger constellation surrounding the smaller Cross on three sides.

Cen X-2 was also observed close to maximum intensity by Brin Cooke and colleagues (University of Leicester) during an April 10 rocket flight to survey the southern sky. The source was absent (below the detection threshold, or 50 times weaker than when at maximum intensity) when FDS' group from Lawrence Livermore Laboratory attempted a repeat observation on September 28. The source was classified as *nova-like*. Cen X-2 has not been seen again and remains unidentified to this day.

The next discovered transient, Cen X-4, flared on July 9, 1969 and was observed throughout its complete outburst by detectors on the two Vela 5 satellites (figure 7.16). A light curve covering 85 days was published by Doyle Evans, Richard Belian and Jerry Connor (Los Alamos National Laboratory) which demonstrated a variation in intensity of a factor of over 300. A second outburst of this source was observed on May 12, 1979 by Lou Kaluzienski, Steve Holt and Jean Swank (Goddard Space Flight Center) with the Ariel V

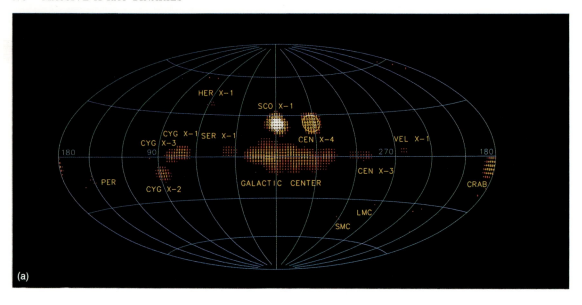

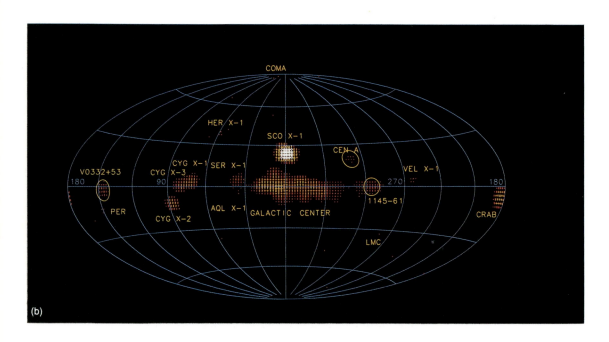

Fig. 7.16 *X-ray transients observed by the Vela 5 satellites. These two pictures of the X-ray sky are in galactic coordinates and show the average brightness of the X-ray sources during the first year of the Vela operation (1969) and in 1973. The bright X-ray transients that occurred in those years are ringed and stand out clearly on comparison of the two maps. Because they monitored the sky all the time, these satellites were ideal for studying the long-term variability of the bright X-ray sources. (Courtesy of Jim Terrell, LANL.)*

all-sky monitor. This resulted in an identification by Claude Canizares and Jeff McClintock (MIT) and Josh Grindlay (Harvard) with a 13th magnitude star which, at the time of the outburst, was at least 6 magnitudes brighter than normal.

The source in quiescence appears to be a late-type star (spectral type K3V), but with a superposed brightness variation that has revealed the binary period to be 15.1 hours. This system, which produces outbursts apparently once every 10 years, is probably a red dwarf and a neutron star bound in an approximately circular orbit. The mechanism which produces these rare episodes of mass transfer onto the neutron star, with consequent strong X-ray emission, remains unknown. Other transients similar to Cen X-4 have become of great interest to astronomers because, during quiescence, it often becomes possible to study the secondary (mass-losing) star in greater detail than in most other X-ray binaries. From this the mass of the compact object can be measured accurately with, as will be shown in chapter 11, quite spectacular results.

Two decades of observing the X-ray sky have produced about 30 transient X-ray sources, and shown that they can be divided into two classes. The first class show very hard (hot) X-ray spectra and are associated with MXRBs, usually *Be* stars, as will be described in the next section. The other class have softer (cooler) spectra and, when identified, are found to be low-mass systems (i.e. the optical companion star is a small, late-type star, usually K–M) which brighten dramatically (usually by 6–8 magnitudes) during the X-ray outburst. This group is called the *soft X-ray transients* and, as will be shown in chapter 11, give us a unique opportunity to directly measure the mass of the compact object in LMXBs.

7.7 *Be* stars and X-ray pulsars

Is there any group of persistent sources which also shows properties in common with the X-ray transients? The answer is yes and they are associated with *Be* stars. In the mid-1970s Laura Maraschi and Aldo Treves (Milan) and Ed van den Heuvel (Amsterdam) noted that the massive X-ray binaries also seemed to divide into two groups: those with OB supergiant primaries showing characteristics of powerful stellar wind mass loss, and those with mid-B giant/main-sequence primaries that were classified as *Be* stars. Now, *Be* stars are a famous class of stars that have been known for more than 50 years and were studied intensively by Otto Struve in the 1930s. The *e* stands for emission lines, which are also variable in both intensity and profile on short and long timescales (hours to years). Their nature has been a subject for speculation for two generations of astronomers. The fact that some of them are in binary systems with neutron star companions (pulsars) has greatly renewed interest in this class of stars.

With the long term X-ray survey by the Ariel V satellite came the discovery that the major outbursts of some of the *Be* systems, which looked just like *normal* X-ray transients, were in fact periodic, but with very long periods. Details of this group are summarised in table 7.5. Perhaps *Be* systems also hold the key to understanding X-ray transient behaviour.

The *Be phenomenon* is almost certainly due to episodes of mass ejection from the star, probably around the equator, so as to form a ring of material. The exact reason for this is unclear, but is probably associated with the fact that *Be* stars are fast rotators and hence are rotating close to their break-up speed.

Any compact object orbiting such a star will almost certainly encounter the ring and likely produce a large increase in its X-ray output. If, in addition, the orbit is eccentric then the X-ray output will be modulated on the orbital period, and by a much larger factor than could be accounted for by a simple stellar wind from the B star. Indeed, in almost all cases the wind from a B

Table 7.5 X-ray emitting Be systems

Source	Spectral type	V (mag)	P_{spin} (s)	P_{orb} (days)	e	L_x(peak) (erg s^{-1})
A0538–66	B2IIIe	15	0.069	16.65	0.7	10^{39}
4U0115+63	Be	16	3.6	24.3	0.34	8×10^{36}
V0332+53	Be(?)	15.5	4.4	34	0.31	?
2S1553–542	?	?	9.3	30.6	<0.09	?
A0535+26	B0Ve	9	104	111	0.3	2×10^{37}
GX304–1	B2Ve	14	272	133	?	3×10^{36}
4U1154–61	B1Ve	9	292	188	?	6×10^{36}
X Per	O9.5III-Ve	6	835	580(?)	?	10^{34}

Fig. 7.17 A plot of the X-ray pulsar periods of the massive X-ray binaries against their orbital periods. The squares are the supergiant binaries in which mass transfer is via a stellar wind, whereas the circles are Be systems. The crosses are supergiant systems also, but they show evidence for an accretion disc as well. The lines labelled POLAR are for a normal stellar wind (uniform in all directions), whereas the EQUATORIAL line is the result expected if the mass loss is via an equatorial ring of material (i.e. in the orbital plane only). (Based on original diagram by Waters and van Kerkwijk, University of Amsterdam.)

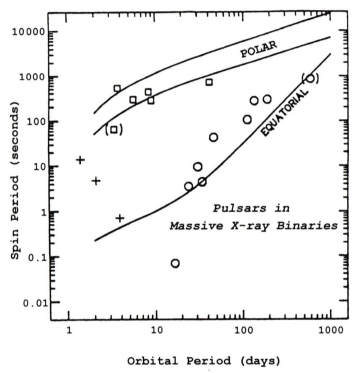

Orbital Period (days)

star could not possibly give the observed X-ray luminosities from these systems. There is strong support for this interpretation from the distribution of X-ray pulsar periods in *Be* systems as a function of their binary periods (see figure 7.17).

The supergiant systems yield a relationship which is almost constant with orbital period and consistent with that expected from the known stellar wind variation of density. The *Be* systems, however, show a strong correlation, first noted by Robin Corbet (Oxford) and one that is very different from that of a stellar wind. Basically, they are consistent with a much *slower* wind, much as would be expected from an equatorial ring around a *Be* star. The larger orbital periods require a greater separation of the neutron star and *Be* companion which implies less matter to spin up the neutron star, and hence a longer spin period.

Of course, because of the episodic nature of the mass ejection, the pulsar spin period will not stay constant. During outbursts it will speed up, and during periods of quiescence it will gradually spin down. But eventually each system will reach a balance between the two and that is assumed to be true for these objects now as they show very little change during outburst.

7.8 An accretion barrier?

One interesting feature of *Be* star outbursts was noted by Luigi Stella and Nick White (ESTEC) and Bob Rosner (CfA) and that was the sudden turn off of X-rays from certain systems as the outburst declined. This was found only in those with the fastest pulsars and implied that some process was stopping the material from accreting even though it was still available. They suggested that a *centrifugal barrier* existed and that this could not be crossed once the mass accretion rate had dropped below a certain level (see figure 7.18).

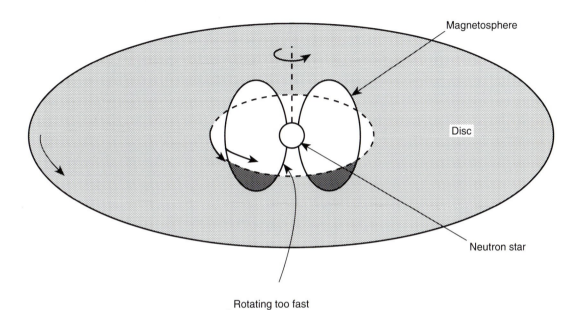

Fig. 7.18 Schematic illustrating the principle of the centrifugal accretion barrier. If the pulsar is spun up sufficiently, or the magnetosphere allowed to grow because of decreased disc pressure, then the material at the inner edge cannot enter the magnetosphere because it would then be travelling faster than its orbital velocity. Hence it would be thrown out instead of being accreted, and the X-ray source turns off.

As the outburst declines the mass transfer rate falls too. This reduces the *pressure* on the magnetosphere of the neutron star, allowing it to expand. Eventually a point is reached at which material entering the magnetosphere would be travelling at a speed greater than its *normal* orbital speed at that radius. It cannot do this and would therefore be ejected from the system by centrifugal forces rather than accreted onto the neutron star. In other words the accretion stops. The X-ray luminosity at which this happens depends of course on the neutron star spin period. The faster the spin period the higher the luminosity at which the system will *turn off*. The observations are shown in figure 7.19.

The maximum luminosities observed for the transient *Be* systems are just above the minimum luminosity line for a 10^{12} gauss magnetic field neutron star. This would have the effect of producing a large orbital variation in an eccentric binary as the X-rays would turn off every time this line was reached. If observations of the precise turn off could be measured then an estimate of the magnetic field of the neutron star could be made.

7.9 A0538–66: the most luminous X-ray transient

From table 7.5 it is clear that A0538–66 is a remarkable object. At its peak it is substantially more luminous than any other stellar X-ray source known. Discovered by Nick White (MSSL) and Geoff Carpenter (Birmingham) in 1977 with Ariel V, and located in the direction of the Large Magellanic Cloud (LMC), its highly variable nature and 16 day peaks during outburst were soon noted. However, its location was not determined well enough for optical identification, because of Ariel V's lack of spatial resolution.

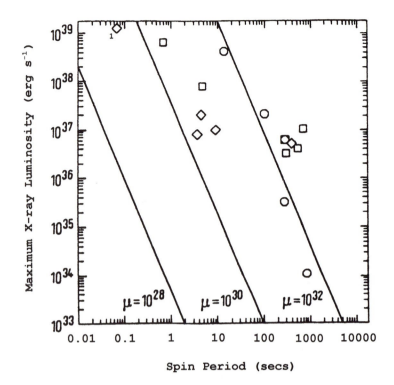

This was soon rectified by HEAO-1 with its scanning modulation collimator (see chapter 2) which provided a ·1 arcminute location. Armed with this information, Mark Johnston (MIT), Richard Griffiths (Harvard) and Martin Ward (Cambridge) immediately found a variable star within the X-ray location on the sky (see figure 7.20). Gerry Skinner (Birmingham) then followed up this identification by searching through Harvard Observatory's celebrated plate library. He found that the optical star also brightens every 16.6 days. What is more, it has been doing so for at least 50 years. However, he also found that there were many occasions when the source was inactive, with no outbursts for very long periods.

The first optical spectrum of this star, obtained by Johnston, Griffiths and Ward in 1980, was at first sight surprisingly ordinary. It showed the hydrogen and helium absorption lines typical of an early B star. There was no sign of the unusual characteristics, such as strong emission lines, found in the optical counterparts of many X-ray sources. However, the large variability range apparent in figure 7.20 was odd. A 2 magnitude change in brightness of an early type star is highly unusual. These massive stars are normally so bright optically that the extra light due to X-ray heating of the surface is actually a rather small effect (usually no more than 0.1~0.2 mag). How then can A0538-66 have an early-type, and hence intrinsically bright, primary and still manage to brighten by at least 2 magnitudes?

It was at this point in 1980 that one of us (PAC), together with John Thorstensen (Dartmouth), recognised the unusual nature of this star and began a series of spectroscopic observations of A0538-66 from CTIO and the AAO. Our optical spectra (figure 7.21) showed remarkable changes between quiescence and outburst. The essentially normal absorption lines of hydrogen and helium were swamped during outburst by enormously strong emission lines of classical P Cygni type. This confirmed that the normal star was then surrounded by a rapidly expanding shell. At times, velocities greater than 3000 km s^{-1} were recorded.

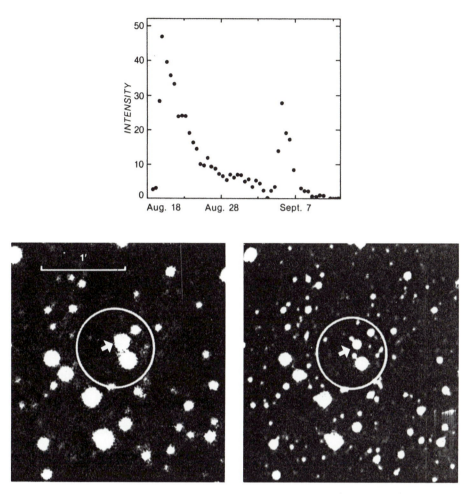

Fig. 7.20 An X-ray outburst of A0538–66 from Ariel V and the corresponding optical behaviour. The X-ray intensity is relative to the Crab Nebula (which equals 1000) and covers a period in 1977. This same outburst was photographed in blue light on September 4, whereas it is significantly fainter a couple of weeks later. The white circle represents the accuracy of the X-ray location and the arrows point to A0538–66. (Optical photographs from the UKSTU.)

The quiescent spectra revealed the velocity of the absorption lines to be 270 km s^{-1} to a high accuracy. This is exactly the recession velocity of the LMC relative to the Sun, and hence establishes beyond doubt that the star is indeed located in the LMC. We therefore know the luminosity of A0538–66 to much greater accuracy than most galactic X-ray sources because we know the distance to the LMC (55 kpc) to about 10%. This meant that, at peak, A0538–66 has an L_x of 10^{39} erg s^{-1}, which is substantially brighter than the next brightest object (which, interestingly enough, is also in the Magellanic Clouds!). This is curious because the 10^{38} erg s^{-1} limit had been thought to have been set by the limiting effects of radiation pressure on the material accreting onto a 1 $M_\odot$ object. The compact object in A0538-66 cannot be much more massive, since the discovery of 69 ms pulsations during an Einstein observation by Marty Weisskopf (MSFC) and Skinner requires that it be a rapidly spinning neutron star.

A key point to understanding A0538–66 is also shown in figure 7.21, and that is the difference between the IUE spectra taken in quiescence and in outburst (by Ian Howarth and Allan Willis of UCL). It is obvious that when A0538–66 is brighter it is redder. This is unlike all known galactic X-ray sources, where increases in X-ray luminosity heat the companion star further and result in hotter, bluer spectra. This physical change is clear in the model atmosphere fits to the spectra summarised in table 7.6.

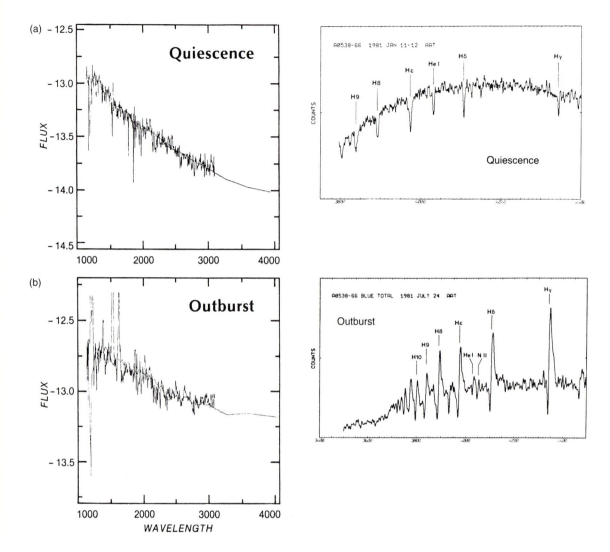

An increase in the size or emitting region of A0538–66 at outburst explains the reddening of its spectrum. During this phase of its 16.6 day cycle, the emitting surface area increases by a factor of almost 16, while the optical luminosity only increases by a factor of 4. Thus less energy is emitted per unit area than at quiescence, and the star appears cooler and redder. Such a large and periodic modulation of the energy output, in both the optical and X-ray regions, indicates that A0538–66 is a compact object in a highly eccentric orbit about a giant companion that is losing mass by the _Be_ mechanism.

The primary (optical) star is shown in figure 7.22 as a _classical Be_ star which is rotating sufficiently rapidly that it is surrounded by an equatorial ring or disc of material. (The widths of the absorption lines in the spectra of figure 7.21 confirm that the star is rotating rapidly.) The companion compact object is in an eccentric orbit that periodically brings it close to the primary and its surrounding ring of material. This matter provides the fuel for each X-ray outburst.

This process occurs so suddenly that the Eddington luminosity limit is temporarily exceeded (see box 7.5). So powerful is the release of radiation that its pressure expands and heats the remaining matter in and around the ring,

Table 7.6.   Physical data for A0538-66

Property	Quiescence	Outburst
Spectral type	B2 giant	B9 supergiant
Temperature (K)	19 000	12 500
Size ($R_\odot$)	12	45
Visual magnitude	15	13
Optical luminosity ($L_\odot$)	12 500	50 000
X-ray luminosity ($L_\odot$)	2 500	250 000

Fig. 7.22 Schematic model of how accretion might proceed in an eccentric, Be binary. The rapid rotation of the B star has created an extended equatorial ring (or atmosphere) which, if the orbit is sufficiently eccentric, the neutron star can pass through or near. If there is enough material and the centrifugal barrier allows, then accretion can take place and an X-ray and optical outburst starts. The enormous X-ray luminosity is then able to drive much of this extended ring material out of the system. For A0538–66 the emitting region during outburst is much larger than during quiescence. (Based on original diagram by Ed van den Heuvel, University of Amsterdam.)

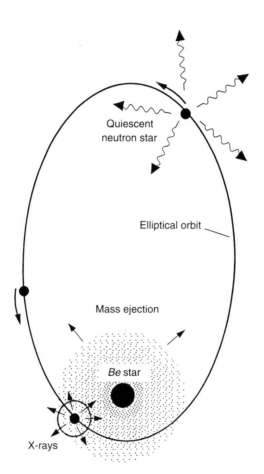

causing much of it to be lost permanently from the system. At the same time the optical brightness increases greatly.

The active period lasted only about 18 months, reaching a maximum in July or August 1981. There have been two further active periods (in August 1983 and January 1988) but they were extremely brief (less than a month or so) and multi-wavelength coverage of these could not be arranged in time. How long it will be until another major activity cycle begins is unknown. Attempts were made by Ginga to search for X-rays during the 1988 optical outburst, but none were detected. Since the optical brightness did not reach anything like the level of the 1981–82 period, this can be explained by the centrifugal barrier model. A0538–66's neutron star has the fastest period of all the Be systems and therefore presents the largest barrier to incoming material. A very large burst of mass transfer will be needed in order to pene-

Box 7.5 The Eddington limit

Although the effect is usually very small, the scattering of electromagnetic radiation off matter exerts a pressure on the material that is proportional to the flux, F, of the incident radiation. The radiation pressure acting on an electron produces a force of magnitude

$$-\frac{\sigma_0 F}{c} = \frac{\sigma_0 L}{4\pi R^2 c}$$

where σ_0 is the Thomson cross-section:

$$\sigma_0 = \frac{8\pi}{3}\left(\frac{e^2}{m_e c^2}\right)^2 = 6.65 \times 10^{-25}\,\text{cm}^2$$

and the flux of radiation is due to a luminosity, L, being generated uniformly over a sphere of radius R. This means that it is possible, if L is large enough, for radiation pressure to overcome the gravitational potential of the compact object and thereby stop the accretion of matter onto it. The luminosity at which this happens is called the *Eddington* luminosity, L_E, and can be estimated from

$$\frac{\sigma_0 L_E}{\left(4\pi R^2 c\right)} = \frac{GMm_p}{R^2}$$

which rearranges to give

$$L_E = \frac{4\pi GM m_p c}{\sigma_0}$$

in which the gravitational force of the compact object of mass M acts on a proton (mass m_p) and the protons and electrons are assumed to be electrostatically bound. Putting in the constants gives

$$L_E = 1.3 \times 10^{38}\,(M/M_\odot)\ erg\ s^{-1}$$

and so a neutron star of mass $1.5\,M_\odot$ cannot produce a steady luminosity greater than about 2×10^{38} erg s^{-1}.

In spite of the simplicity of this calculation (in particular the assumption of spherical symmetry in the mass infall onto the compact object) this effect must be important in the generation of X-rays by neutron stars and black holes since there are no *steady* X-ray sources brighter than about 5×10^{38} erg s^{-1}. In addition there appears to be a maximum X-ray luminosity associated with the production of X-ray bursts on the surface of a neutron star which can also be attributed to the Eddington limit.

trate the magnetosphere by reducing the corotation radius. The small outbursts of 1983 and 1988 were unable to do this and so no X-rays were produced. In fact, A0538–66 is entry number 1 in figure 7.19 and requires a significantly weaker magnetic field than in other *Be* systems in order to be able to produce this prodigious X-ray luminosity. This suggests that A0538–66 may be much older and more evolved than those *Be* X-ray transients we have been studying in our own galaxy.

Nevertheless, this system holds, by a significant factor, the title of the most powerful stellar X-ray source. What is more, it is not located in our own galaxy. This must surely indicate that A0538–66 is in a short-lived phase of its evolution, since any comparable object in the Milky Way would have been discovered even during the early days of X-ray astronomy. Observation of future outbursts will serve to answer many of the remaining questions. In particular, if the centrifugal barrier is indeed the limiting factor in the X-ray behaviour of A0538–66, then this barrier should reduce as the neutron star spins down, thus allowing X-ray emission at lower levels. In addition, the eccentric orbit can be verified by observing the X-ray pulsar for a much larger fraction of the 16.6 day orbit than Einstein was able to do. We also have little idea of what drives the evolution of the *Be* star in replacing the material lost during the accretion passages of the neutron star. Is it purely rotation that replenishes the ring or does the atmosphere of the giant star have to re-expand through normal stellar evolution?

In addition, tidal forces will cause the very eccentric orbit to circularise on the astronomically brief timescale of about 100 000 years, and yet it is thought to be an old system. In such a case, some process must be maintaining the orbit's eccentricity against the circularising forces. But, for the moment, A0538-66 presents us with two remarkable opportunities: first, to study the X-ray transient phenomenon in a *controlled environment*, and second, to watch the extraordinary forced evolution of the B giant's atmosphere as it and its neutron-star companion regularly, but temporarily, become a common-envelope binary producing 10 times the energy normally possible in such a system.

7.10 Evolution of a massive X-ray binary

It is a well-known phenomenon in stellar evolution that the more massive a star the faster it consumes its fuel and the shorter is its life as a *normal* star. This goes against one's intuition that a star which has more fuel will last longer. In fact a $10\ M_\odot$ star will last only roughly $1/100$ as long as our Sun! However, in massive X-ray binaries we find a neutron star (virtually the end-point of a star's life) orbiting a much more massive companion star. How can this be? Surely the more massive star would be expected to have evolved first?

The answer to this paradox is that the neutron star started out as the more massive of the pair, but that it transferred the greater part of its mass to its companion during its very short lifetime. Computer codes have been generated by a number of groups (especially at MIT and Amsterdam) to follow the evolution of such a system, and an example is given in table 7.7.

The simulation starts at time 0 with a massive pair $(13 + 6.5\ M_\odot)$ in a 2.6 day binary assumed to have been formed together (stage *a*). The larger star evolves rapidly through normal hydrogen burning, generates a helium core and then within 12 million years expands to fill its Roche lobe. Further expansion will transfer matter from its envelope (unburnt hydrogen) onto the less massive star and the first stage of mass transfer begins (*b*). Within a mere 60 000 years over $10\ M_\odot$ of matter has been transferred onto the originally less massive star, now weighing in at around $17\ M_\odot$! Only a $2.5\ M_\odot$ helium star remains of the originally more massive component (stage *c*). However, conservation of mass and angular momentum has widened the binary orbit to just over 20 days and left an equatorial ring or disc around the newly massive star because it is rotating so rapidly. It is this ring that imparts the P Cygni profiles to the hydrogen lines which is the signature of a *Be* star.

Further evolution of the helium star procedes rapidly. Its energy source is now helium fusion (to form a carbon core) and within 3 million years it will again have expanded to fill the Roche lobe, thus initiating a second mass transfer stage (*d*). This is completed within 3000 years and only transfers 0.3 $M_\odot$ of mostly helium onto the *Be* star, leaving behind a carbon core (*e*). Within another 50 years this carbon star will collapse and detonate as a supernova to form a neutron star which is now in an even wider (31 day)

Table 7.7 Evolution to a Be X-ray Binary

	Time (yrs)	Orbital period (days)	Stage
(a)	0	2.58	Formation
(b)	12 000 000	2.58	First mass exchange starts
(c)	12 000 000 + 60 000	20.29	Mass transfer stops
(d)	14 800 000	20.29	Second mass exchange starts
(e)	14 800 000 + 2 900	28.09	Mass transfer stops again
(f)	50 yrs after e	30.63	Helium star detonates as a supernova, leaving a neutron star

orbit about the *Be* star (f). An asymmetry in the supernova explosion imparts an eccentricity to the orbit (see figure 7.7) and thus provides the path for the neutron star to pass regularly through or close to the equatorial ring, thereby producing the powerful X-ray outbursts that bring these systems to our attention.

8 Low-mass X-ray binary stars

8.1 Introduction

As the first extrasolar X-ray source to be discovered, and to this day the brightest X-ray object regularly visible in the sky, Sco X-1 was not surprisingly the system that observers and theoreticians alike tried hardest to explain. The binary nature of the massive systems just described was often obvious in the X-ray data alone, and could easily be determined from optical observations as well. However, this other group of X-ray sources, also very bright, gave little clue to their identity from their X-ray variations and behaviour. Perhaps a clue could be obtained from looking at their distribution in the sky?

Figure 8.1 shows that the massive systems (the bright sources away from the galactic centre region) are confined to the plane of our galaxy and, more importantly, are associated with its spiral arms. This is not surprising, for this is exactly where we would expect to find the young stars and early-type associations that contain the most massive (and of course short-lived) stars. However, the figure also shows a grouping of X-ray sources in and around the centre of the galaxy. These have become known as the *bulge sources*, since they have the same distribution as the galactic bulge. The stars in the galactic bulge are very old and of Population II, indeed they are some of the oldest stars in the galaxy. They must therefore be less massive than the Sun, since any higher mass ($> 1\ M_\odot$) stars will have evolved and disappeared long ago. Perhaps the bulge X-ray sources are associated with the same kind of object?

This is substantiated by the optical appearance of Sco X-1 (see figure 7.1). Although Sco X-1's spectrum is very odd, it is certainly not that of a massive star. Depending on the (unknown) distance, any 'normal' star present in the system had to be quite faint in order for it not to be seen, and this meant it had to be of relatively low mass. However, this still begged the question of what was actually producing the X-rays. The fundamental model for objects such as Sco X-1 was developed by Cambridge theoreticians Jim Pringle and Martin Rees in the early 1970s.

This basic idea (figure 8.2) will crop up again and again as different aspects of it are demonstrated by different sources. Essentially, a (slightly) evolved late-type star is transferring material through its inner Lagrangian point into the gravitational field of a compact object, which is either a white dwarf, neutron star or black hole. Because of the angular momentum of the material as it leaves the star it cannot fall directly onto the compact object, but must instead go into orbit around it, thereby forming an *accretion disc*. The formation and evolution of this disc forms one of the most important current

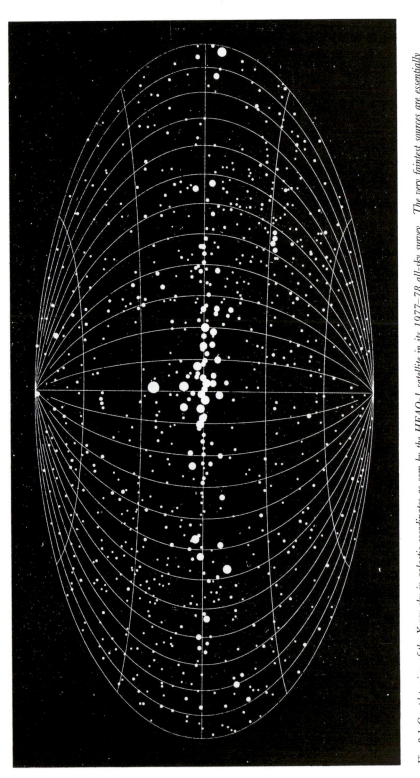

Fig. 8.1 Complete view of the X-ray sky in galactic coordinates as seen by the HEAO-1 satellite in its 1977–78 all-sky survey. The very faintest sources are essentially uniformly distributed over the sky and are therefore mostly extragalactic. The brightest sources, however, are concentrated in the plane of the galaxy. The group around the centre are the galactic bulge sources, whereas the bright sources elsewhere in the plane are mostly massive X-ray binaries associated with young OB stars in the spiral arms of the galaxy. (Courtesy Naval Research Laboratory.)

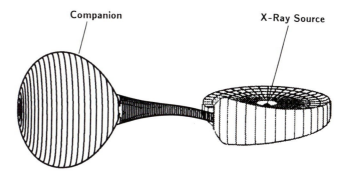

Fig. 8.2 Basic model of a low-mass X-ray binary. The mass-losing star is of low mass and has evolved so that it fills its Roche lobe, thereby transferring mass that, because of its angular momentum, forms an accretion disc around the compact object, usually a neutron star. The accretion stream impacts the disc, forming a bulge which, if the inclination is appropriate as here, can occasionally obstruct our view of the X-ray source (producing dips). (Sketch by Keith Mason, MSSL.)

branches of astronomy with applications ranging from the formation of stars and planetary systems to distant galaxies containing the most powerful processes known in the universe.

The major difference between this model and that for massive systems is in how the matter is transferred from the normal star onto the compact object. A stellar wind carries very little angular momentum and a (large) accretion disc is unlikely to form. The wind from a low-mass (late type) star, however, is feeble (as is the Sun's!), and would lead to completely undetectable levels of X-ray emission. Such a star must therefore *direct* its mass into the gravitational influence of the compact object in order to yield detectable X-rays. But the

Table 8.1.   *Properties of low-mass X-ray binaries*

Source	Period (hrs)	X-ray type	Visual magnitude	Optical modulation	Companion star
4U1820–30	0.19	Burster	–	–	White dwarf
4U1626–67	0.7	Burster	19	Yes	Degenerate
A1916–05	0.83	Burster	21	Yes	Degenerate
X1323–619	2.9	Burster, dipper	–	–	
MXB1636–536	3.8	Burster	17	Yes	
EXO0748–676	3.8	Burster, dipper, transient	17		
4U1254–69	3.9	Burster, dipper	19	Yes	
4U1728–16	4.2	ADC[a] ?	17	Yes	
X1755–338	4.4	Dipper	18.5	Yes	
MXB1735–444	4.6	Burster	17	Yes	
Cyg X-3	4.8		(IR)	Yes(IR)	
4U2129+47	5.2	ADC	16	Yes	
2A1822–371	5.6	ADC	16	Yes	
MXB1659–29	7.2	Burster, dipper	19		
A0620–00	7.3	Transient	12–19	Yes	K
LMC X-2	8.2?		19	Yes	
4U2127+11	8.5	ADC	16	Yes	
4U1956+11	9.3		18	Yes	
CAL 87	10.2	ADC	19	Yes	
GX339–4	14.8	Multi-state	15–21	Yes	
Sco X-1	19.2	Prototype LMXB	12–14	Yes	
4U1624–49	21	Dipper	–	–	
CAL 83	25	ADC	17	Yes	
Her X-1	40.8	Dipper	15	Yes	F
GS2023+338	155	Transient	12–19	Yes	K0
2S0921–630	216	ADC	16	Yes	
Cyg X-2	235	Dipper	15	Yes	F giant

[a] ADC = accretion disc corona

intermediate step for this is through the accretion disc, formed by the angular momentum of the transferred material, and it is the disc which is responsible for much of the wide variety of behaviour seen in galactic X-ray binaries. Table 8.1 lists some properties of low-mass X-ray binaries.

But where does our information come from about the various components of figure 8.2? As a summary of what is to come in this chapter, let us look at each part of a low-mass X-ray binary (LMXB) in turn:

1 The compact object.
If the source is weak ($<10^{33}$ erg s^{-1}) and close to us, then the compact object is almost certainly a white dwarf. Above this luminosity, it has to be a neutron star or black hole. But if X-ray bursts are seen, then it must be a neutron star. If it is luminous and has a soft X-ray spectrum, then it *may* be a black hole.

2 The accretion disc.
Sources at high inclination enable the structure of the disc edge to be seen by its erratic absorption effects (*dips*) on the X-rays at certain times. The inner part of the disc is very hot and is bright in the ultraviolet. If the inclination prevents us seeing the compact object at all then some systems show hot, X-ray emitting coronae extending above and below the disc.

3 The secondary star.
Usually very faint and cool, they are impossible to see in many X-ray sources. However, by searching in the infrared and by looking during eclipses it is possible to show that the mass-losing star is of low mass (usually much less than the Sun) and sometimes evolved.

8.2 Searching for binary behaviour

Surely if the model described above were correct then it would be relatively easy to find direct evidence for such binaries through the observation of eclipses, or at least orbital modulation? In fact, Sco X-1 and the other bulge sources observed by Uhuru and other satellites in the 1970s showed no such behaviour at all. Indeed, it took more than 10 years of careful study of Sco X-1 before the binary period was discovered, and even then it was as a result of optical spectroscopic observations, not X-ray. By the mid-1970s there were enough sources known for Mordy Milgrom (Weizmann Institute) to realise that nature could not have placed us in just the right position that we would see no eclipsing sources. There must be a selection effect that prevents us from seeing high inclination systems. This is demonstrated in figure 8.3.

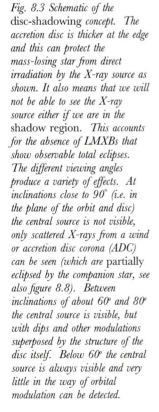

Fig. 8.3 Schematic of the disc-shadowing concept. The accretion disc is thicker at the edge and this can protect the mass-losing star from direct irradiation by the X-ray source as shown. It also means that we will not be able to see the X-ray source either if we are in the shadow region. This accounts for the absence of LMXBs that show observable total eclipses. The different viewing angles produce a variety of effects. At inclinations close to 90° (i.e. in the plane of the orbit and disc) the central source is not visible, only scattered X-rays from a wind or accretion disc corona (ADC) can be seen (which are partially eclipsed by the companion star, see also figure 8.8). Between inclinations of about 60° and 80° the central source is visible, but with dips and other modulations superposed by the structure of the disc itself. Below 60° the central source is always visible and very little in the way of orbital modulation can be detected.

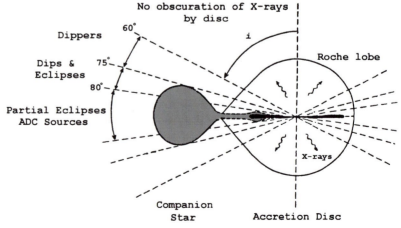

Fig. 8.4 Relative sizes of the orbits and companion stars of a selection of LMXBs from (almost) the shortest to the longest periods. The bar at the top indicates the size corresponding to 50 light seconds (remember that the Earth is about 8 light minutes from the Sun!). To put this into perspective we also show below the size of one of the LMXB orbits (4U1626−67) relative to an image of the Sun. (From diagrams originally by Saul Rappaport (MIT) and Jeff McClintock (CfA).)

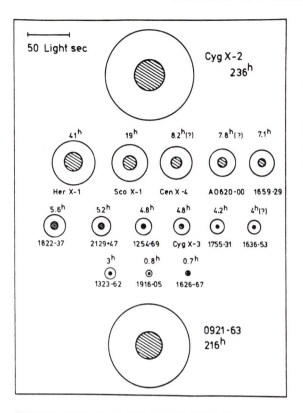

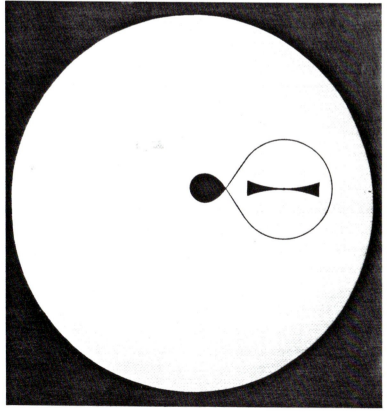

When the LMXB is viewed in cross-section it becomes clear that the disc itself is the source of the problem. The X-ray eclipse would be expected in cases where the mass-losing companion star comes between us and the compact object. However, it is at just these inclinations that the thickness of the disc itself prevents us from seeing any of the X-rays at all! Such systems obviously exist but we cannot find them by their X-ray emission. The binary's inclination to us must be such that we can just see over the edge of the disc down into the central region where the X-rays are produced. This explains why early X-ray searches for binary behaviour were so unsuccessful.

In fact, much of the information about LMXBs contained in table 8.1 has come from observations (usually CCD photometry) of the *optical* counterparts. It is their *intrinsic* faintness that makes this possible (compared to the massive systems) as the effects of X-ray heating (of the disc or companion star) are then more readily apparent. One of us (PAC) has been active in this area and has exploited the new generation of high sensitivity CCDs (charge coupled devices) that enable accurate time-series photometry of faint stars to be performed. Orbital periods are now known for 26 LMXBs, and the relative sizes of a selection of these are shown in figure 8.4. To put these into perspective, figure 8.4 also shows a schematic representation (to scale) of the 42 minute binary 4U1626–67 superposed on an image of the Sun!

8.3 Partial and total eclipses, and X-ray dippers

But there are other reasons why none of the low-mass systems showed binary evidence from early X-ray observations. Look again at figure 8.3. It is possi-

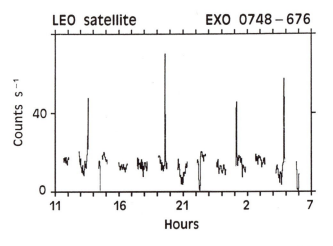

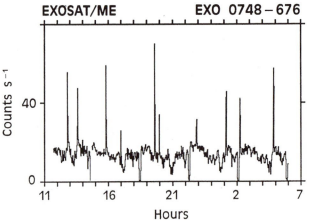

Fig. 8.5 A single 20 hour EXOSAT observation of the transient LMXB EXO0748–676 is shown in the lower panel. This source was discovered by EXOSAT during its manoeuvres around the sky (between other observations) and was found to exhibit dips, eclipses and bursts! For comparison the top panel shows this same light curve as it would have appeared to a conventional low-Earth orbit satellite. (Courtesy of EXOSAT Observatory, ESA.)

ble to imagine a system in which the companion star is just larger than the projected *flare* of the disc, and if we were at just the right inclination the top of the companion star would very briefly eclipse the central X-ray source. Unfortunately, that means the eclipse would occupy a very small fraction of the total binary period and hence be very difficult to catch, unless it were possible to monitor the X-rays from these objects without *any* break at all. All X-ray satellites of the 1970s suffered greatly from the frequent breaks in observing that were a natural consequence of their low-Earth orbits. This greatly hindered the chances of catching any brief eclipse.

With the arrival of EXOSAT in 1983 it was possible to observe all these objects for long intervals without any breaks at all. A new transient X-ray source, EXO0748–676, that was discovered by EXOSAT demonstrates superbly the power of EXOSAT's extended monitoring capability. A continuous 20 hour observation is shown in figure 8.5 in which a wide variety of behaviour is contained in this one object. The most important feature is the eclipse, that lasts for just *8 minutes*! And that is out of a total binary period of almost 4 hours.

The upper part of this figure shows what the data would have looked like if it had been observed from an earlier, conventional-orbit satellite. The gaps destroy most of the fine temporal detail and, depending on the phase at which the observations start, it is possible to miss the eclipses altogether. The other main feature of this new source is the X-ray bursts that permeate the light curve, which will be described fully in the next section. This indicates that the compact object in EXO0748-676 is a neutron star (see section 8.6). Removing these bursts from the data, and then folding it on the binary period of 3.8 hours gives the light curve shown in the lower part of figure 8.6.

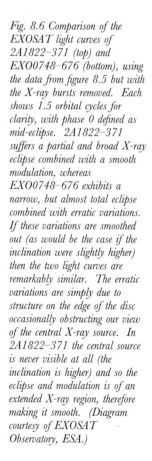

Fig. 8.6 Comparison of the EXOSAT light curves of 2A1822–371 (top) and EXO0748–676 (bottom), using the data from figure 8.5 but with the X-ray bursts removed. Each shows 1.5 orbital cycles for clarity, with phase 0 defined as mid-eclipse. 2A1822–371 suffers a partial and broad X-ray eclipse combined with a smooth modulation, whereas EXO0748–676 exhibits a narrow, but almost total eclipse combined with erratic variations. If these variations are smoothed out (as would be the case if the inclination were slightly higher) then the two light curves are remarkably similar. The erratic variations are simply due to structure on the edge of the disc occasionally obstructing our view of the central X-ray source. In 2A1822–371 the central source is never visible at all (the inclination is higher) and so the eclipse and modulation is of an extended X-ray region, therefore making it smooth. (Diagram courtesy of EXOSAT Observatory, ESA.)

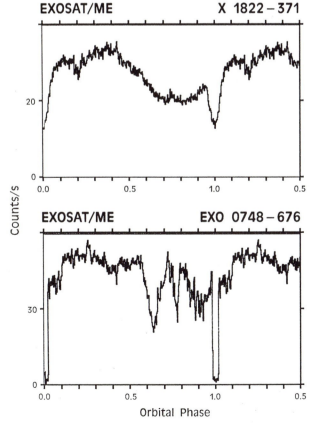

The 8 minute total eclipse is now readily visible. Its sharpness of ingress and egress indicates that the X-ray emitting region must be very small, as expected for a compact object such as a neutron star. However, this light curve shows other interesting features at phases between 0.6 and the eclipse. They look like erratic *dips* and that is exactly what they are. They are a result of the line of sight to the X-ray source passing very close to the edge of the accretion disc (as demonstrated in figure 8.3).

But why are they confined to just part of the binary orbit? The answer is to look back to figure 8.2. There has to be a region of the disc where the accretion stream from the companion star impacts the disc, and probably penetrates in to quite some distance. Theoretical calculations also suggest that the stream is thicker than the disc itself, thereby easily penetrating the inner regions of the disc. The result of this is that the disc is not of uniform height around its periphery. When observing a source such as EXO0748-676 at a high inclination there is a chance that the variable height of the disc will obscure the neutron star from time to time. This idea is vindicated by examining the X-ray spectra of the dips. The only thing that varies in the spectrum is the amount of absorbing material (the column density), exactly as expected. Because of the orbital motion of the two stars (of which the mass-losing star is corotating) the mass transfer stream will follow the path shown in figure 8.2. Since phase 0 is the time of X-ray eclipse by the companion star, the *bulged* part of the disc will only absorb X-rays in the period immediately preceding the eclipse.

Although X-ray dipping behaviour was actually discovered from Einstein observations of A1916–05 (by Fred Walter and colleagues from Berkeley), it was not until the launch of EXOSAT that they could be studied in detail. In fact, A1916-05's behaviour was only found at all because of its exceptionally short period (50 minutes) and hence the regularity of the dips was recognised immediately (see figure 8.7).

A1916–05 demonstrates very clearly that such structure at the edge of the disc is *not* stable over long periods of time. Although the 50 minute period is always visible (and is marked), the degree of obscuration presented by the disc edge can vary dramatically on a timescale of months. However, the dipping source most difficult for early satellite observations to have recognised is 4U1624–49. With its 21 hour binary period and quite spectacular light curve, its discoverers (Mike Watson and colleagues at Leicester) have given it the name *The Big Dipper*! Note, however, the common feature of all of these light curves, which is that there is no total eclipse present. In most cases then our line of sight to the X-ray emitting object just misses the companion star but intercepts from time to time part of the edge of the accretion disc. The orbital inclination is slightly lower than in the case of EXO0748–676.

8.4 Accretion disc coronae

What happens if the inclination is slightly higher? (The line of sight would be slightly closer to the plane of the binary orbit.) If our interpretation of figure 8.3 is correct, then we should see no X-rays at all. This is true if the X-ray emitting region is confined to the compact object only. However, if there were an *extended* emitting region of X-rays then it might be possible to see this if the bright main source were permanently occulted from our view. There is a small group of X-ray sources that display such behaviour, of which one of the best examples is 2A1822-371, whose light curve is displayed in the upper part of figure 8.6. There is an X-ray eclipse here but, it is *partial*! Note also the gradual ingress and egress as the companion star passes across our line of sight. This immediately tells us that we are observing an extended X-ray source. Since we cannot be observing the compact object itself, but the timing of the partial eclipse tells us that the X-rays are somewhere near where

Fig. 8.7 Montage of EXOSAT light curves of X-ray dippers. The orbital period is immediately evident from the repetitive nature of the dips. The structure is more complex in some sources (e.g. XB1254–69 and 4U1624–49 compared to X1755-33) and this may indicate slight differences in viewing angle (see fig. 8.3). However, the disc structure in individual sources can evolve substantially with time as shown in the montage of EXOSAT observations of A1916–05 (bottom panels), which are separated in time by many months. In the central panel the disc has clearly expanded to a height that creates substantial dipping structure for almost the entire orbital cycle. The very regular dips that are always present on the 50 minute orbital period are marked with a series of vertical dashes. Note also that the X-ray bursts have been truncated here so as to make the dip structure stand out more clearly. (Diagrams based on EXOSAT Observatory originals combined with data from Mike Watson, Leicester and Alan Smale, MSSL.)

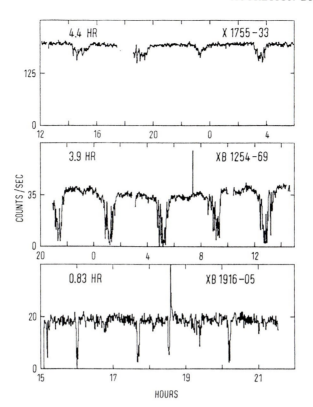

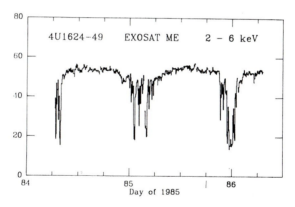

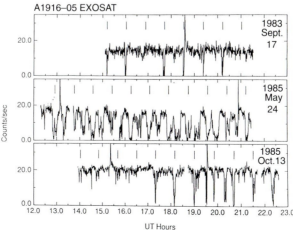

we would expect the compact source to be, then this extended source must be a kind of *corona* which extends above and below the disc. 2A1822–371 is actually quite a weak X-ray source as the LMXBs go. This is because it is only X-rays scattered into our line of sight from the corona that we manage to see.

The folded light curve of 2A1822–371 plotted immediately above that of EXO0748–676 in figure 8.6 brings out their striking similarity. Admittedly the dipping structure of the transient is very ragged, but if it were smoothed out it would follow the same general shape of that of 2A1822–371. Both are the result of a modulation of the X-ray source by obscuration due to the varying height of the edge of the disc. Because the transient's source is compact (i.e. point-like as far as we are concerned) then the dipping behaviour will be highly erratic, since there is no reason why the disc edge itself should be smooth and well-defined

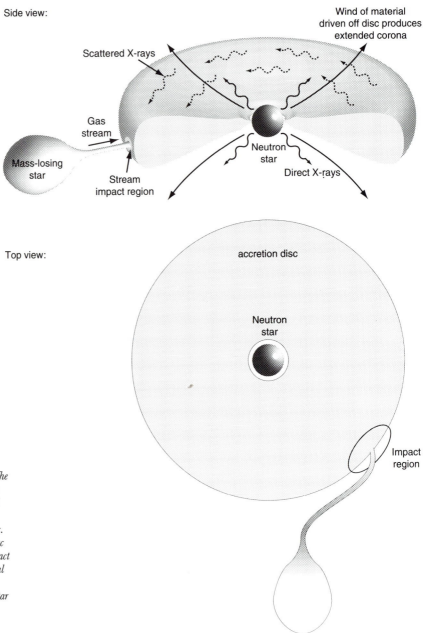

Fig. 8.8 Schematic of an accretion disc corona system. The neutron star is not visible to the observer, who only sees scattered X-radiation from the extended corona above and below the disc. The bulge on the edge of the disc (where the accretion stream impact occurs) gives rise to a substantial modulation of the X-rays observed, and the mass-losing star (partially) eclipses the coronal region.

like that of a star. However, if the source itself is extended, then the disc irregularities will be smoothed out by the large object beyond it (see figure 8.8).

Almost all that has been learned so far about LMXBs has come from observations of *periodic* phenomena. The orbit of a binary adds the signature of a precise clock to the data, which immediately tells much about the object itself. But in the mid-1970s and again in the mid-1980s, two types of X-ray variability were discovered which were to revolutionise our understanding of compact objects and the process of accretion. These were X-ray *bursts* and *quasi-periodic oscillations*.

8.5 X-ray bursters

Their very name conjures up an accurate picture of this X-ray event. First discovered by Josh Grindlay (CfA) and John Heise (Utrecht) in 1975 from observations with the Astronomical Netherlands Satellite (ANS), and by Dick Belian and colleagues (LANL), the X-ray sources responsible were initially all located in globular clusters. The next chapter is devoted to the globular cluster sources, but for now let us concentrate on just the X-ray bursts. As the reader will probably have noticed, examples have appeared in figure 8.5, which shows the EXOSAT observations of the transient EXO0748-676. These dramatic X-ray spikes, called bursts, are over in about 10 seconds and contain typically 10^{39} ergs of X-ray energy. (For comparison, it would take the Sun's X-ray emitting corona about 3000 years to radiate as much energy!)

X-ray bursts could have been discovered much earlier from the pioneering Uhuru satellite. The problem was that the appearance of an X-ray burst is similar to a *non*-astronomical phenomenon, charged particles in the Earth's upper atmosphere. Unfortunately, Uhuru did not have the sophistication of a separate charged particle detector that would have enabled the discrimination to be made between real X-ray events and those that originated in the atmosphere. [In fact, X-ray bursts had been discovered already by Richard Belian, Jerry Conner and Doyle Evans (Los Alamos) from observations made with the Vela satellites. These satellites were designed to look for bursts of high energy radiation, but not from the sky! The principal role of the Vela satellites was to search for evidence of atmospheric nuclear explosions. Unfortunately, their discovery of cosmic X-ray bursts could not be announced until much later because of the classified nature of the mission.]

The clues to the nature of X-ray bursts were contained in some of the earliest observations (see figure 8.9). The start of the burst is extremely rapid, lasting about a second, and is simultaneous at all energies. However, the decay of the burst is very different in the various energy bands. The lowest band, corresponding to the coolest material, displays a distinct *tail*, whereas the higher energies show a sharper decay. In other words, the hardest X-rays last for a shorter length of time than the softer X-rays. This means that the temperature of the burst, which starts off high, drops as the burst progresses. It changes typically from about 30 million degrees at peak to about 15 million degrees 10 seconds later.

A model requiring a black hole was developed by Grindlay and Herb Gursky (CfA), to explain this behaviour. If a massive black hole (say, 100 $M_\odot$) were residing at the centre of the globular cluster then it could accrete material from the winds of the red giants in the cluster, thereby producing observable X-rays. A temporary increase in the amount of material accreted could produce an extremely hot primary X-ray burst which would then be scattered in the surrounding gas cloud giving the X-ray burst observed. Unfortunately, the burst profiles are too short for this. The X-ray emitting region is smaller than the event horizon of such a massive black hole. Another explanation is required.

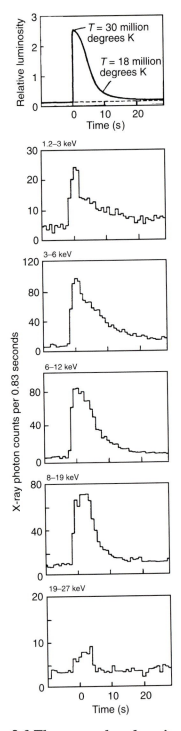

Fig. 8.9 X-ray burst profiles observed by SAS-3 in different energy bands from MXB1728–34. Note how in the lowest energies the burst persists for much longer (it has a tail), which does not happen at higher energies. This is because the radiating material cools significantly during the burst, as indicated in the theoretical burst profile shown at top. (Original diagram by Walter Lewin, MIT.)

8.6 Thermonuclear burning on the neutron star

Using observations by the Goddard detector on board OSO-8, Jean Swank and colleagues (at GSFC) noted that the X-ray spectrum of one particular X-ray burst was remarkably similar to that of a cooling blackbody. A blackbody is a perfect absorber and radiator (see chapter 3) and can be described by just two parameters, the temperature and the radiating area. Apart from being associated with the globular clusters, the X-ray bursters were generally distrib-

Fig. 8.10 Evolution of the neutron star parameters (temperature and radius) during an X-ray burst from MXB1608–52 observed by EXOSAT. After the initial expansion of the radiating area early in the burst, it then cools over about 1 minute and settles back to its original radius. (Diagram courtesy of EXOSAT Observatory, ESA.)

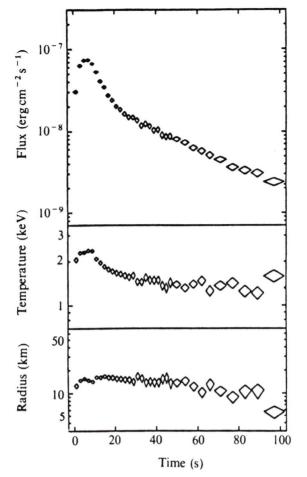

uted amongst the galactic bulge, suggesting that bursting was an intrinsic property of bulge sources and that they must be one and the same population. This infers a distance for most sources of 5 to 10 kpc. Armed with this assumption, Swank's analysis (see box 8.1) showed that, after an initial expansion, the radius of the sphere required to radiate the observed blackbody X-rays in the cooling tail was always in the range 10–15 km, exactly the expected size of a *neutron star*. With its high sensitivity, EXOSAT demonstrated this effect very nicely when it observed a strong burst from X1608–52 (see figure 8.10).

But what is the physical origin of the burst itself? The solution to this problem was first proposed by Italian theoreticians Laura Maraschi (Milan) and Alfonso Cavaliere (Rome), and independently by Stan Woosley (University of California at Santa Cruz) and Ron Taam (Northwestern). They suggested that unstable nuclear burning on the surface of a neutron star could lead to a thermonuclear flash of radiation, with the peak energy occuring at X-ray wavelengths. The first detailed numerical computations of such an event were produced by Paul Joss of MIT, and are illustrated in figure 8.11.

The accretion of hydrogen onto the neutron star from the accretion disc will produce a surface layer of hydrogen which will burn (in the sense of nuclear fusion) steadily, producing an underlying layer of helium. Eventually the density and temperature in the helium layer reaches a critical point and it too starts burning, fusing into carbon. However, in the conditions prevailing, the helium burning process is *unstable* and all the helium will be rapidly consumed producing a thermonuclear flash that we see as an X-ray burst. The steady accretion of *fresh* hydrogen from the disc then continues, and after a similar, but not

Box 8.1   Blackbody radiation from a neutron star

If the observed X-ray flux, F_x, during an X-ray burst from a source at distance d is due to blackbody radiation from a sphere of radius R and temperature T then

$$4\pi\, d^2\, F_x = 4\pi\, R^2 \sigma T^4$$

where the right-hand side is merely Stefan's Law. From this equation the radius can be calculated as

$$R = d/T^2\,(F_x\,/\,\sigma)^{0.5}$$

where everything, except the distance d, can be measured directly from the X-ray spectrum. However, regardless of whether d is known or not, it was clear from observations of X-ray bursts that F_x/T^4 was a constant during the cooling tail of the burst and, for sensible estimates of d, was consistent with an R of about 10 km. The compact object must therefore be a neutron star, and the X-ray burst arises from physical processes occurring on its surface.

X-ray bursts as a standard candle?

After a dozen or so bursters had been discovered, it was realised that there appeared to be a maximum burst luminosity for all sources that was never exceeded. This value was in the region of 1.8×10^{38} erg s^{-1} and is exactly what is expected for a $1.4\ M_\odot$ neutron star. This is set by the *Eddington limit*, and is due to the effects of radiation pressure, as described in box 7.5. A heavier object (which would have a higher limit) would be a black hole and hence incapable of producing X-ray bursts. Therefore the maximum burst flux seen from any burster is independent of distance and can therefore be used as a *standard candle* to provide an estimate of the distance to these sources. This is a particularly valuable diagnostic for investigating the enigmatic bulge sources, and has been used by Jan van Paradijs (Amsterdam) to estimate the distance from the Earth to the centre of our galaxy as about 6–7 kpc (assuming that the bulge sources are uniformly distributed about the centre). This is significantly (30%) less than previously accepted values and, if correct, has considerable implications for the entire cosmic distance scale.

exact, amount of time another burst will occur. However, if a longer period of time elapses before the next burst, then more helium will have likely been produced and hence a larger burst will be expected. This is demonstrated in figure 8.12 in which the recurrence times for the bursts from the EXOSAT transient EXO0748–676 (figure 8.5) are plotted against the total energy in the burst.

It is worth pointing out again that this diagram could not have been produced before EXOSAT, because of the gaps in the coverage of earlier satellites. One would never know if a burst had been missed! This behaviour is as predicted by the thermonuclear flash model, although in the ideal case it is expected that the linear relation would go through zero. That it does not indicates that not all the fuel appears to be consumed in the burst, and this can lead to some very strange bursts (figure 8.13)!

Apparently normal bursts are sometimes followed a short time later by another burst. The waiting time is too short for material to accrete, fuse into helium and ignite again. The only explanation can be that all the material in the previous burst was *not* consumed, and hence could act as part of that to be burnt in the next burst.

The success of this model is graphically illustrated by the theoretical calculation of an X-ray burst (top of figure 8.9). It correctly reproduces the rapid rise, the temperature at peak and its subsequent cooling. The length of time between bursts depends on the rate at which matter is being accreted, but for typical values this is about 3 hours. Since the entire process takes place on and in the surface layers of a neutron star, it is clear that the observation of bursts from an X-ray source rules out the possibility of the compact object

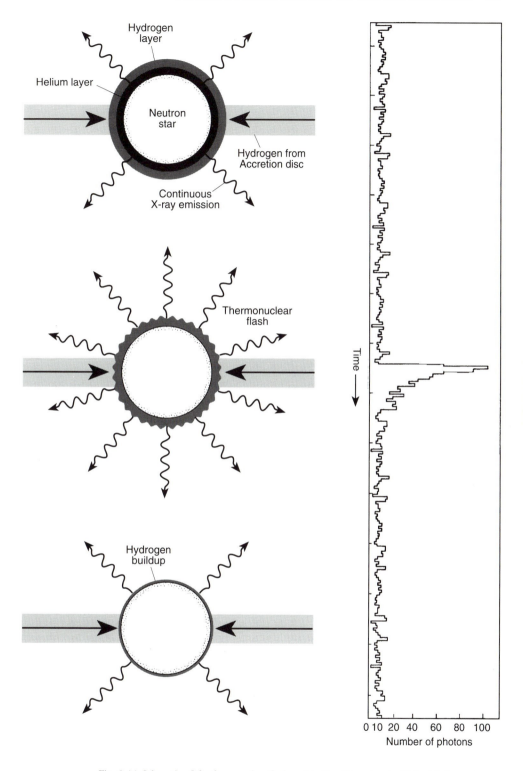

Fig. 8.11 Schematic of the thermonuclear flash model of an X-ray burst. At the top the neutron star is accreting hydrogen from its accretion disc, forming a layer typically 1 m thick. This hydrogen burns steadily into helium, forming a layer of comparable thickness. Eventually the conditions in the helium layer go critical and a thermonuclear flash takes place (centre panel). The process then begins again. (Diagram by Walter Lewin, MIT.)

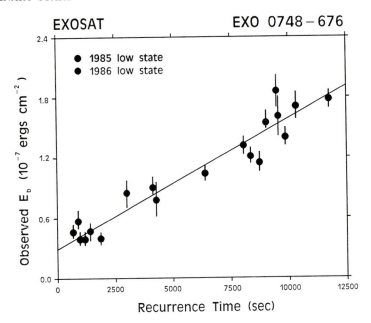

Fig. 8.12 Relationship between total burst energy and the burst recurrence time, for EXO0748–676. Clearly the longer the recurrence time, the more energy there is in the burst. This is because more fuel has accumulated during the longer intervals. (Diagram courtesy of EXOSAT Observatory, ESA.)

being a black hole. Accreting material onto a black hole would pass through the event horizon, and thereafter be unobservable. A *surface* is needed for the fusion to take place.

The thermonuclear flash model also accounts nicely for the observed ratio of the energy released in steady accretion to that released in X-ray bursts. An atom of hydrogen (actually it would be a proton because of the high temperatures) falling in the potential well of a neutron star would gain about 100 MeV of kinetic energy, which it would eventually release (mainly as X-rays) when it hits the neutron star surface. However, when that same atom (after a short wait) takes part in the nuclear flash that we see as the burst, it releases about 1 MeV of energy. In other words we should see about 100 times more energy in *steady* emission (averaged over time) than in *burst* emission. This is exactly the ratio observed in many of the X-ray bursters.

The brightest bursts (which occur after the longest waiting times) show a double peaked structure that supports the basic idea of a hydrogen-burning layer accumulating above a layer of helium. Figure 8.14 shows the EXOSAT observations of 4U1820–30, the X-ray source at the centre of the globular cluster NGC 6624 (which also happens to be the source from which bursts were first discovered).

The X-ray spectrum shows how the blackbody temperature varies during the burst and indicates that the radius *changes*. There is an expansion of the neutron star's atmosphere (usually called photosphere) followed by a contraction. The X-ray luminosity reaches the Eddington limit and the radiation pressure then forces the hydrogen layer to expand outwards. The luminosity can then increase again as the Eddington limit for helium is higher by a factor 1.75. (See also the discussion in box 8.1). A globular cluster such as NGC 6624 is an excellent location to test these ideas because its distance is much more accurately known than that of most other galactic bulge sources. Obtaining direct evidence for such rapid motions during an X-ray burst will be difficult. Figure 8.15 shows Tenma observations of the X-ray spectra during bursts from X1608-52 and X1702–43, in which there is an apparent emission feature near 5 keV.

This is attributed to red-shifted iron emission (which should normally be at an energy of around 6.7 keV). But the redshift is not due to motion away from us, it is explained as a *gravitational* redshift because of the very high sur-

Fig. 8.13 Examples of double bursts from EXO0748–676. In the standard model described here they should not happen because there has been insufficient time for new material to accrete and produce fresh helium. These are explained by not burning all the available fuel during the first (larger) burst, thereby leaving helium ready for another burst to follow more rapidly (only minutes later). (Diagram courtesy of EXOSAT Observatory, ESA.)

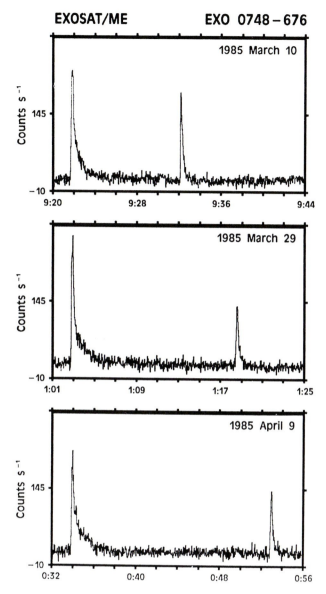

face gravity of the neutron star. Further observations by larger instruments with better time resolution will be needed to confirm or deny this interpretation of a remarkable feature.

The complexity of the nuclear reactions involved is nicely demonstrated in a series of EXOSAT observations of 4U1705–44 (figure 8.16). As the average strength of the source varies, the character and frequency of the bursts changes. (This was first noted in 4U1820–30. Bursts were not observed when the source was X-ray bright, only when it was in its low state.) The average brightness of the source is a measure of the rate at which matter is accumulating on the surface of the neutron star. At very high rates the surface fusion of hydrogen cannot keep pace with the rate at which new material is arriving. Consequently the hydrogen-burning shell is pushed into the layer of helium that is accumulating beneath it, thereby creating a mixture of helium and hydrogen. This inhibits the thermonuclear burning of helium, thereby changing the burst behaviour of the source.

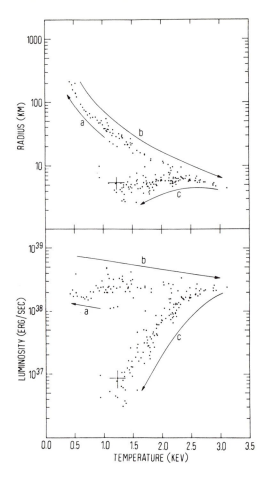

Fig. 8.14 Evolution of double-peaked bursts from 4U1820–30 (NGC 6624). Three phases are marked: (a) expansion, (b) contraction, and (c) cooling, with the arrows indicating the direction of time. Initially the radiating area expands to almost 200 km whilst the luminosity remains constant. This is due to the Eddington limit being reached, and the radiation pressure driving the material out to a larger radius. (Diagram courtesy of EXOSAT Observatory, ESA.)

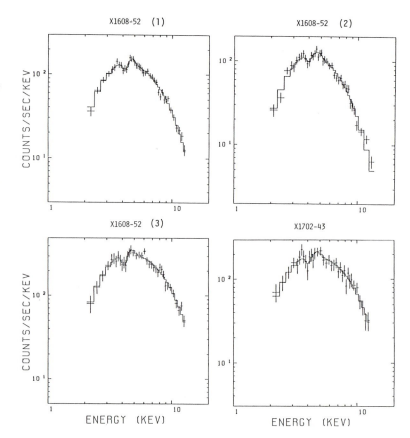

Fig. 8.15 X-ray spectra of three X-ray bursts from X1608–52 and one from X1702–43, as observed by the Japanese Tenma satellite. These spectra are unusual compared to the quiescent source behaviour in exhibiting an apparent emission feature near 5 keV. This is attributed to an iron emission line at 6.7 keV which has been gravitationally red-shifted by the intense gravitational field near the surface of the neutron star. (Courtesy of the Tenma group, ISAS.)

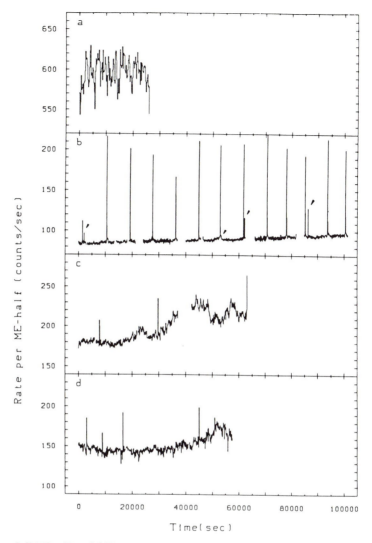

Fig. 8.16 Dependence of burst properties on X-ray brightness in 4U1705–44. These EXOSAT observations nicely demonstrate how, when the source is at its brightest (a), no X-ray bursts are seen at all, whereas when it is faint (b) it bursts frequently. The other panels are at intermediate brightness levels which show bursts, but less frequently. (Diagram courtesy of EXOSAT Observatory, ESA.)

8.7 The Rapid Burster

Early on in the study of X-ray bursters, an object was found that at first seemed to threaten the well-developing foundations of the thermonuclear flash model. This object was called the *Rapid Burster*, for reasons that are evident in figure 8.17. These bursts look like astronomical *machine-gun fire*, recurring as rapidly as every 10 seconds. Sometimes there is a large burst that is followed by a longer gap until the next burst. The strongest bursts contain as much as 1000 times the energy of the weakest. This was at first a blow to the model involving thermon clear fusion because it cannot explain the behaviour of the Rapid Burster. However, on careful examination, it became clear that these rapid bursts had a different character to those emitted from other X-ray bursters and therefore probably had a completely different origin.

Examine figure 8.17 carefully. The time between bursts is not random, but depends directly on the strength of the *previous* burst. A big one is always followed by a long gap; a small one is followed rapidly by another burst. This is illustrated in figure 8.18 where the time to the next burst is plotted as a function of the energy in the burst.

Also, the X-ray spectra of these bursts showed no evidence for the cooling in the tail that is the hallmark of thermonuclear fusion; the temperature was essentially constant. As is usual in astronomy, when confronted with

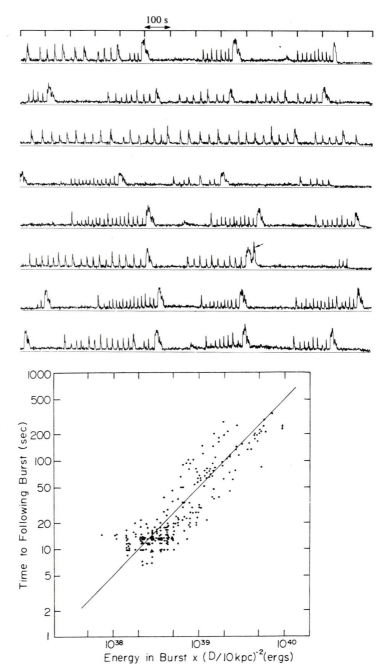

Fig. 8.17 The Rapid Burster. These eight orbits of SAS-3 observations (each panel covering 24 mins) show its extraordinary bursting behaviour that is classified as type II, together with a single, special burst that is in fact an ordinary type I burst (i.e. a thermonuclear event). Note how the smallest bursts are very close together, whereas a big one is always followed by a long gap. (Diagram by Walter Lewin, MIT.)

Fig. 8.18 Burst properties of the Rapid Burster. As is qualitatively clear from figure 8.17, the time to the following burst depends simply on the energy in the previous burst. (Diagram by Walter Lewin, MIT.)

different types of behaviour, we classify them. The normal (or *flash*-type) bursts are called Type I, whereas the rapid repeating ones from the Rapid Burster are Type II. To this day, the Rapid Burster is the only object known to emit type II bursts. A clue to the nature of the Rapid Burster is contained in figure 8.17. There is one burst, quite a large one, that occurs at an unexpected time given the relation of figure 8.12. This burst is marked with an arrow and was called a *special burst* by its MIT discoverers Walter Lewin and Herman Marshall. On close examination it was seen to be nothing more than the common-or-garden variety type I burst. Even the Rapid Burster sometimes undergoes thermonuclear flash events.

Fig. 8.19 Magnetospheric gate *model of the Rapid Burster. Material accreting from the disc is held back (top panel) by the neutron star's magnetosphere. When enough material has built up outside this gate, the magnetosphere can no longer hold it and it ruptures (middle panel), thereby allowing it to fall onto the neutron star, producing a type II burst. With the material gone, the gate re-forms and the process starts again. (Diagram by Walter Lewin, MIT.)*

This requires a neutron star which, in this case (see figure 8.19), has a strong enough magnetic field to produce a magnetosphere which acts as a *gate* that the accreting material has to pass through. Matter builds up around this barrier until it can no longer support the pressure. Suddenly, part of the magnetosphere gives way for a moment (the gate opens and then immediately closes), and matter can then fall directly onto the neutron star surface producing an X-ray burst. If a large amount of matter manages to get through then it will take some time for more to build up sufficient strength to force the gate (the magnetosphere) to open again. However, if only very little material gets through, then it is likely to open again within a short space of time. In physics this process is described as a *relaxation oscillator*. It accounts for the temporal behaviour of type II bursts very well indeed.

This accreted material is hydrogen that is passing through the gate onto the neutron star. If it were not for the magnetosphere this would appear as *steady* accretion. This accreted hydrogen fuses steadily into helium, eventually producing a thermonuclear flash when the helium ignites, which we observe as a *special burst*. The special bursts are emitted every 3 to 4 hours, exactly as for other type I bursts. So, after an initial scare, the thermonuclear flash model emerged from the Rapid Burster discovery fully established as the basis for explaining X-ray bursts.

8.8 Optical bursts, too?

Within a year or two of the discovery of X-ray bursts, campaigns were organised to search for an optical counterpart to the X-ray flashes. However, the technical difficulties were immense. There were two problems that could not be avoided. Firstly, the sources were faint. The brightness of the steady optical counterpart of even the brightest X-ray burster had a visual magnitude of 17! Secondly, the exact times of X-ray bursts could not be predicted. Indeed, sources would often stop bursting altogether. This was compounded by, once again, the limitations of near-Earth orbit satellites whose observational coverage was usually only about 40%. Add to this the vagaries of weather for Earth-bound observatories and the chances of success become slim. The first season in which this was attempted (summer 1977) produced no positive detections of optical bursts.

In the following year the project moved to the southern hemisphere (Cerro Tololo) and thereby gained access to the brightest (optically) X-ray bursters, MXB1636–536 and MXB1735–44. The MIT group led by Jeff McClintock

Fig. 8.20 Simultaneous optical and X-ray bursts observed from MXB1636–536 in June 1979. The X-ray burst profile, which is typical, was observed by the Japanese satellite, Hakucho, but note the several second delay between the X-ray and optical peaks. This is due to the time taken for the X-ray burst to irradiate the disc which then produces the optical burst. (Diagram by Jan van Paradijs, Amsterdam.)

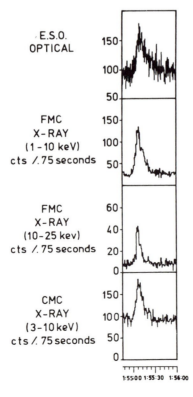

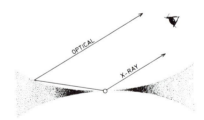

were successful, and the first simultaneous optical and X-ray bursts were observed. The next two or three years saw further intensive campaigns of this type, and figure 8.20 shows one of the best results to be obtained.

The optical burst is delayed by 1–3 seconds with respect to the X-ray burst, and, for a given object, this delay does not vary. The optical burst cannot therefore be caused by the X-ray burst heating up the surface of the companion star. If that were true, the delay would vary as the stars orbit each other. Also the level of *smearing* of the optical burst is much as would be expected if the X-ray burst heated up the surface of the accretion disc itself, which then reradiated the energy at optical wavelengths. This model, which is favoured, is shown schematically in figure 8.20.

Future observations of optical bursts are an excellent project for very large telescopes. This would enable optical colours and possibly even spectra of the bursts to be determined as a function of time through the burst and allow us to infer directly the geometry of the region of the accretion disc visible to the neutron star. But this is a challenging task even for a very large telescope!

8.9 Why don't X-ray pulsars burst or X-ray bursters pulse?

Two ways of using X-rays to study neutron stars and their immediate vicinity have been discussed in some detail – the X-ray bursts that arise from nuclear reactions occurring close to the neutron star's surface, and the pulsations that show the presence of a very powerful magnetic field. It would seem there might be a mixture of these two properties in most sources. The Rapid Burster must have a magnetosphere (which acts as the gate), so why are there no X-ray pulsations as the material flows down the magnetic field lines onto the poles of the neutron star? Also the matter flowing onto the poles in the bright X-ray pulsars must be hydrogen rich, so why are X-ray bursts not seen from time to time? For that matter, why are X-ray bursts not a characteristic of *all* X-ray sources in which the compact object is a neutron star?

The answer to the latter question has already been hinted at. It has been observed that the X-ray bursters do not burst all the time, and that their bursting behaviour is directly related to the overall X-ray brightness level. At high mass transfer rates it is found that the helium burning instability does not occur, and so bursting behaviour is suppressed. Almost all of the X-ray pulsars are in highly luminous massive X-ray binaries. This leads to a high mass transfer rate in which X-ray bursts are inhibited.

However, the principal difference between the pulsars and the bursters is *age*. The pulsars all have massive early-type stellar companions. Such systems must be *young* because an OB star has a lifetime on the main sequence of only 1–10 million years (and a lot less off the main sequence!). They must have formed very recently. The bursters on the other hand are all associated with LMXBs and are likely to be very *old*. Their mass-losing companions are mostly less massive than the Sun and could have existed essentially in their present form since the formation of the galaxy, approximately 10 billion years ago.

Although both contain neutron stars it is the age that is critical in determining the strength of the magnetic field. Frozen in during the initial collapse of the star that formed the neutron star, this field gradually decays over time. Since a strong field is needed to channel the accreting matter onto the poles of the neutron star, X-ray pulsations are only expected in young neutron stars, i.e. those associated with massive stellar companions. The weak field of the bursters gives way rather than direct incoming material to the magnetic poles. Hence, regular X-ray pulsations will not be observed from X-ray bursters.

The characteristics of the Rapid Burster indicate it has a field that *is* weaker than in most of the pulsars. The pulsars' fields never give way to the accreting matter to allow the type II bursts to occur. A more sensitive X-ray detector may yet reveal the presence of pulsations in the Rapid Burster as a result of the remnant magnetic field.

8.10 Quasi-periodic oscillations

The neutron star in the majority of LMXBs is sufficiently old that its magnetic field will have decayed significantly. However, for a very long period of time, the neutron star will have been accreting material from its companion. In the young OB X-ray systems, the accretion process can spin up the neutron star noticeably, even during the short times in which X-ray sources have been observed. Since accretion through a disc can transfer much larger amounts of angular momentum than that acquired from a stellar wind, the neutron stars in old LMXBs should be spinning very rapidly indeed. Perhaps as fast as a thousand times a second or so. However, the weakness of the field would mean that any modulation at this period would be of very low amplitude and hence very difficult to detect – even more so given the fact that these objects vary erratically anyway, thus adding considerable *noise* to the data.

The best way to search for such an effect is to use an X-ray detector with a large collecting area (allowing a significant number of X-rays to be collected on a timescale of milliseconds). Since the detector should be pointed at the target for as long a period of time as possible, EXOSAT was well suited for

Fig. 8.21 The discovery of QPOs in GX5–1. The upper curve shows the X-ray intensity as seen by EXOSAT during 8 hours of observations of this very bright source near the galactic centre. This curve is superposed on the power spectrum of the same data, and which has been derived as a function of time during the observation. If there is an oscillation present in the data it will show up as a peak in the power spectrum. For example, at the beginning of the observation (time 0) there is an oscillation present at about 19 Hz. As the source varies in brightness, the frequency of the oscillations change, generally going to higher frequencies when it is brighter, but disappearing altogether above a certain brightness. This correlation is shown more clearly in the lower diagram in which the frequency of the oscillation is plotted against the brightness of the source. They are linearly related until GX5–1 is at about 2900 counts s[1], when the oscillations disappear. (Diagram courtesy of EXOSAT Observatory, ESA.)

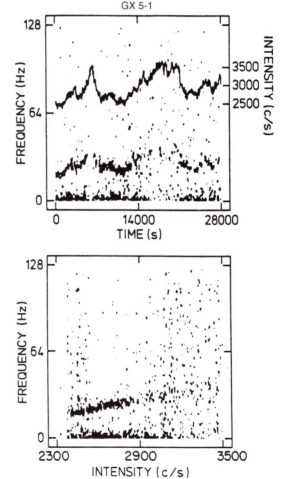

Box 8.2 Fourier analysis and temporal variability

The basic idea behind Fourier analysis is quite simple. In the 1800s the French mathematician Fourier proved that any function (or series of data points) could be represented by just the sum of a set of sines and cosines. This is demonstrated in figure 8.B1 using a square wave as the function to fit, and alongside it is the sum of three appropriately chosen sine functions. As more and more (smaller) terms are added together, a closer and closer approximation can be obtained to the square wave. In the limit of an infinite number of terms, the fit is exact.

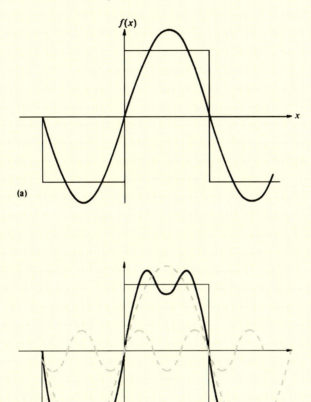

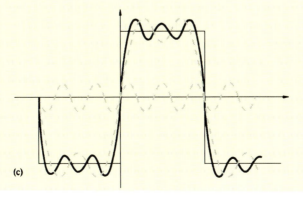

Fig. 8.B1 Synthesis of a square wave by the superposition of a Fourier series of sine waves. The pure sine wave at the top is a poor representation of the square wave, but the addition of each addition term (plotted in the centre) produces a closer approximation. If an infinite number of terms could be employed then an exact square wave could be reproduced.

But how does this result help in the search for periodic signals in data? Usually the periodicity searched for will appear as something fairly close to a sine wave. Therefore, in the Fourier series that represents the data the dominant term will be that sine wave corresponding to the frequency of the periodicity. The noise on the other hand will be spread out over all frequencies and thereby its effect will be greatly reduced. This is demonstrated using sample data in figure 8.B2.

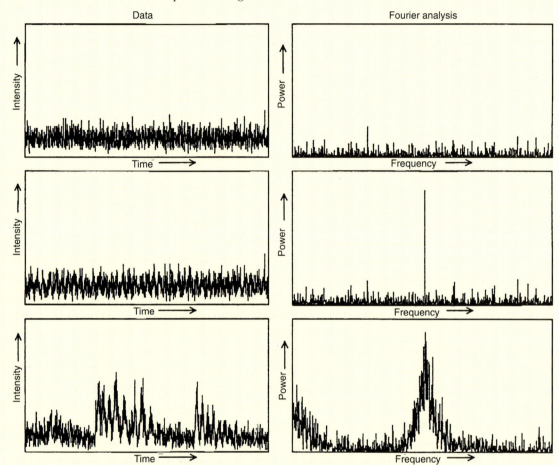

Fig. 8.B2 Demonstration of the ability of Fourier analysis to detect periodic signals in noisy data. There are three examples of such data here, shown in the left hand panels simply as counting rate against time. The top one is in fact pure noise, with no periodic signal present at all; the middle one, although noisy, contains a regular pulsation; while the bottom dataset contains pulsations that cover a range of frequencies (quasi-periodic). The Fourier analysis gives us information about how much power is present in the data at particular frequencies. Not surprisingly, the top power spectrum shows essentially a random scatter amongst all frequencies. However, the regular pulsation present in the middle panel stands out dramatically in this type of analysis (indeed, very much weaker signals can be extracted than that in the data example here). The reason is that the noise in the data is uniformly distributed over all frequencies, whereas the regular pulsation is only present at one frequency where its power is correspondingly greater. In the lower panel, the peak has been spread out because of the range of periods present in the data. (From a diagram by Michiel van der Klis, Amsterdam.)

The left panels represent three different sets of data. The top one is random noise only. The centre panel contains a regular periodic signal, whereas the bottom panel is of a QPO. The right panels show how much of each sine term (for each frequency) is needed in order to represent the data on the left.

Such plots are termed *power spectra*, because they show how much power is present in the data at each frequency. Not surprisingly, the noise plot requires essentially a random amount of each sine term. There is no dominant frequency. However, the central panel shows the value of this technique. The regular variation is visible to the eye in the data, but the power spectrum has a dramatic spike at the frequency that corresponds to the periodicity. It could have been detected in data with noise much worse than that shown.

The bottom panel is the one relevant to this section. Apart from irregular variations, it is also clear to the eye that there is some kind of periodic phenomenon present, but it is not so clear as in the centre panel. The power spectrum shows why this is. Instead of one particular frequency being present, there is a spread of frequencies, hence we say that the data are quasi-periodic.

these observations. In 1984 and 1985 Michiel van der Klis and his colleagues at the EXOSAT Observatory observed one of the brightest of the galactic bulge sources, GX5–1 (so-named because of its position relative to the centre of our galaxy, 5° longitude and −1° latitude). The resulting light curve revealed its remarkable property of *quasi-periodic oscillations*, or *QPOs* (figure 8.21).

The source was obviously varying in an irregular manner, but was there any evidence of variations on the shortest timescales? The answer to this is given by Fourier analysis of the data (see box 8.2, and in particular look at figure 8.B2 which illustrates the basic technique). This analysis shows that there are regular variations present, but the period of the modulation *varies* with the intensity of the source. This effect can be seen in the figure where the peak of the power spectrum is plotted below the source intensity curve. The period varies over the frequency range 20–40 Hz (i.e. 50–25 milliseconds), with the frequency increasing as the source brightens. Because the period is always changing it is called a *quasi*-periodic oscillation. However, at the brightest levels the QPOs disappear completely. Although these oscillations had been observed from a LMXB, they obviously do not directly represent the rotation of the neutron star because it is not possible to change the spin of such a massive object so quickly. They were also much slower than the expected period of rotation.

8.11 The magnetospheric model

The discovery of QPOs caused great excitement in X-ray astronomy, particularly the behaviour relating their frequency to the brightness of the source. The oscillations, although slower than the expected spin period of the neutron star, are still fast enough that they have to be originating from the inner part of the accretion disc where it interacts with the (remnant) magnetosphere. Because of the weak magnetic field, the accretion disc will reach down almost all the way to the neutron star's surface and therefore completely surround the magnetosphere (figure 8.22). When the source's X-ray output increases, the accretion rate has increased. More matter flowing through the disc will lead to greater pressure on the magnetosphere which will therefore shrink. Because it is then closer to the neutron star, the matter immediately above the shrunken magnetosphere will orbit faster, thereby producing oscillations at a higher frequency, exactly as observed.

As so often happens, this beautiful explanation did not stand up to careful examination. The observed relationship between the QPO frequency and source intensity (as demonstrated for GX5–1 in figure 8.23) could not be reproduced by the model. However, the basic idea of figure 8.22 was rescued

Fig. 8.22 The beat-frequency
*model of QPOs. This shows the
inner part of the accretion disc
surrounding the neutron star's
magnetosphere. Both are rotating
rapidly, but accretion of* plasma
clumps *onto the neutron star can
only take place when a clump
finds itself over the pole of the
neutron star. This occurs at a
frequency which is the* difference
*between the neutron star's spin
frequency and the orbital frequency
of the inner disc. (Based on a
diagram by Michiel van der Klis,
Amsterdam.)*

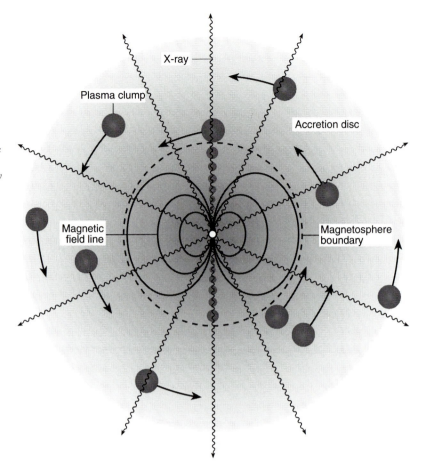

by theoreticians Ali Alpar (Illinois) and Jacob Shaham (Columbia), together
with Fred Lamb and Noriaki Shibazaki (also at Illinois). They proposed that
the X-ray modulation is not due directly to orbital effects of the matter
immediately outside the magnetosphere, but instead is the result of the *difference* in frequency between this orbital period and the as yet unknown spin
period of the neutron star.

Returning to figure 8.22, imagine a large clump of matter orbiting just
outside the magnetosphere, which is rotating rapidly beneath it at the neutron star spin period. The magnetosphere can be imagined as a barrier that
will stop the clump accreting unless it finds itself above the poles, which look
like an *open door*. When this happens, part of the clump falls onto the neutron
star, releasing a short flash of X-rays. It was shown that such a model could
reproduce the spread of QPO frequencies and, more importantly, how they
change with intensity. It is also necessary to assume that, because of the
weakness of the field (compared to Class I pulsars), the hot spots on the neutron star surface (at the poles) are actually quite large and therefore smear
out any periodic signal produced from the rapid spin. Otherwise we would
have expected to see the basic spin period itself in the data too.

8.12 Slow QPOs and the inner structure of the accretion disc
After the discovery of QPOs from GX5–1, the announcements of QPOs
being found in other bright LMXBs and bulge sources came thick and fast.
As Michiel van der Klis pointed out at the time, QPOs could have been dis-

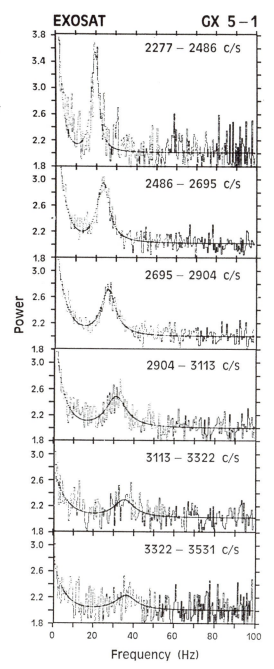

Fig. 8.23 The linear relation of QPO frequency with X-ray intensity is evident here in these series of power spectra of GX5–1 which have been divided into different intensity bands (faintest at the top, brightest at the bottom). Note also that, at higher intensities, the QPOs are not just shifted to higher frequencies, but the peak broadens as well. (Diagram courtesy of EXOSAT Observatory, ESA.)

covered much earlier, but the astronomers were searching for a strictly periodic signal, not a broad peak in the power spectrum. Unless a sufficiently wide range of frequencies is displayed in the power spectrum it is quite easy for the broad peak to go unnoticed! Once this was realised QPOs were hunted down and the variety of behaviour that they exhibited was bewildering (see table 8.2).

QPOs were found in Sco X-1 itself by Bill Priedhorsky and John Middleditch (Los Alamos) using EXOSAT. The QPOs not only varied over a wide frequency range, but they also depended on the form of the X-ray spectrum at the time in a very complex way that is still very poorly understood.

Table 8.2 Quasi-periodic oscillations in LMXBs

Source	Low freq. QPOs (Hz)	High freq. QPOs (Hz)	Erratic QPOs (Hz)
GX5–1	~5	20–36	—
Sco X-1	5.7–6.4	9–22	6-20
Cyg X-2	5.3–5.9	18–55	—
GX17+2	7.0–7.4	24–28	—
GX349+2	4.6–6.0	~11	—
Rapid Burster	2.1–6.0	—	0.4–4.5
4U1820–30	~7	—	15–30
GX3+1	7–9	—	—
Cir X-1	~1.4, ~5.6	8-18	—
GX340+0	5.3–5.9	—	—

Fig. 8.24 A model to explain the slow QPOs. This was suggested by Luigi Stella and Nick White (ESA) and is based on the intense radiation pressure from the X-ray source creating a bloated *inner disc whose size varies with X-ray brightness. If the orientation of the observer is appropriate then this inner disc can obscure the view of the X-ray source in an erratic way that is related to the X-ray intensity, thereby producing the QPOs. (Based on a diagram by Michiel van der Klis, Amsterdam.)*

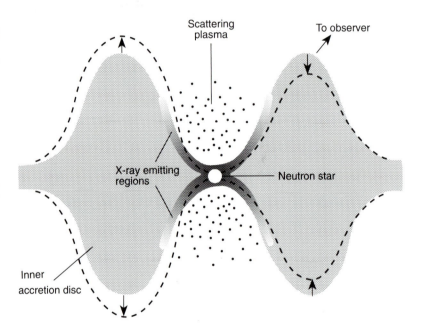

More importantly, there is a time when Sco X-1 (and many other sources) exhibit *slow* QPOs, 5 Hz or so, compared to the 20–40 Hz of the *normal* QPOs. These are difficult to account for in the magnetospheric beat-frequency model described above, and alternative models have been proposed. Luigi Stella and Nick White (EXOSAT Observatory) have suggested that slow QPOs arise because of an inner bulge in the accretion disc (figure 8.24).

All the QPO sources are very bright in X-rays and therefore radiation pressure has a significant effect close to the neutron star. It is possible perhaps to produce a bloated inner disc with size directly controlled by the strength of the X-ray emission. Depending on the angle at which the neutron star is viewed, variations in the size of this *torus* could intersect the line of sight, thereby imposing a modulation on the X-ray signal that is seen as a slow QPO. Some support for the presence of the scattering cloud is revealed by the fact that there is a delay of a few milliseconds between the hard and soft X-rays of the QPOs. This is explained as due to scattering close to the neutron star. The hard X-rays will suffer more scatterings, thereby taking longer to exit the emitting region relative to the softer X-rays.

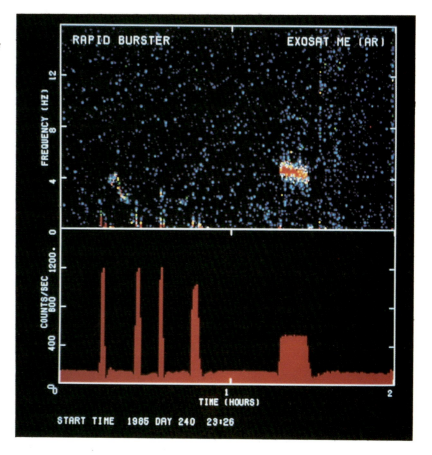

Fig. 8.25 QPOs in the Rapid Burster. The lower panel contains the EXOSAT light curve during a time that rapid bursts were occurring. The upper panel shows the power spectrum as a function of time (as in figure 8.21). The QPOs (at frequencies around 5 Hz) occur during the bursts, and are most evident during the long flat-topped bursts; but even slower (below 1 Hz) QPOs sometimes occur between bursts. (Diagram courtesy of EXOSAT Observatory, ESA.)

8.13 QPOs and the Rapid Burster

Perhaps one of the most bizarre, but potentially illuminating, X-ray observations in the QPO field has been the discovery that the Rapid Burster also exhibits QPOs, but the strongest QPOs are *only* seen during the bursts themselves! This phenomenon was found originally with the Hakucho satellite, but in figure 8.25 we show the observations of the Rapid Burster with EXOSAT.

The 4 Hz QPOs turn on dramatically at the start of the long flat-topped burst and end in the same way. But the most recent observations with Ginga by Tadayasu Dotani (ISAS) and his colleagues show this effect in very great detail (figure 8.26).

The large area of Ginga enables the individual cycles of the QPOs to be seen. The long bursts actually show three distinct (and fairly flat) maxima in which the QPOs are present in the first and third peaks, but *not* the middle one! This remarkable result was not a fluke, as they saw it happen twice. But what does it mean? First of all, it has already been shown that the bursts in the Rapid Burster are a magnetospheric phenomenon and not the thermonuclear flash which is confined to the surface of the neutron star. And it is in the region of the magnetosphere that we believe the processes are occurring that give rise to QPOs. This is confirmed by the association of the QPOs with the bursts.

But if this were true, then all of the bursts should display QPOs, not the alternating *odd–even* effect seen. Dotani's group suggest an explanation for this, in which the *reservoir* of material accumulating above the magnetosphere can move up and down. When it is down, or deep in the magnetosphere,

Fig. 8.26 Detail of QPOs from the Rapid Burster observed by the Ginga satellite. These bursts are not nuclear explosions on the neutron star surface (as in all other burst sources), but are related to the behaviour of the magnetosphere as evidenced by their appearance primarily during the bursts. The large area of Ginga enables a significant number of X-rays to be detected within 20 milliseconds, which means that the individual QPOs are observable. Panel (a) covers 40 seconds of a burst (with a blow-up of the first 10 seconds of the burst), panels (b) and (c) each cover 15 seconds of subsequent bursts. One of the most extraordinary properties of these QPOs is demonstrated in the triple-peaked burst of panel (c). QPOs are only discernible in the first and third peaks, not the second one! Explanations for this effect are highly speculative at present. (Diagram by Tadayasu Dotani, ISAS.)

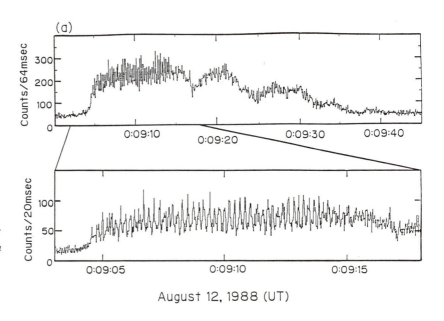

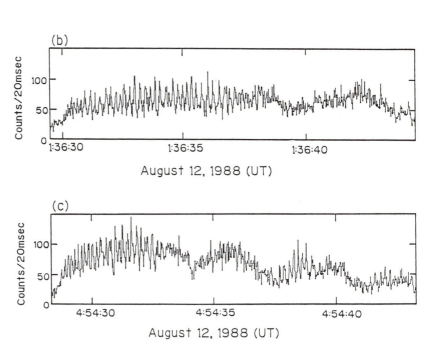

accretion occurs predominantly by leakage through the magnetosphere (and QPOs will be seen). However, when it is up, then material can spill over the *top* of the reservoir and accrete onto the poles of the neutron star. This change in accretion geometry leads to the suppression of the QPOs. It is a speculative idea at present, but demonstrates why the Rapid Burster is such an important object in the study of the accretion structure of LMXBs.

The future should see more research on QPOs, both from further analysis of EXOSAT data and new information from the XTE satellite which was designed specifically for timing studies of bright X-ray sources.

9 X-ray binaries in globular clusters

9.1 Introduction

The first all-sky surveys conducted by the Uhuru and OSO-7 satellites revealed a handful of sources that appeared to be associated with globular clusters. Although there was a large positional uncertainty associated with those surveys (as described in chapter 2), these sources were at high galactic latitudes where the star density is low and there were no other obvious candidates. Hence, by 1975 there were five bright X-ray sources believed to be located within globular clusters. They were NGC 1851, NGC 6441, NGC 6624, NGC 6712 and the well-known M15 (NGC 7078). This result attracted considerable attention because this was a large number of X-ray sources relative to the total number of stars within the clusters. The calculation is simple. There are $\sim 10^{11}$ stars in our galaxy which contains about 100 bright X-ray sources. So for the $\sim 10^7$ stars in all the globular clusters in our galaxy, there should be only 0.01 sources, whereas there were five! (We now know, as a result of more thorough surveys, that there are nine bright X-ray sources associated with globular clusters).

But what are globular clusters? They are almost perfectly spheroidal aggregations of stars (hence their name) which are amongst the oldest objects in our galaxy. Beautiful and impressive to observe (even through modest sized telescopes) they were formed from the original proto-galactic gas cloud *before* the main disc of our galaxy.

This is evident from the distribution of globular clusters in the sky, which shows a uniform spread about the galactic centre, with no concentration in the plane or disc. The clusters travel in orbits about the galaxy which, of course, take them through the plane. During such passages some loosely-bound stars are lost together with any cluster gas which is *stripped* by the higher gas pressure in the plane. This process helped ensure that all the stars in a given globular cluster were formed at about the same time and from the same mix of chemicals (mostly hydrogen, with very little in the way of heavy elements in those early days of the galaxy). Hence a cluster's H–R diagram consists of a truncated main sequence together with red giant and horizontal branches containing evolved stars.

This well-known situation occurs because of the more rapid evolution of the heaviest stars. Those heavier than about 0.8 $M_\odot$ have exhausted their original nuclear fuel and have evolved off the main sequence. This *turn-off* point depends principally on the age of the cluster and it is clear that these

Fig. 9.1 Colour photograph by the US Naval Observatory of the globular cluster M13, a naked-eye object visible in the constellation of Hercules. Note how the density of stars increases rapidly towards the centre of the cluster, there being several hundred thousand in all. All of these stars are less massive than our Sun, but the brightest have evolved to become red giants. However, the low metal abundance in globular clusters (due to their age) means that the stars will be much bluer than their galactic counterparts.

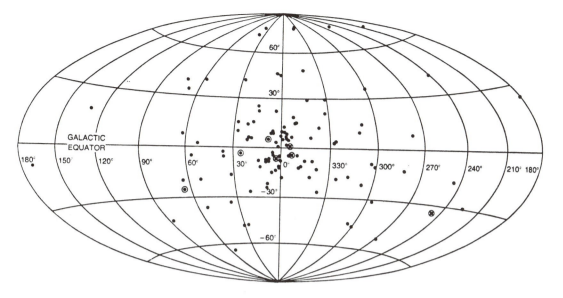

Fig. 9.2 Distribution of globular clusters on the sky, in galactic coordinates. They are not confined to the galactic plane at all (as are most of the visible stars of our galaxy), but instead are essentially uniformly distributed around the galactic centre. The X-ray emitting clusters are circled. (Diagram by George Clark, MIT.)

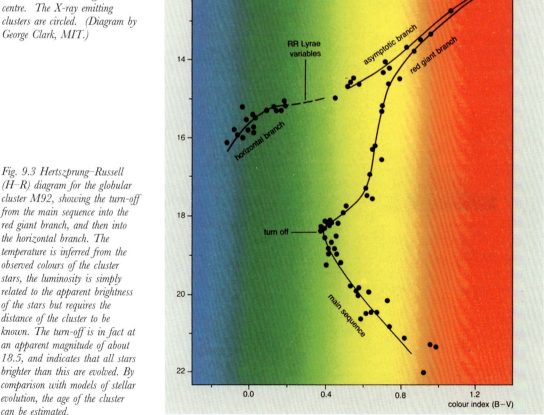

Fig. 9.3 Hertszprung–Russell (H–R) diagram for the globular cluster M92, showing the turn-off from the main sequence into the red giant branch, and then into the horizontal branch. The temperature is inferred from the observed colours of the cluster stars, the luminosity is simply related to the apparent brightness of the stars but requires the distance of the cluster to be known. The turn-off is in fact at an apparent magnitude of about 18.5, and indicates that all stars brighter than this are evolved. By comparison with models of stellar evolution, the age of the cluster can be estimated.

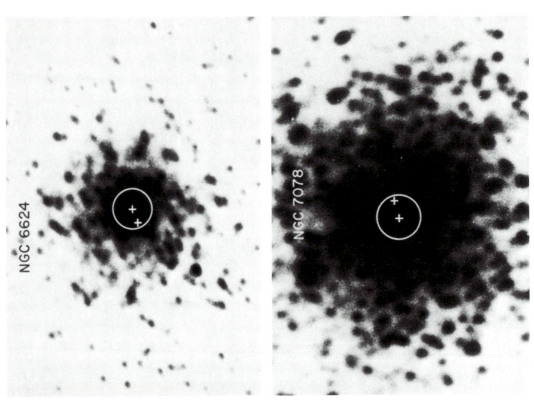

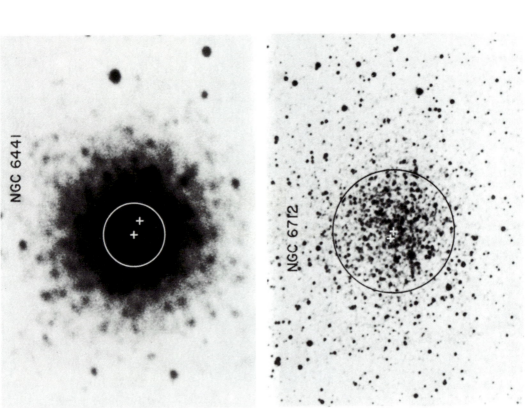

(a)

Fig. 9.4 (a) Montage of optical photos of centres of some of the bright globular cluster X-ray sources. The circle is not, in fact, the X-ray source location, but encircles the core of the cluster as defined by the radius at which the brightness has fallen to half of its peak value. The centre of the circle therefore represents the best estimate of the optical centre of the cluster. The X-ray position is the offset cross as determined by Einstein HRI images, and its uncertainty is the size of the cross. A compendium of these positions for all the X-ray clusters is also shown (b), in which the core radius for each cluster is set to unity and the relative position for each source then plotted. Deep X-ray observations of globular clusters that do not contain bright X-ray sources have revealed much fainter sources that could be cataclysmic variables (X-ray binaries in which the compact object is a white dwarf; see chapter 10). (c) Four, perhaps five, low-luminosity sources are revealed in this ROSAT HRI picture of the central 128 × 128 arcsecs of the cluster NGC 6397. These sources are likely to be cataclysmic variables (see chapter 10). (Optical photos by Josh Grindlay, CfA; X-ray image courtesy of Adrienne Cool and Josh Grindlay.)

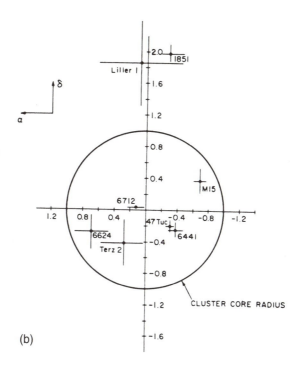

(b)

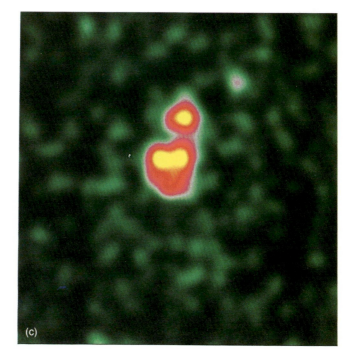

(c)

clusters are indeed extremely old (typically older than our Sun). More importantly, the more massive original members of the cluster will have left behind stellar remnants after supernova explosions in the form of white dwarfs, neutron stars and black holes. Thus the clusters contain just the objects needed to form X-ray sources. [The reader who is familiar with the physics of the formation of neutron stars in supernova explosions may challenge this assertion on the grounds that radio pulsars are known to have quite large space motions, acquired during the supernova explosion itself, which are typically much larger than the escape velocity from a globular cluster. Whilst this is true, there are enough pulsars observed with low space velocities to suggest that a fraction of new neutron stars formed (believed to be between 1/10 and 1/5) will be retained by the parent globular cluster.]

Long-studied because of their importance to theories of stellar evolution, globular clusters are now found to display new properties, such as the frequent occurrence of X-ray sources and millisecond radio pulsars. Clearly the process by which X-ray sources are formed in globular clusters is greatly enhanced. In order to attack this problem we need more details about the nature of the X-ray sources themselves. They cannot be massive binaries like Cyg X-1 (because of the age of the clusters), but are they low-mass systems similar to Sco X-1, or perhaps lone compact objects accreting material lost by the many stars packed into the very dense cluster core? One of the first models proposed was built upon the presence of the red giants and their winds in the cores of the clusters. This material could accrete onto a massive ($\sim 10^3 \ M_\odot$) black hole, thereby liberating the observed flux of X-rays. However, the discovery of X-ray bursts from virtually all the globular cluster X-ray sources, and their subsequent accurate positioning soon disposed of this idea.

9.2 X-ray bursts and precise X-ray locations

The first X-ray burster was discovered in 1975 in the globular cluster NGC 6624 (see chapter 8). Within a year, bursting behaviour had been found in several of the other cluster sources, as well as other galactic X-ray sources. As we have seen, X-ray bursts are now well explained by the model of thermonuclear flashes on the surface of a neutron star. This identification constrains the mass of the compact object to be in the range 1 to 2 $M_\odot$. With virtually all of the globular cluster sources displaying bursting behaviour, something must make the conditions for the formation of bursters ideal within clusters.

The mass range was further constrained when the Einstein Observatory HRI obtained accurate (< 3 arcsecs) X-ray positions showing that all the globular cluster X-ray sources were located very close to the *centre* of the cluster. Indeed, this work by Josh Grindlay and his co-workers at CfA, showed that the X-ray sources were all within or close to one core radius of each cluster (defined for a cluster as the radius within which a certain fraction of stars are contained).

This could not happen by chance. The X-ray objects had to be different from the other stars in the cluster. They had to be *heavier*. A detailed statistical analysis showed that the X-ray sources were between two and three times as heavy as the average cluster star. Since this average is about 0.5 $M_\odot$ the cluster X-ray sources have an average mass of about 1.5 $M_\odot$. Note that this assumes that all the cluster X-ray sources are similar, and gives no information about individual sources, but does confirm that the X-ray sources are neutron stars and not the black holes that had been proposed earlier.

In addition to locating the bright globular cluster X-ray sources, the Einstein HRI could survey for the presence of hitherto unknown weaker sources in the clusters. Several such objects were found, and a complete summary of the X-ray sources detected in globular clusters is given in table 9.1.

9.3 Are the cluster sources X-ray binaries?

The answer to this is certainly yes, but it is worth noting that more than 10 years after the discovery of globular cluster X-ray sources no direct evidence (observation of periodic variability or eclipses) was available for their binary

Table 9.1 Properties of X-ray emitting globular clusters

Source	Cluster	Log L_x (erg s^{-1})	Distance (kpc)	Bursts	Period	r_{core} (arcsecs)	$\Delta r_x/r_{core}$ [b]
(a)	*BRIGHT*	$>10^{36}$ erg s^{-1}					
4U1820–30	NGC 6624	38.0	8.0	Yes	11.4 min	5	0.75
4U1746–37	NGC 6441	36.8	11.7	Yes	—	8	0.45
MXB1730–335[a]	Liller 1	36.8	10.0	Yes	—	4	1.9
X1732–304	Terzan 1	36.8	10.6	Yes	—	6	—
4U2127+11	M15	36.7	9.7	Yes	8.5 hrs	6	0.8
X1724–308	Terzan 2	36.7	10.0	Yes	—	6	0.5
X1850–086	NGC 6712	36.4	6.2	Yes	—	49	0.13
X0513–401	NGC 1851	36.1	12.0	Yes	—	6	2.0
MX1746–20	NGC 6440	Transient	7.1	—	—	12	<5
(b)	*DIM*	$<10^{36}$ erg s^{-1}					
E0021–722	47 Tuc	34.6	4.6	No	—	24	0.4
	NGC 5824	34.3	23.5	—	—	4	15
	M79	33.9	13.0	—	—	16	<4
	M3	33.6	10.4	—	—	29	<2
	NGC 6541	33.3	7.0	—	—	34	<2
5 sources	ω Cen	32.6–32.9	5.2	—	—	144	
4 sources	M22	32.0–32.6	3.1	—	—	114	

[a] Also known as the *Rapid Burster* (see chapter 8).
[b] This gives the position of the X-ray source relative to the cluster's core radius, r_{core}.

nature. Other models were discarded (such as the single massive black hole accreting from red giant winds, referred to above) once their low mass had been established. There was no alternative but to think of them as LMXBs in which a neutron star (or perhaps a black hole) is accreting material from a low mass companion. Nothing else was capable of delivering the observed X-ray power, and in any case they were otherwise very similar (X-ray spectrum, X-ray bursts, distribution) to the X-ray sources of the galactic bulge.

But how are so many binaries formed inside globular clusters? Obviously there are lots of stars in a small region of space, but what is the process by which a compact object (neutron star or black hole) acquires the companion it needs to start life again as an X-ray source? There are two mechanisms that have been examined in detail (see figure 9.5).

(a) Compact object captures single star.

Unless the two stars physically collide (which is unlikely), they have too much kinetic (motion) energy to be trapped into a stable binary orbit. However, as pointed out by Andy Fabian, Jim Pringle and Martin Rees (Cambridge) in the mid-1970s, there is a mechanism by which some of this excess kinetic energy can be dissipated, and that is *tides*. As the compact object approaches the normal star it distorts it substantially (raises tides) which absorbs energy (in exactly the same way that our Moon raises tides in the Earth's oceans). If this happens quickly enough then the two stars become bound.

(b) Compact object interacts with a wide binary, a triple-star event.

Two normal stars exist in a wide (non-interacting) binary. This is approached by the compact object, and a complex triple-star interaction takes place. This results in the compact object replacing the lower mass star in the binary and ejecting it. This is akin to the exchange collisions seen in nuclear particle reactions.

Fig. 9.5 Schematic of the processes of binary formation thought to be applicable in globular clusters. In the tidal interaction of two stars (a), some of their relative kinetic energy must be lost in order for them to become bound. This is done through the tides that are raised on the normal star by the compact object. Since the initial orbit in such a capture is eccentric, these tides continue to operate at each close approach, thereby causing the orbit to rapidly circularise. The alternative mechanism depicted (b) involves a three-body interaction (compact object approaching an existing wide binary) and is more complex. Eventually one of the normal stars in this temporary triple system is ejected, thereby leaving the compact object with the other as a companion. (Based on original diagrams by George Clark, MIT.)

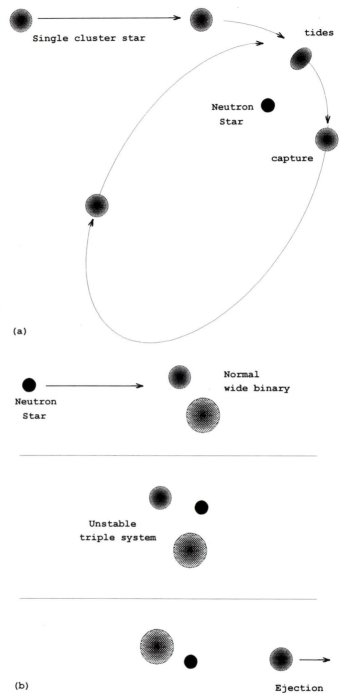

Although the cross-section (likelihood of encounter) is much greater for the triple-star event (the compact object only has to *hit* somewhere between the two normal stars), the problem is that such wide binaries appear to be extremely rare in globular clusters. Calculations by Frank Verbunt (Utrecht) show therefore that process (a), the *tidal capture* mechanism, is by far the more dominant. It is also clear in that case that the capture is more likely the larger the size of the normal star. This means that an evolving star (onto the red giant branch) has a greater chance of being captured, a point we shall return to later.

In the search for evidence to support the binary interpretation, although substantial X-ray variability was present in all of these objects, *none* exhibited any regular modulations that could be attributed to the binary period. This greatly restricted further interpretation and, in particular, any attempts to establish how they had formed and evolved.

9.4 Search for optical identifications

Perhaps the study could be continued at other wavelengths. If an optical counterpart could be established then, as in the case of Sco X-1, it might be possible to measure a period on the basis of photometric or spectroscopic variations. However, there is a serious problem which is obvious upon examination of figure 9.4. All of these objects are located in the most crowded star fields known to ground-based astronomy. In addition, by comparison with known LMXB counterparts (such as Sco X-1) the brightness of these objects is expected to be in the range 17th to 18th magnitude, or some four magnitudes fainter than the brightest stars in the pictures. Searching for such a star presents enormous technical difficulties and is possible only at sites on Earth which harbour the finest *seeing* properties (i.e. exceptionally stable atmospheric conditions).

9.5 M15 (4U2127+11)

In 1984, as a result of the pioneering work of Michel Aurière at the Pic-du-Midi Observatory, an extremely ultraviolet and variable star was found in the core of the globular cluster M15. This star, labelled AC211 in the catalogue that Aurière and Cordoni prepared of stars in M15, was also found to lie about 3 arcseconds from the Einstein HRI X-ray location. This combination of very unusual behaviour and proximity to the X-ray position strongly supported its association with the X-ray source. But at 15th magnitude it was much brighter than expected! Its ultraviolet colour also makes it stand out superbly from the other stars in the *Hubble Space Telescope* image (figure 9.6).

However, it was the brightness of this star that attracted the attention of one of us (PAC), and prompted the use of the re-furbished Isaac Newton Telescope (INT, newly moved to La Palma in the Canary Islands) to obtain the first optical spectrum of AC211. This immediately confirmed the identification through the discovery of very strong emission by ionised helium. This is a characteristic feature of LMXBs and one requiring extremely high temperatures or the presence of high energy ionising radiation (i.e. X-rays). The large optical and UV variability combined with the apparent weakness of its X-ray output also suggested that AC211 was an *accretion disc corona* system (see chapter 8) in which the X-ray source is not viewed directly, but only through X-rays scattered from above and below the accretion disc (which shadows the X-ray source itself). However, this explanation has been shot down with the discovery by the Ginga satellite (Tadayasu Dotani and colleagues at Tokyo's ISAS) of a giant X-ray burst from AC211, one of the largest bursts ever detected. This immediately tells us that the compact object is a neutron star, and the luminosity means that we are seeing it directly! The cause of AC211's optical brightness therefore remains a mystery.

A spectroscopic study (by PAC and graduate student Tim Naylor) in 1986 revealed the binary period of about 9 hours through the radial velocity variations of lines in the spectrum. This period was subsequently found photometrically (see figure 9.7) in data accumulated at the Canada–France–Hawaii Telescope (on Mauna Kea) and Pic-du-Midi by Aurière and Sergio Ilovaisky (Observatoire Haute Provence). It was also quickly confirmed by the analysis of archival HEAO-1 X-ray data. (This had been masked in the first analysis by large intrinsic *erratic* variability). The star AC211 is still the *only* confirmed optical identification of a globular cluster X-ray source and hence the only

Fig. 9.6 (a) CCD images of the core of M15 obtained in July 1985 with the Pic-du-Midi 2 m telescope by Michel Aurière and colleagues. The set of two images in the ultra-violet (U) band (top) shows the large amplitude (>1 mag) variability of AC211 on a timescale of only an hour, and by comparison with the blue (B) and visual (V) band images (below) then the very blue colour of AC211 becomes clear. These excellent ground-based images should be compared with the Hubble Space Telescope U band image obtained with the Wide Field Planetary Camera (b). This covers a region of approx. 5 × 7 arcsecs of the core of M15 (but is rotated approx. 20° clockwise compared to (a)). AC211 is the bright star marked with a pointer. The diffuse light in the core is not real. It is a result of HST's spherical aberration of the primary mirror.

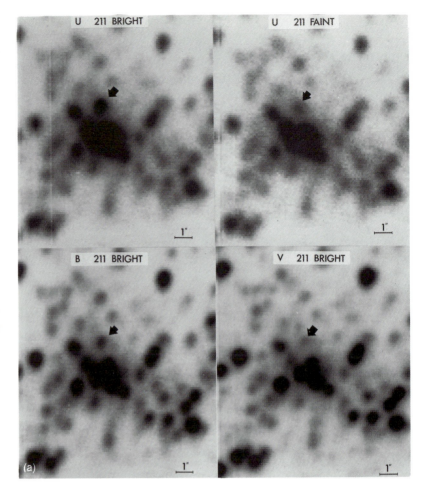

one with a spectroscopically determined binary orbit. (Interestingly enough, AC211 is still the only spectroscopic binary known in any globular cluster!) A search for similar objects in the other X-ray globular clusters has thus far revealed only two faint candidates (in NGC 6712 and 47 Tuc), but they are much fainter than AC211 and little is known about them at present.

Fig. 9.7 (a) The optical spectrum, radial velocity curve and (b) light curve of AC211. The spectroscopy was obtained by PAC and Tim Naylor with the 2.5 m INT on La Palma, and shows the HeI absorption (at 4471 Å) and HeII emission (4686 Å) that are characteristic of high temperature gas. The HeI line was used to produce the radial velocity curve, with which the binary period of 8.5 hrs was first discovered. The light curve (in ultraviolet light) is from joint CFHT and Pic-du-Midi observations by Aurière and Ilovaisky and was used to accurately define the orbital period.

(a)

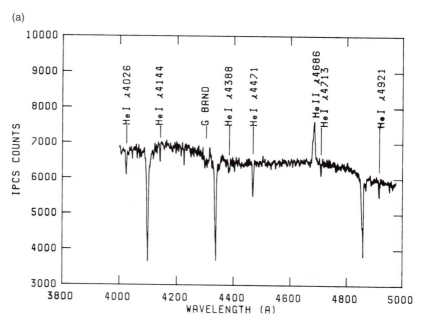

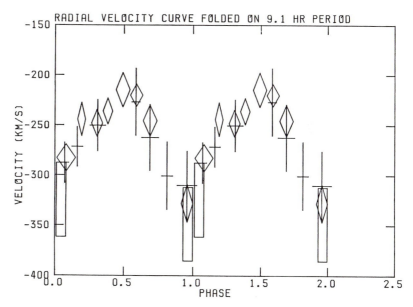

9.6 Is AC211 being ejected from M15, or ejecting matter from itself?

The most extraordinary fact to emerge about AC211 is illustrated in the radial velocity curve of figure 9.7. The average of the curve represents the velocity of the object with respect to us, and that value is -250 km s^{-1}. The velocity of M15 as a whole, however, is very accurately known to be -100 km s^{-1}. In other words AC211 appears to be travelling towards us at 150 km s^{-1} with

Fig. 9.7 (b) For legend see page 229.

(b)

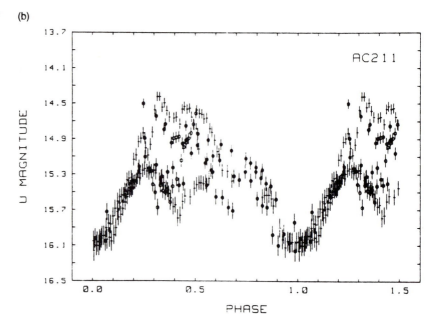

respect to the cluster. This is truly extraordinary since the escape velocity of the cluster is only about 30 km s^{-1}. This would imply that AC211 is rapidly leaving the cluster and must also have been formed in this state very recently (since it is so close to the core now), at least within the last 10^4 years.

Alternatively, the material that is producing the absorption line (from which figure 9.7 is derived) could be moving towards us at this velocity. This would require a large shell of material to be blown off the disc or companion star, presumably driven by the radiation of the X-ray source itself. It would also require that this shell be within the binary system so that it can display the orbital motion that is seen. Recent observations suggest that this explanation is closer to the truth since the average radial velocity has been seen to vary, an impossibility in the ejection hypothesis. But how could the source have reached this remarkable phase and how will it evolve from here? To answer this question we must first look in a little more detail at AC211 itself.

9.7 Mass estimates

Even though the actual locations of the line-forming regions in the binary are unclear, it is still possible to use Kepler's Laws to estimate the masses of the individual stars in binaries. To do this unambiguously requires a knowledge of the period, the inclination of the orbit with respect to us and the radial velocity curve for each component of the binary (see box 7.3). Such knowledge is rarely complete and that is especially true here! For AC211 the period and radial velocity curve are known (but what does it represent?). What else can be inferred?

The light curve in figure 9.7 shows a very large amplitude and the minimum looks suspiciously like an eclipse. This suggests that the inclination is close to 90° (i.e. its orbital plane is close to our line of sight). The neutral helium, usually an indication that hot material is present, is believed to be intimately associated with the compact object and the accretion disc. With these assumptions, the data yield the mass of the compact object for any given mass of the companion star. Here are four possible solutions:

Companion star mass, M_c	1.05	0.8	0.65	0.4	$M_\odot$
Compact object mass, M_x	21	14	10	3	$M_\odot$

which clearly indicate that something exciting is happening. The first three masses chosen as possibilities for the companion star correspond to particular stages in its evolution. The first (and largest) mass is the solution if the companion star were a main sequence star filling its Roche lobe and transferring mass. This cannot be the case for M15 because it is too old. In fact, the most massive stars still just on the main sequence (and hence at the *turn-off* point, where they are just about to leave the main sequence and evolve into a giant, see boxes 4.3 and 9.1) have a mass of 0.8 $M_\odot$ which is the second solution. The lightest stars known in M15 are those which have evolved onto the *Horizontal Branch* in the H–R diagram. Their mass has been measured from the pulsation properties of stars in this region (called RR Lyrae stars) to be 0.65 $M_\odot$, the third solution.

But these all yield very high masses for the compact object (indicating that it is a black hole), contrary to the statistical determination of the mean masses of the X-ray systems in globular clusters of about 2–3 $M_\odot$. Only if the companion star were 0.4 $M_\odot$ or less, would the compact object be less than 3 $M_\odot$, and therefore reconcilable with being a neutron star as required by the discovery of the X-ray burst and the results of the Einstein *average* mass measurements. But how could such a system be formed? The answer has come from recent observations of the brightest of all the X-ray globular clusters, NGC 6624.

9.8 NGC 6624 (4U1820–30), the shortest orbital period known

As well as being the brightest X-ray source in a globular cluster, 4U1820-30 was also the first X-ray burster to be discovered, and is therefore certainly a neutron star system. As part of a study of X-ray *QPOs* (chapter 8) in globular cluster sources, Luigi Stella and Nick White (ESA), and Bill Priedhorsky (Los Alamos) obtained extensive observations of this source with EXOSAT. They had been searching for variations which were quasi-periodic and changed with the intensity of the source. To their amazement they discovered a weak (few per cent amplitude) but significant *regular* modulation at a period of 11.4 minutes (see figure 9.8).

At first sight such a discovery does not appear to be remarkable. Neutron stars with strong magnetic fields give rise to X-ray pulsations as they rotate, with periods in the range seconds to minutes, as described in chapter 7. However, one important characteristic of X-ray pulsars is that the rotation period changes with the luminosity of the source (more accreting material spins the neutron star up faster, less material and it spins down; see figures 7.9 and 7.10). In the case of 4U1820–30 the periodicity was found to be present in earlier observations, with *exactly* the same value, even though this object is extremely bright in X-rays and must therefore be accreting at a very high rate. In fact, the limit on any change in the 11.4 minute period was soon lowered considerably when the pulsations were also found in SAS-3 data from 1976–77, almost eight years earlier.

The only solution was then to assume that the periodicity was not due to rotation but instead was *orbital* in origin. If confirmed, and no other explanation appears viable, 4U1820-30 has the shortest orbital period known for any astronomical object. The immediate consequence of this deduction is that the system must be small in size. So small in fact that no normal star could fit within it as the mass-losing object. Only a degenerate object is possible, and the current model consists of a 0.05 $M_\odot$ helium white dwarf (which is only 0.03 $R_\odot$ in size) filling its Roche lobe and transferring material to a 1.5 $M_\odot$ neutron star.

Fig. 9.8 The shortest orbital period known. The EXOSAT discovery of this 685 second period of 4U1820-30 (NGC 6624) is revealed in these power spectra (a) of the 1–3 keV X-ray data from three observations in 1984 and 1985. The folded light curve for each observation shows the low amplitude of this modulation, normally less than 3%, which is why it was not discovered by earlier observations. The fact that the modulation does not change frequency between observations indicates that the periodicity is orbital, and not due to the simple rotation of the neutron star. The schematic (b) demonstrates how compact this entire system must be in order to have such a short period. The mass-losing star is itself a white dwarf, transferring mass into the accretion disc around the 1.5 $M_\odot$ neutron star, only 80 000 miles away. For comparison, this system is shown relative to a prominence on the solar rim and with a projection of the Earth's size just above it. (Diagrams courtesy of EXOSAT Observatory, ESA.)

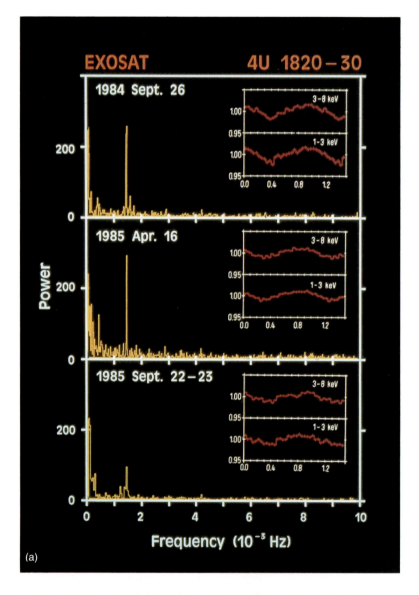

(a)

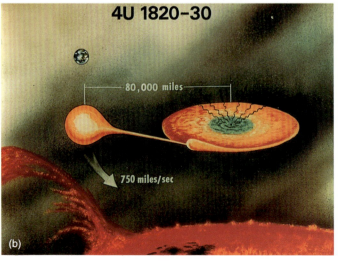

(b)

9.9 Origin and evolution of M15 and NGC 6624 X-ray sources

These two systems are the only ones for which there is now sufficient physical information for study. Both are remarkable when compared with the LMXB population as a whole. M15/AC211 is optically the brightest LMXB in our Galaxy, and 4U1820–30 is an exotic double degenerate contained within an 11 minute binary. How did they form and reach the stage that we see them at now? Verbunt has done considerable work in this area, and his ideas are illustrated graphically in figure 9.9.

Fig. 9.9 Evolutionary scenarios for different ways of capturing a neutron star in the crowded core regions of globular clusters. Most encounters involve main-sequence stars, the majority of which will form typical LMXBs. Alternatively, a very short period LMXB can be formed by a direct hit with a red giant star in which the neutron star spirals in through the red giant's extended atmosphere. The end results of all these scenarios have already been observed. (Based on a diagram by Frank Verbunt, Utrecht.)

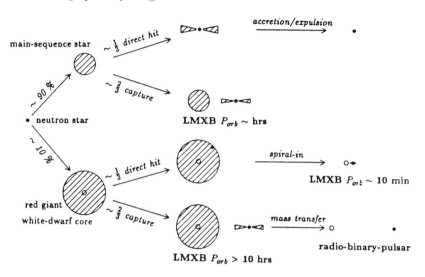

Obviously compact objects are needed to form X-ray binaries. As mentioned earlier, these will have evolved from some of the earlier, more massive stars in the cluster. The clusters are very old and we only see low-mass stars now. When originally formed there must have been stars of all spectral types (and hence a wide range of masses). The most massive will have gone supernova, leaving behind neutron stars and black holes. Admittedly some of these will have been ejected from the cluster in the explosion, but some will have remained. And being heavier than the average star in the cluster (typically $0.5\,M_\odot$), they will eventually settle in the *inner* regions of the cluster.

There is therefore a pool of compact objects circulating within the highest density region of the cluster. What interactions are expected? The most likely events involve main-sequence stars (upper part of figure 9.9). In 1/3 of the collisions the neutron star actually scores a direct hit on the star and numerical simulations indicate that the normal star will be totally disrupted, forming a massive disc around the compact object. Eventually this disc will be accreted (or expelled by radiation pressure) leaving behind the lone neutron star. In the other 2/3 of the interactions, the star is captured by the neutron star, forming an LMXB with an orbital period of several hours, like the source AC211 in M15.

But what happens when a neutron star meets a red giant? There are fewer of these stars, but they are more likely to be found in the core *and* are much larger objects (and hence better targets for compact objects to hit). About 10% of collisions are thought to involve red giants. A direct hit (again expected in 1/3 of the events) does not disrupt the star, because a red giant is physically very different from a normal star (see box 9.1).

The neutron star barely notices the presence of the atmosphere of the red giant, and settles into orbit about the compact centre of the red giant (which will be between 0.1 and $0.5\,M_\odot$ in size, and composed almost entirely of heli-

Box 9.1 The structure of a red giant

A red giant forms when a normal star has consumed a significant fraction of its main nuclear fuel (hydrogen). The central helium core, no longer supported by internal nuclear energy, then begins to contract under its own weight. Hydrogen burning continues, however, in a shell around the core (see figure 9.B1).

Fig. 9.B1 Internal structure of a star on the main sequence, and then as it evolves into a giant. Although the hydrogen in the core has been fused into helium, hydrogen-burning continues in a shell surrounding this core, but at a higher rate because of the increased density of the core.

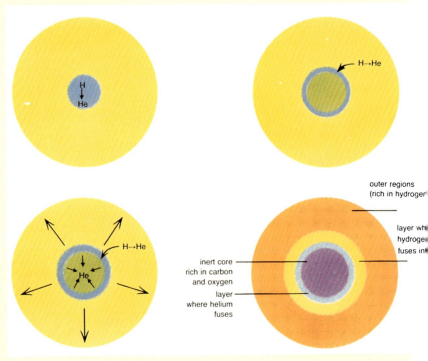

As the core contracts, the surface gravity in the shell increases. This in turn increases the rate of energy generation in the shell. The increased luminosity causes the outer parts of the star to expand and become less dense. Because the radiation we see from the star is being emitted from a larger area (than a normal star) the surface appears to be cooler or redder. Hence, the name *red giant*. In fact, the centre of a red giant is much hotter and denser than a normal star. The diagram shows that the star is becoming a double entity: a compact central core (the progenitor of a white dwarf) surrounded by a large, extremely diffuse envelope. One day our Sun will suffer this fate.

um, the nuclear *ash* from when the red giant was a normal star). However, the atmosphere will cause this orbit to spiral in quickly (in the same way as our Earth's atmosphere causes drag on a satellite's orbit, and eventually it too spirals in unless rockets are fired to raise it again). During this process the neutron star accretes material, releasing copious X-rays that disperse the red giant's atmosphere. It is possible that M15/AC211 is in exactly this phase now. If this were the case, then the dispersing (expanding) atmosphere could give rise to the 150 km s^{-1} offset observed in the radial velocity curve. At the end of the spiral-in phase the neutron star and helium white dwarf would be left in a tight orbit, with a period of about 10 minutes. This would correspond to the state of 4U1820−30 that is observed now.

Alternatively, Charles Bailyn and Josh Grindlay (CfA) have suggested that a neutron star may capture a main-sequence star that is just at the *turn-off* point. If this happens, the subsequent mass-transfer will be unstable (i.e. the loss of a small amount of material from the companion will lead to the star further overfilling its Roche lobe, hence transferring more mass, and so on) and the outer envelope of the star will rapidly expand to encompass the neutron star as well. At this point, which is again suggested to be the M15/AC211 case, the situation is almost indistinguishable from that postulated by Verbunt for the neutron star – red giant collision, and the subsequent evolution is the same.

In a grazing collision, the neutron star will capture the red giant. They will form a relatively wide binary, with a period greater than 10 hours. Again this would appear as an LMXB. However, mass transfer will widen the binary, eventually leaving a non-interacting white dwarf – neutron star pair. This happens quite quickly because the red giant is itself evolving rapidly. For that reason, we would not expect to be lucky enough to find a source in this state.

9.10 Formation of millisecond radio pulsars

The story is not yet over. What happens when the mass transfer phase is completed for either of the extreme scenarios (neutron star hits a red giant or captures a main-sequence star) just described? (Both are rapid processes.) A lone, or at least detached binary, neutron star will be left. But before the binary was formed, the neutron star was already old, as it was created in the early days of the globular cluster from a massive star. Its magnetic field would have decayed and would therefore have had relatively little effect on the accretion process after the capture of another star. The accretion disc that was formed thus extended down virtually to the surface of the neutron star, and eventually the transfer of angular momentum from the accreted material (requiring about 0.1 $M_\odot$ of matter) would have spun up the neutron star to the speed of that of the innermost orbits of the accretion disc. Thus a spin period of a few *milliseconds* was created.

At such a period the neutron star is *re-born* as a radio pulsar. It was proposed by Julian Krolik (Harvard) that the best place to look for very rapid radio pulsars is in globular clusters, and indeed that is where many are now being found (M15, M28 and 47 Tuc for example). All these pulsars have rotation periods shorter than 10 milliseconds. Some are single objects, others are binary. Interestingly enough, the fastest of this new breed of pulsars (at 1.6 ms) is represented by two objects, one single (PSR1937+21), the other binary (PSR1957+20). This new binary is proving to be a wonderful source of physical information, because it is *eclipsing*. It has a period of 9.2 hours and the radio pulsations are eclipsed for a total of 50 minutes per cycle.

The eclipse is *not* sharp and the pulsar timing residuals indicate the presence of diffuse plasma around the eclipsing star. The inclination of the orbit must be near 90°, and it can be assumed that the neutron star mass is 1.4 $M_\odot$. The pulsar's radial velocity curve (from the pulse period change around the orbit) then gives the companion star's mass as only 0.02 $M_\odot$. The diffuse matter around the star is apparently driven off the surface by the enormous flux of relativistic particles from the pulsar. The Roche lobe of such a low mass star is three times smaller than the size of the eclipsing object, so the matter *outside* the Roche lobe will be rapidly lost from the system and must be continuously replenished.

Although the pulsar was discovered in late 1986 at Arecibo by Andy Fruchter, Dan Stinebring and Joe Taylor, an accurate radio position was only acquired in spring 1988 from the VLA. Armed with this position, one of us (PAC) used the newly commissioned William Herschel Telescope (WHT) on La Palma, a telescope which points to an accuracy of 1 arcsecond, to obtain direct CCD images of the field. A variable star was quickly found at the radio position, as is evident from the pictures at maximum and minimum

Fig. 9.10 Montage of radio and optical observations of the binary, millisecond radio pulsar PSR1957+20. (a) The pulsar radial velocity curve and its residuals show the extended plasma surrounding the star (the pulses do not disappear and reappear sharply during eclipse). (b) The WHT optical CCD images were taken at maximum (left) and minimum (right) light in the 9.2 hour binary cycle. (c) The complete optical light curve is also shown. It is almost certainly asymmetric. (Optical images by PAC, Paul Callanan and Jan van Paradijs, radio data from Andy Fruchter and colleagues at Princeton.)

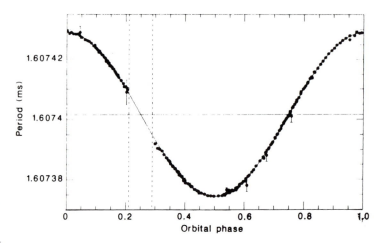

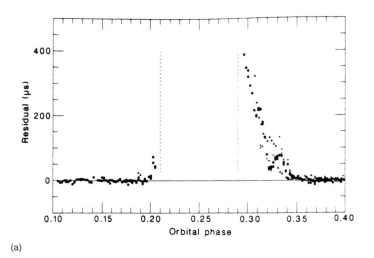

(a)

(see figure 9.10). With La Palma's superb seeing it became clear that the optical counterpart of PSR1957+20 was contaminated by an unrelated field star barely 0.7 arcseconds away. In addition, the pulsar varied from 21st magnitude at maximum, to below 23rd magnitude at minimum. Only the heated side of the low-mass star is seen. The other (normal) side of such a low-mass object is, so far, completely undetectable. A schematic of this system is shown in figure 9.11, which illustrates the heating by the pulsar and the presence of the wind being blown off the star.

Extrapolations of the current behaviour of PSR1957+20 suggest that the companion star will be completely *evaporated* within 10 million years, leaving only an isolated radio pulsar which will gradually spin down. It is likely that PSR1937+21, the other fast but single pulsar, has already gone through this phase and evaporated its companion.

Such objects must have gone through an LMXB stage at some point (figure 9.9) and have been spun up by the mass lost as the companion star evolved. It has been suggested by van den Heuvel and van Paradijs that binary millisecond pulsars also offer an explanation of why there are so

Fig. 9.10 (b, c) For legend see opposite page.

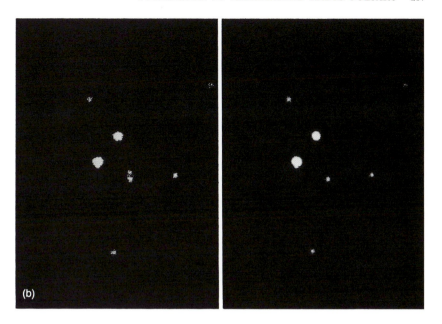

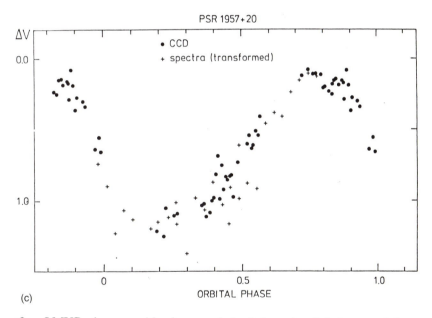

(c)

few LMXBs known with short periods (below the 2–3 hour *period gap* that is observed for cataclysmic variables, see chapter 10).

In the case of cataclysmic variables the mass transfer is driven (for periods below 10 hours) by loss of orbital angular momentum (due to either gravitational radiation or magnetic braking or both). This reduces the period gradually until it reaches about 3 hours. At this point magnetic braking ceases (probably due to structural changes inside the star that greatly reduce the strength of the magnetic field and hence the braking effect) and mass transfer stops, thereby rendering the system very difficult to observe since it would be extremely faint. Only when the period has reduced to 2 hours through the effects of gravitational radiation will mass transfer start again, making the system observable once more.

All this can also be applied to LMXBs, but with one fundamental differ-

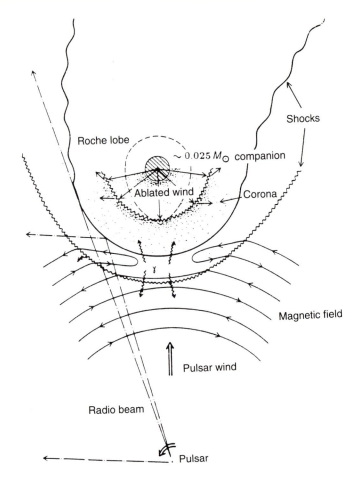

Fig. 9.11 Schematic of the PSR1957+20 binary system. The extremely fast pulsar generates an ultra-high energy wind that irradiates the companion star and boils material off its surface. This material interacts with the pulsar wind at an extended shock front that is responsible for the extended region seen close to pulsar eclipse (figure 9.10). (Based on a diagram by Phinney and colleagues at Cal-Tech.)

Roche lobe

Shocks

$\sim 0.025\,M_\odot$ companion

Ablated wind

Corona

Magnetic field

Pulsar wind

Radio beam

Pulsar

Fig. 9.12 Outline of the evolution of a low-mass interacting binary with period below 10 hours for the two cases of a neutron star and white dwarf compact object. The initial formation (a) from a much wider binary star system is probably common to both types of compact object. The key point is the common envelope phase which allows the lower mass normal star to spiral in towards the core of its companion red giant. This disposes of the orbital angular momentum, resulting in a short-period binary (f). But the subsequent evolution (opposite page) depends on the nature of the compact object. The spun-up neutron star becomes a relativistic pulsar which dramatically changes the evolution of the system when its period has reached the edge of the period gap. This does not happen in the case of the white dwarf as it cannot generate the high energy wind that a pulsar can. ((a)–(f) are based on originals by R. Webbink and M. Politano.)

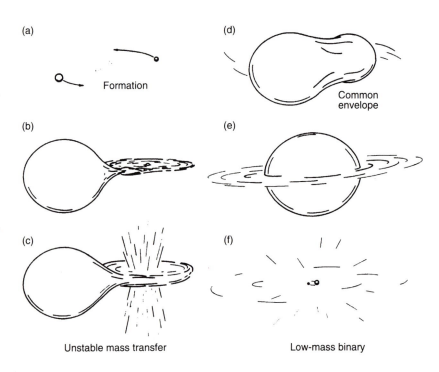

(a) Formation

(d) Common envelope

(b)

(e)

(c)

(f)

Unstable mass transfer

Low-mass binary

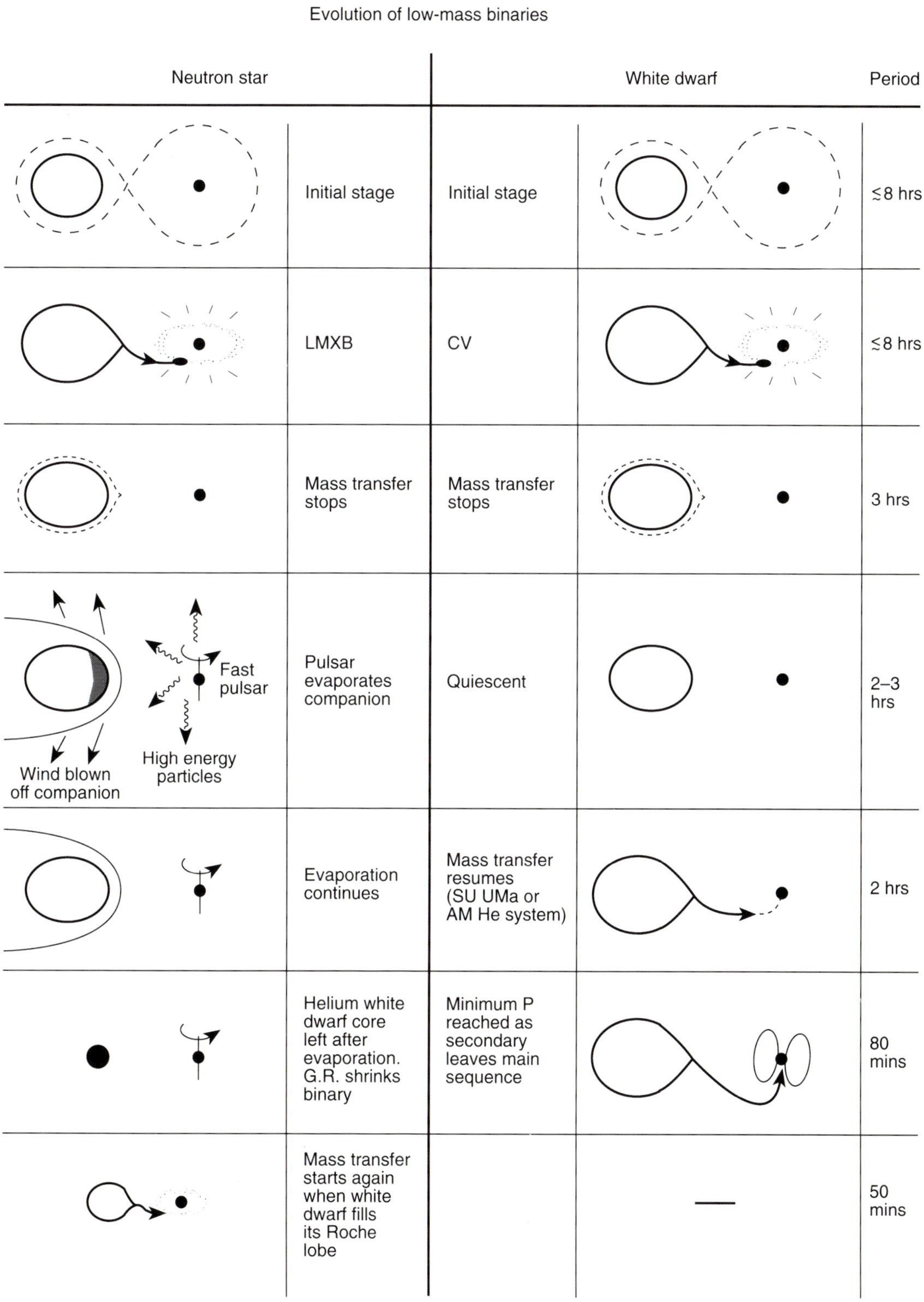

Evolution of low-mass binaries

Neutron star			White dwarf		Period
	Initial stage	Initial stage			≤8 hrs
	LMXB	CV			≤8 hrs
	Mass transfer stops	Mass transfer stops			3 hrs
Fast pulsar / High energy particles / Wind blown off companion	Pulsar evaporates companion	Quiescent			2–3 hrs
	Evaporation continues	Mass transfer resumes (SU UMa or AM He system)			2 hrs
	Helium white dwarf core left after evaporation. G.R. shrinks binary	Minimum P reached as secondary leaves main sequence			80 mins
	Mass transfer starts again when white dwarf fills its Roche lobe		—		50 mins

Fig. 9.12 For legend see opposite.

ence, the compact object is now a neutron star. At the point where mass transfer stops (at the 3 hour period) the neutron star is likely to have already been spun up and be capable of switching on as a powerful radio pulsar. It will then start evaporating its companion and, if successful, will never show up below the period gap! However, if the secondary star already has a helium core of sufficient size, this will survive the evaporation stage, leaving a neutron star – helium white dwarf binary again. But this time, gravitational radiation will be driving them closer together, and when the period reaches about 50 minutes, the white dwarf will start to fill its Roche lobe and, once more, provide material to accrete onto the neutron star, re-initiating it as an LMXB. There are two such systems, A1916–05 (an X-ray dipper already mentioned in chapter 8) and 4U1626–67, which come into this category and may have formed in this way (see figure 9.12).

Hence, the globular cluster sources and the millisecond radio pulsars have provided us with ideas to unify the evolution of X-ray binaries, and give clues to the understanding of the end points of stellar evolution itself.

10 Cataclysmic variable stars

10.1 Dwarf novae and novae

Although remarkably similar to the low-mass X-ray binaries already described, cataclysmic variables (CVs) are significantly fainter in X-rays, particularly at the higher X-ray energies of the early generation of X-ray satellites. But they have a long history that goes back well before the era of X-ray astronomy, as they include among their number both dwarf novae and novae. Curiously enough only one previously known cataclysmic variable, EX Hya, is to be found in the Uhuru survey of the X-ray sky. It should be noted that a supernova event results from a fundamentally different process (as described in earlier chapters) which is *irreversible*. The progenitor star collapses rapidly under gravity and then explodes catastrophically. For any given object it happens once, whereas nova and dwarf nova explosions can and do recur. Indeed, it is hypothesised that all novae recur, it is just that the typical recurrence time is believed to be about 10 000 years which is difficult to monitor!

Novae have been known to mankind throughout history, as for a few weeks the nearest and brightest examples become important naked eye objects. The most recent nova to attract popular attention was Nova Cygni 1975, which at its brightest reached 2nd magnitude and for a time dramatically changed the appearance of the constellation of Cygnus, the Swan. Novae are now known to be the result of unstable thermonuclear burning on the surface of a white dwarf, and accretion energy contributes little to the outburst. Also, such outbursts have not been seen at X-ray wavelengths.

What makes them of interest to us in this book is the fact that all novae are binaries, a discovery only made in the 1960s! The material that is burnt during a nova outburst is provided by a companion star, thereby creating a system remarkably similar to an LMXB. But the compact object in a nova outburst is a white dwarf, not a neutron star. The mass transfer occurs gradually over a long period of time between nova outbursts, and that low level of mass transfer does accrete onto the white dwarf, liberating energy that produces weak X-ray emission.

Dwarf novae, on the other hand, have been known only since the invention of the telescope. The brightest of these only reaches about 8th magnitude during an outburst. Because of the accessibility of a significant number of such objects to binoculars and small telescopes, they are popular targets for observations by amateurs. Indeed, this is one field of astronomy where, even today, amateur groups play an extremely important role in supporting and stimulating professional observations.

Table 10.1 Properties of cataclysmic variables

	Quiescent magnitude (visual)	Outburst magnitude (visual)	P_{orb} (hrs)	Notes
Novae	~15–20	~0–3		
Dwarf novae	~12–16	~8–13		
Examples:				
Novae				
Nova Her 1934 (DQ Her)	14.6	1.5	4.6	71 s oscillations
Nova Cyg 1975 (V1500 Cyg)	>17	2.2	3.4	AM Her type
Dwarf Novae				
U Gem	14.5	8.9	4.2	Prototype
Z Cha	15.2	12	1.8	SU UMa system (has superoutbursts)
Z Cam	14.5	10.2	7.0	Standstills
VY Scl	~13	~17		Anti-dwarf nova
Nova-like				
UX UMa	13.5	–	4.7	
AM Her	12.9	15.2	3.1	Prototype polar
AO Psc	13.3	–	3.6	Intermediate polar (DQ Her type)

Although the dwarf nova outburst is much less dramatic than a typical nova explosion, they occur more frequently (some objects producing outbursts every week or two) and last for much shorter periods of time (usually less than 10 days or so). Table 10.1 summarises the properties of typical nova and dwarf nova outbursts.

There is a great variety in the nature of dwarf nova outbursts as is indicated by the sample shown in figure 10.1. A classical dwarf nova, such as U Gem or SS Cyg, increases in brightness by 3–5 magnitudes within a day, and then decays on a timescale of about 10 days. There are typical times between outbursts for any particular object, but these do vary considerably thereby making their study through conventionally assigned telescope time very difficult.

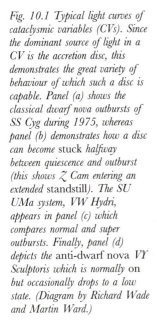

Fig. 10.1 Typical light curves of cataclysmic variables (CVs). Since the dominant source of light in a CV is the accretion disc, this demonstrates the great variety of behaviour of which such a disc is capable. Panel (a) shows the classical dwarf nova outbursts of SS Cyg during 1975, whereas panel (b) demonstrates how a disc can become stuck halfway between quiescence and outburst (this shows Z Cam entering an extended standstill). The SU UMa system, VW Hydri, appears in panel (c) which compares normal and super outbursts. Finally, panel (d) depicts the anti-dwarf nova VY Sculptoris which is normally on but occasionally drops to a low state. (Diagram by Richard Wade and Martin Ward.)

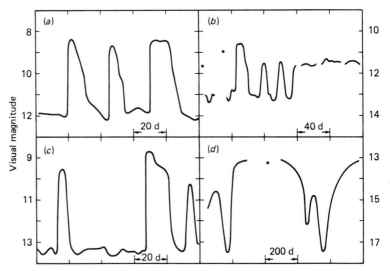

The figure also indicates the behaviour of other types of dwarf novae. The short period SU UMa systems appear at first sight to be like other classical dwarf novae in their outburst patterns. But, much more rarely than their normal outbursts, they deliver an exceptionally long outburst which has become known as a *superoutburst*, an explanation for which we will present later on.

The Z Cam stars also appear to erupt normally like other dwarf novae, but occasionally they too do something very different. They get stuck in *standstills*, sometimes for months on end, at a brightness level below their normal outburst maximum, but well above normal quiescence. Finally, there are the *anti-dwarf novae*, or VY Scl stars, which spend most of their time in an *outburst* state, and then occasionally dip into quiescence for a few days.

There are also objects known as *nova-likes*, because they look like a nova would long after its eruption, but in this case they have not actually been observed in a nova state. Nor do these objects display typical dwarf nova outbursts, as they appear to be in a permanently high state of mass transfer. The classical example of this is UX UMa.

For a long time this wide variety of behaviour was considered to be an impenetrable forest of different types which may or may not relate to one another. However, observations at X-ray and UV wavelengths have enabled major inroads to be made into the puzzles of these low-mass interacting binaries. At the same time, more can be learned about the end-points of stellar evolution which all stars must eventually go through.

10.2 Presence of accretion discs and UV/X-ray emission

It was only during the 1960s that the work of Bob Kraft (Lick Observatory) and co-workers demonstrated that dwarf novae contained accretion discs. The eclipsing binaries (in which we happen to be in the orbital plane of the two stars) gave the opportunity for time-resolved optical spectra to demonstrate the presence of an *extended object*, in this case the accretion disc (see figure 10.2)

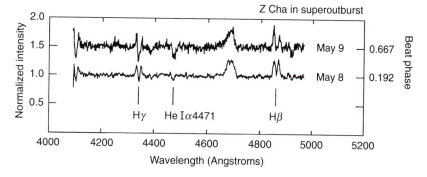

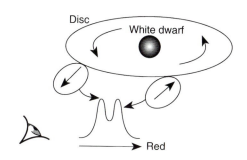

Fig. 10.2 Optical spectra of Z Cha obtained at SAAO show the classical double-peaked emission lines of an accretion disc viewed at high inclination (the plane of the disc is close to our line of sight). The schematic (below) shows how this arises due to the rotation of the disc about the white dwarf. The projection on the sky leads to the dominant emission coming from each side of the white dwarf, one of which is approaching us (blue-shifted), whilst the other is receding (red-shifted), thereby giving rise to the two peaks.

The clearly double-peaked structure arises due to the motion of material in and from the disc itself as shown schematically in the figure. This interpretation was verified by observing the spectrum through eclipse. As the companion star occults one side of the disc, one part of the emission line (in this case the part coming towards us, or the blue-shifted component) disappears. Then as the occulting star moves in its orbit this section of the emission line starts to reappear just as the red-shifted component starts to be occulted. We can actually *see* the accretion disc.

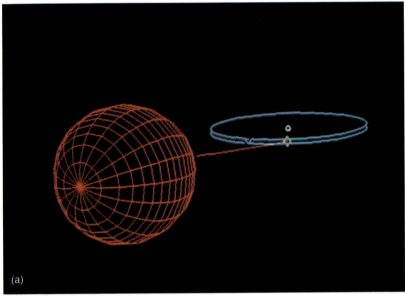

(a)

Fig. 10.3 (a) Canonical model of a CV showing the gas stream from the secondary impacting the accretion disc and producing a bright (or hot) spot. This spot is seen in CVs in quiescence only, and is swamped by the light of the disc itself during outbursts. A disc can only form if the white dwarf has a weak magnetic field. (b) The presence of a strong field disrupts the disc and allows material to accrete directly onto the white dwarf (lower diagram), forming a polar. *The accretion stream becomes* threaded *onto the magnetic field of the white dwarf, prior to accretion onto the magnetic pole caps. The region where the accretion column impacts the white dwarf is shown more clearly in the inset. A shock forms due to the very high velocity of the impact and this produces the hard X-rays (radiated essentially as a fan-beam). These hard X-rays heat the immediate area of the white dwarf's surface producing a* hot cap *which radiates predominantly in soft X-rays and the ultraviolet. (Based on diagrams by Mark Cropper, MSSL and Mike Watson, Leicester.) (c) The accretion geometry of an* intermediate polar, *however, combines properties of both disc and magetic binaries. The outer regions are remarkably similar to a normal cataclysmic variable, but the inner regions are disrupted by the magnetic field of the white dwarf in much the same way as an AM Her system. (Diagram by Keith Mason, MSSL.)*

The main component observed in a U Gem type dwarf nova is the accretion disc itself. Matter is gradually transported through the disc onto the white dwarf, in the process being heated to temperatures that release large amounts of UV and X-ray emission. This radiation then heats up the inner part of the disc so that its temperature decreases towards the outside. This means that a different *picture* of the disc is obtained for each wavelength region or colour observed. The infrared region is dominated by the extended, cooler outer regions of the disc (and frequently the mass-losing star itself), whereas the UV emission is predominantly from the inner disc.

The source of matter for the disc, which must be continuously replenished, is the companion star. It is usually not too different in size or appearance from that of a main-sequence star of comparable mass, except for the fact that it is transferring material to the compact object that it is orbiting. This is happening because the star is filling its Roche lobe and matter flows through its inner Lagrangian point (figure 10.3). As this matter falls in the potential well of the white dwarf, it supplies energy to the disc which enables it to radiate. It is a remarkable fact in these systems that the luminosity of the disc far exceeds that of either the normal mass-losing star, or the white dwarf itself. And because the inner disc is very hot, the overall colour of the system is very blue. It is this characteristic above all others that is used most frequently to distinguish cataclysmic variables from other objects.

Given that virtually all cataclysmic variables have binary periods of less than 1/2 day and that the compact object must be a white dwarf, it is easy to show that the companion star must be of low mass. There just is not room enough in the system for a massive star.

Because of the analogy with LMXBs, CVs were anticipated to be strong

Fig. 10.3 (b, c) For legend see opposite.

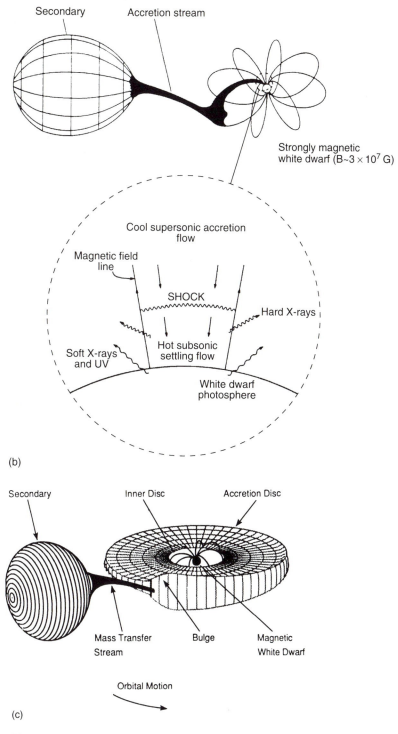

(b)

(c)

X-ray sources. But this was not confirmed by the Uhuru survey, with only one CV being detected. However, early rocket flights which were sensitive only to very soft X-rays (with energy below that of the Uhuru mission) suggested that during outburst CVs may be strong soft X-ray sources, a result that was confirmed by subsequent satellites such as HEAO-1 and then Einstein. But what happens during a CV outburst and what causes it?

10.3 Outburst

A typical CV outburst involves the system brightening by as much as a factor of 100 within a day, remaining at that brightness for a day or two and then returning gradually to its previous level over a period of about a week. The only component of the system that can vary in this way is the disc. There is no known mechanism by which the mass-losing star can brighten significantly, and the white dwarf presents too small an area to radiate this much luminosity. Since the early 1970s there have been two competing models to account for the origin of the dramatic change in brightness of the accretion disc.

The first links the outburst to the behaviour of the mass-losing star. Since it is transferring matter, then a straightforward increase in the amount of matter lost from the secondary must result in a corresponding increase in the luminosity of the disc. This is called the *mass-transfer* model.

The other explanation for CV outbursts assumes that the mass transfer from the secondary is roughly constant with time. The disc accumulates more and more material whilst changing little in appearance until, suddenly, the internal structure of the disc changes (its viscosity is believed to increase) and it brightens dramatically. This is the *disc-instability* model.

Much observational and theoretical effort has gone into attempting to distinguish between these two models. However, it is worth pointing out that in both models the disc is provided with material by the secondary star. The difference between them is in the manner of the mass transfer (steady versus irregular) and the resultant behaviour of the disc. It has proved frustratingly difficult to devise observations that can clearly distinguish between these models.

10.4 Magnetic binaries

X-ray astronomy, although not playing a significant role in the early understanding of dwarf novae, has been responsible for the discovery of two new types of white dwarf binaries. These are the *polars* and *intermediate polars*, which are often referred to as *AM Her* and *DQ Her* systems respectively.

As the prototype, AM Her was the first magnetic binary to be discovered, and is in the Uhuru catalogue, designated as 4U1809+50. But interest in the star was not generated until an accurate position was found and intense soft X-ray emission was discovered by SAS-3. 4U1809+50 was highly variable and brighter in soft X-rays than any other *stellar* source. Then in 1976 came the discovery by Santiago Tapia (Steward Observatory) of extremely high

Box 10.1 Polarisation and magnetic fields

Chapter 3 contains a description of the process of synchrotron radiation, where high energy electrons spiral in a magnetic field and emit radiation. If the magnetic field lines are ordered in some way, then the radiation will be *aligned* so that it will appear to us to be polarised. The case of synchrotron radiation involves such high energies that the electrons are relativistic. However, what happens if there is a strong magnetic field but the electrons do not have such high energies? In this case they will emit *cyclotron* radiation. (So called because it was first found as radiation by charged particles in a cyclotron, an early form of particle accelerator.) The physics is essentially the same as for synchrotron radiation, but the relativistic calculations are unnecessary. For magnetic fields of 20 or 30 megagauss, the fundamental cyclotron frequencies (the rate at which the electrons circle the field lines) lie in the optical or near IR region. Such radiation is polarised and if the cyclotron emission is a significant fraction of the total radiation from the system then there will be large values for the polarisation observed.

(and variable) circular polarisation in the light from AM Her. The variations repeated with a period of 3.09 hours which was the binary period. The polarisation values were greater than any measurements previously recorded and indicated the presence of a powerful magnetic field (see box 10.1).

Optical spectroscopy revealed strong emission lines which exhibited remarkable radial velocity curves (very large velocity amplitudes that could not possibly be indicative of matter in a *normal* binary orbit). But the X-ray light curves were to prove an equally strong probe of these systems.

These variations indicated that the X-rays were eclipsed for part of the cycle and suggested that the emitting region was confined to a small region of the surface of the white dwarf (see figure 10.3). Once again a red star is losing mass to an orbiting white dwarf, but now the white dwarf has such a strong magnetic field that it takes complete control of the matter after it leaves the red star. The matter is therefore not allowed to form an accretion disc, but instead forms an extra long mass transfer stream or accretion column directed along the magnetic field lines straight onto the magnetic pole (or poles) of the white dwarf. This demonstrates the origin of the name of these systems as *polars*. The matter in the column is in almost *free fall* and this accounts for the large velocities seen in the optical spectra.

When the matter impacts the surface of the white dwarf, a shock is formed that emits the hard X-rays seen originally by the Uhuru satellite. This is immediately above the actual surface of the white dwarf and so some of the hard X-rays are absorbed by the white dwarf surface which is heated up to much higher temperatures than its surroundings. This heated area emits the intense soft X-rays that were seen by SAS-3, but represents only a small part of the surface area of the star. As the white dwarf rotates it is possible for the emitting region to be occulted from our view by the white dwarf itself, thus accounting for the form of the X-ray light curves. Note in figure 10.4 the dip in the hard X-rays that is seen when the soft X-rays are at a maximum (pole cap pointing towards us). This is also the time when it is possible for the accretion stream to absorb some of the hard X-rays as we are *looking down* the column.

The strength of the field is such that the rotation period of the white dwarf is locked to that of the orbital period of the system. Hence the modulation of the X-ray flux by geometrical effects (the area of the white dwarf surface which can be seen) occurs at the same period as indicated by the optical spectroscopy and the variations of polarisation. The complexity of this process is also demonstrated by figure 10.4 in the light curves of E2003+225. During three months in 1985 the soft X-ray light curve changed dramatically. One possible explanation for this remarkable *about-turn* is that there is another magnetic pole cap which has *turned on* for some reason. This may be the case, but it is also possible to explain this effect with complex changes in the accretion geometry near a single pole cap. Only further observations will resolve this issue.

Now the X-rays are intercepted by the inner face of the mass-losing star and this too will be heated. The effect of this can be observed in the optical spectra and is valuable for outlining the motion of the secondary star. Together with studies of the un-heated side of the red star it has been possible to estimate accurate masses for both components in several of the AM Her systems. The white dwarfs are very close to 1 $M_\odot$ and are essentially indistinguishable from single white dwarfs that have been found.

Table 10.2 collects together the properties of the currently known set of AM Her systems. One of the most curious of these is that a significant number of them have periods in the range 112–113 minutes, an incredibly narrow band! We shall return to this point after looking at how short period binaries evolve with time.

The long-term behaviour of AM Her systems demonstrates *off* states when

Fig. 10.4 X-ray light curves of two AM Her systems. AM Her itself (top) displays a strong, variable soft X-ray flux in this EXOSAT observation which is anticorrelated with the hard (higher temperature) X-rays. The eclipse of the soft X-rays for a large fraction of the 3.1 hour binary cycle is due to the occultation of the X-ray emitting pole cap by the white dwarf itself, as shown schematically below. The hard X-rays are produced by a shock within the accretion column and are only modulated when the column cuts across our line of sight (see also figure 10.3(b)). However, another long-period AM Her system, E2003+225, has a soft X-ray light curve (lower panels) that changed dramatically over several months in 1985. One possibility is that another magnetic pole turned on and dominated in the later observation. (Diagrams by Julian Osborne, EXOSAT Observatory.)

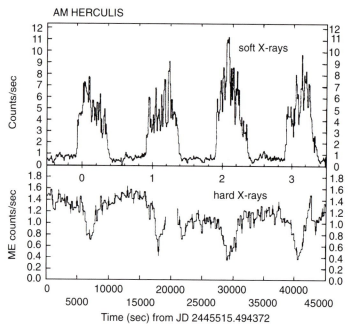

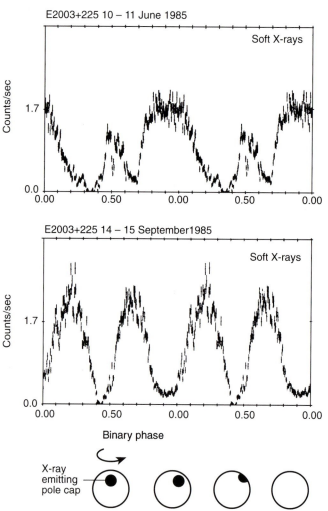

Table 10.2.   Magnetic variables (AM Her systems)

Source	Variable name	Period (mins)	Visual magnitude	Distance (pc)	Magnetic field (Mgauss)
H0139–68	BL Hyi	113.6	14.9		30
EXO0234–52		114.6	18.8		—
2A0311–227	EF Eri	81.0	13.7	>100	—
H0538+608		186?	14.6		41
	VV Pup	100.4	14.5	155	32
1E1048.5+5421		114.5	18.0		—
	AN UMa	114.8	14.5	>300	36
CW1103+254	ST LMi	113.9	15.0	130	30
1E1114+182	DP Leo	89.8	17.5		32
E1405–451	V834 Cen	101.5	14.2	90	20
PG1550+191	MR Ser	113.6	14.9	140	25
4U1809+50	AM Her	185.6	12.0	>70	20
E2003+225	QQ Vul	222.5	14.5	450	—

the star is much fainter (several magnitudes) than normal, and there is no evidence of an accretion column (the emission lines disappear or greatly weaken). Unlike a dwarf nova system, which has a disc as the dominant optical light source (and which will take some time to decay if mass transfer *switches off*), an AM Her system is powered by the energy released directly in the accretion column, and if this ceases there are only the white dwarf and the red star to look at. However, at such times it can be possible to measure the magnetic field of the white dwarf directly. At these fields of tens of megagauss the *Zeeman effect* becomes important and splits the normal spectral lines (usually hydrogen) into several components (as a result of the energy levels, see box 12.4, being split by the strong field into multiple levels very close together). By measuring this effect (only visible when there is no accretion column present) the strength of the magnetic field can be estimated accurately.

As a final note on AM Her systems, consider again Nova Cygni 1975. Many of you reading this book must be able to remember this event, which, although bright, was essentially like any other nova explosion. During its long decline from maximum (3 years after the outburst it had reached 14th magnitude) Joe Patterson (Columbia University) found that it had a 3.35 hour orbital period. However, in 1987 when the star had faded to 17th magnitude, a significant polarisation was found by Harvey Stockman (STScI), Gary Schmidt (Steward Observatory) and Don Lamb (University of Chicago), but at a period of 3.29 hours! In other words, Nova Cygni 1975 was a hitherto unknown AM Her star, and is the first such system known for certain to undergo a nova outburst. But this should not be surprising. The material being accreted onto the white dwarf is almost certainly hydrogen rich, and a thermonuclear event will eventually happen. In this case it was the mechanism by which our attention was drawn to this system. What is interesting in this case is the small (1.8%) difference between the white dwarf rotation period (as revealed by the polarisation) and the orbital period, since all other AM Her systems are orbitally locked. It is speculated that it was as a result of the nova explosion itself, with the loss of perhaps as much as 0.001 $M_\odot$ of material from the system, that the synchronisation has been temporarily lost.

10.5 Intermediate polars

The AM Her binaries are clearly very different from other cataclysmic variables in one important respect. They do not have accretion discs. It is this that makes them observationally distinct. As figure 10.3(b) shows, the powerful magnetic field of the white dwarf completely dominates the process of mass transfer between the two stars. What would happen if the binary were a little wider, or the magnetic field were not quite so powerful? In this case the magnetic field would not have the strength at the distance of the red star to control the motion of the matter as it left the star. This would lead to the formation of an accretion disc, but at some point the inner part of the disc would be disrupted by the magnetic field and matter would accrete onto the white dwarf only at its poles in the manner of an AM Her system.

What would such a system look like? With a disc it might already have been found and catalogued optically as a cataclysmic variable. But the polar accretion would lead to the formation of hard X-rays and it might have been seen by the Uhuru satellite survey. It would combine the properties of a dwarf nova and an AM Her system. Such *hybrid* systems have been found and are called, not surprisingly, *intermediate polars*. In fact, although it was not the first to be recognised as such, the CV EX Hya is a member of this class (see figure 10.3), but prior to the recognition of its X-ray output it had been thought to be a fairly normal CV.

The first object to be placed into this class distinct from the AM Her systems was H2252-035 (AO Psc), a new variable X-ray source discovered by the HEAO-1 survey. The property that sets *intermediate polars* aside from other CVs is the presence of several periodicities. AO Psc exhibits an X-ray period of 805 s (see figure 10.5), an optical period of 858 s and a longer optical period of 215 minutes. The longer period is the orbital period, whereas the X-ray pulsation is due to the spin of the white dwarf. How then do we make sense of the optical pulsation? Using figure 10.3 as a guide it is possible to see that if the rotating X-ray beam from the white dwarf sweeps across the disc *hump* or secondary star (or anything rotating at the *orbital* frequency) then there will be a modulation at a period that is different from that of the white dwarf spin period. (In fact, they are related by the equation $1/P_{opt} = 1/P_x - 1/P_{orb}$, and the observed optical period is said to be the *beat* period of the other two.)

It is the weakness of the magnetic field, or separation of the stars, that prevents the white dwarf being synchronised with the orbital period as in AM Her stars. Table 10.3 gives the properties of the known intermediate polars,

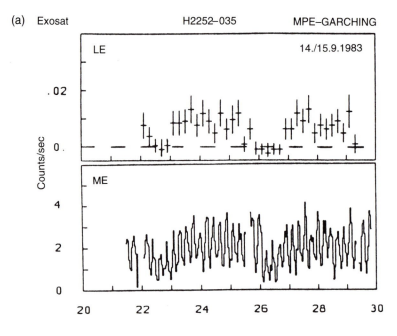

(a)   Exosat   H2252–035   MPE–GARCHING

Fig. 10.5 Observations of the spin periods in three intermediate polars. (a) EXOSAT reveals both the 13.4 minute white dwarf spin period and the 3.6 hour orbital period of AO Psc (H2252–035). The spin period is very clear in hard X-rays (ME), which originate close to the white dwarf.

Fig. 10.5 (b) A 7 hour EXOSAT observation of the CV with the longest orbital period, GK Per, shows the 6 minute spin period (above), whilst the lower panel (c) indicates how these effects can be observed by ground-based optical and infrared photometry, in this case the 21 minute spin of the white dwarf in FO Aqr. Other periods are also often present as a result of reprocessing of the X-ray energy in other parts of the binary. (Diagrams by Julian Osborne, EXOSAT Observatory, Mike Watson, Leicester and Graham Berriman, University of Arizona.)

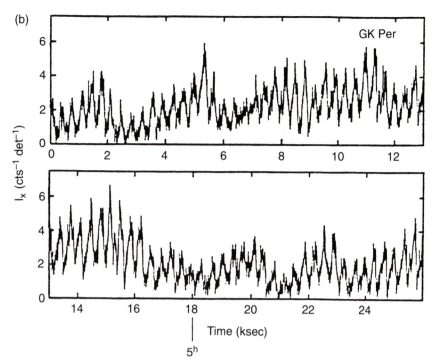

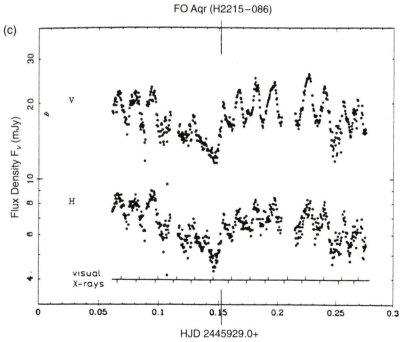

from which it is clear that, as a group, they have orbital periods that are longer than AM Her stars. For comparable masses they indeed have wider separations. As to whether their magnetic fields are also weaker, that debate is still not resolved.

These objects are sometimes referred to as DQ Her systems. DQ Her was discovered through its nova outburst (Nova Her 1934) and subsequently was

Table 10.3.   Properties of intermediate polars

Source	P_{spin} (mins)	P_{orb} (hrs)	Visual magnitude	X-ray flux (μJy)	Distance (pc)
AE Aqr	0.55	9.88	11	0.5	100
V533 Her	1.06	6.7	14	<0.01	1200
DQ Her	1.18	3.67	14	<0.01	420
GK Per	5.85	47.9	10–13	0.5–9	480
V1223 Sgr	12.4	3.37	13	2	600
AO Psc	13.4	3.59	14	3	<750
BG CMi	15.2	3.24	15	0.7	?
SW UMa	15.9	1.36	12–17	<0.3	140
FO Aqr	20.9	4.85	13	0.5	250
TV Col	31.9	5.49	14	2	<500
H0542–407	32	6.2	16	1.5	?
V426 Oph	60	6.0	12–13	4	100
EX Hya	67.0	1.64	10–13	5	125

found to exhibit optical pulsations with a period of 71 seconds. There are others which behave similarly and only AE Aqr is an X-ray source. It is possible then that these very short period variables may be somewhat distinct as a group from the intermediate polars described above, but this is also an area which is currently being actively researched.

10.6 Formation of a cataclysmic variable

The general idea of how low-mass interacting binaries (such as CVs) were formed was put forward by Bohdan Paczynski (Princeton) in the mid-1970s (see figure 9.12). Double stars are normally formed with wide separation (several AUs) and periods of years, a remarkably different system from that of a CV! It is virtually impossible to create a double star system in which the two components are close because of the problem of disposing of the angular momentum of the two stars (which is inherited from the initial gas cloud out of which they formed). However, it is normal stellar evolution that will radically change the picture.

The two stars are formed on the main sequence (like the Sun), but the more massive component will evolve (burn its fuel) more rapidly and start to expand into a red giant. The expansion is so great that the atmosphere of the more massive star actually swells to embrace the other star, which then spirals in towards the core of the red giant as a result of *atmospheric drag* (with the atmosphere carrying off the angular momentum of the binary). You can almost imagine the core of the red giant (which itself is collapsing towards the white dwarf stage) and the companion star orbiting each other actually *inside* the overall envelope of the red giant (but it must be admitted that this envelope is very tenuous). The envelope dissipates rapidly as a result of the spiral in, and a short period binary is left with the companion star orbiting the white dwarf. Eventually the companion star will evolve to fill its own Roche lobe, and mass transfer onto the white dwarf will then turn on. A CV is born.

10.7 Evolution to short periods

Interestingly enough there are virtually no CVs known with orbital periods between 2 and 3 hours, a range which has come to be called the *period gap* (see figure 10.6). There are now a sufficient number of CVs with known orbital periods that this diagram cannot be substantially distorted by selection effects. The *gap* must be due to a ʳeal physical effect. The secondary star is losing mass because it is filling its *Roche lobe*, the equipotential gravitational surface that surrounds the binary. It might be thought that it is the *normal* stellar evolution of the red star that causes it to expand and hence overflow its Roche lobe. For low mass stars, however (as is the case for virtually all CVs), this cannot come close to accounting for the observed luminosities of CVs. Another mechanism is required that causes the Roche lobe to shrink. Although not well understood, this mechanism is believed to be *magnetic braking.* Just like the Sun, the CV's red star will have a stellar wind. If the star also has a magnetic field, this wind will be forced to corotate out to some distance from the star. (In the Sun's case this distance is about 100 solar radii, or about half way to the Earth.) The energy required to do this will slow down the spin rate of the red star. (Imagine spinning around with arms outstretched. Then consider doing it whilst holding a long pole outstretched in each hand. The latter case will require a lot more effort and will tend to slow you down.) But the spin rate of the red star is locked gravitationally and precisely to the orbital period, and so the effect of the drag imposed by the stellar wind will be to reduce the orbital angular momentum of the binary and hence cause the binary period to reduce. This is the mechanism that drives the mass transfer rate in CVs and LMXBs above the period gap.

Fig. 10.6 Distribution of orbital periods amongst cataclysmic variables. This clearly shows the absence of any CV with period in the range 2 to 3 hours (the period gap), and also the minimum orbital period of about 80 minutes. The only objects outside the two main groups are the long period GK Per (which has a giant secondary) and the three double degenerates (in which the mass-losing star is also a white dwarf). (From a diagram produced by Hans Ritter.)

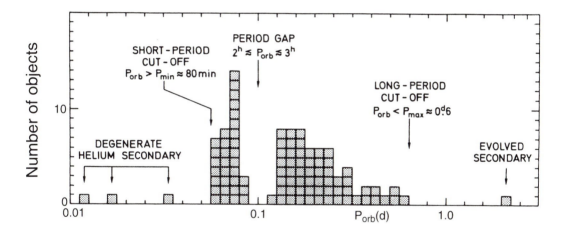

However, at periods of about 3 hours something happens (perhaps the magnetic field of the red star is dramatically reduced) and the mechanism is *switched off.* The red star shrinks slightly and the mass transfer stops. Without its main power source such a system would be almost impossible to find because it would no longer contain the accretion disc that provides its distinguishing characteristics of blue colour, variability and emission lines. Orbital angular momentum is then lost by the process of gravitational radiation and, eventually, the Roche lobe shrinks again to catch up with the star's surface and re-initiate mass transfer, at a period of around 2 hours. The system is *reborn* and can be seen again as a CV (see figure 9.12).

It has already been noted that the period distribution of the magnetic binaries is odd and appears to cluster around 113 minutes. Indeed, there are now six AM Her systems known with periods within 1 minute of this value! A detailed evolutionary model has been developed by Andy King (University of Leicester) and colleagues which can account for this if the white dwarf masses all have similar values (0.6–0.7 $M_\odot$). This model received striking support when EXOSAT observations revealed a new AM Her system (EXO033319–2554.2) with the unusually long period of 126 minutes (just inside the bottom end of the period gap). To account for this in their model required a more massive white dwarf (1.2–1.3 $M_\odot$). Since this system was also eclipsing it was possible to follow the motion of the secondary star and there-by determine the mass of the white dwarf. The value so obtained was exactly that predicted!

How do such short-period systems evolve from this point? The binary is so small that gravitational radiation is now the main driving force for the mass transfer as the period continues to reduce. However, this mass transfer rate is much smaller than it was when the binary was *above* the period gap. If that is the case, though, how then is it possible to detect X-rays from them that are comparable to longer-period systems? The answer is that, when mass transfer is restored on emerging from the period gap, there is a temporary enhance-ment in the mass loss as the secondary star settles down. This enhancement will eventually die away and the source will become a much weaker X-ray emitter. Hence, we would expect X-ray surveys to select those systems with larger mass transfer rates which are those that have just emerged from the period gap. i.e. most AM Her systems will have the *same* orbital period, which is exactly what we observe!

After this initial activity on emerging from the period gap, gravitational radiation will act to reduce the orbital period as mass continues to be lost. The source will become much fainter and hence more difficult to detect. But the period does not decrease indefinitely. Models constructed by, amongst others, Saul Rappaport (MIT), Frank Verbunt (Utrecht) and Hans Ritter (Munich), show that the orbital period goes through a minimum at about 80 minutes (see figure 10.7). This is because, as the mass of the secondary star reduces, its nuclear evolution slows down, but it is still losing mass. The star therefore departs from thermal equilibrium and gravitational radiation now drives it to longer periods, but with no mass transferred. The ultimate fate of the system is a double degenerate binary consisting of a white dwarf and a very low-mass black dwarf. Such systems would be extremely difficult for us to find.

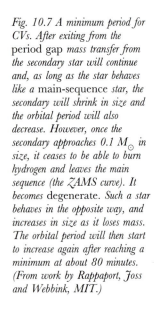

Fig. 10.7 A minimum period for CVs. After exiting from the period gap mass transfer from the secondary star will continue and, as long as the star behaves like a main-sequence star, the secondary will shrink in size and the orbital period will also decrease. However, once the secondary approaches 0.1 $M_\odot$ in size, it ceases to be able to burn hydrogen and leaves the main sequence (the ZAMS curve). It becomes degenerate. Such a star behaves in the opposite way, and increases in size as it loses mass. The orbital period will then start to increase again after reaching a minimum at about 80 minutes. (From work by Rappaport, Joss and Webbink, MIT.)

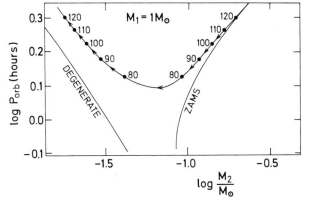

10.8 Superoutburst

Below the period gap (i.e. orbital periods < 2 hours) there appear to be only two types of CVs: the magnetic AM Her systems and the SU UMa systems, the subject of this section. The principal feature that distinguishes SU UMa systems from other CVs is the nature of their outbursts. Even classical dwarf novae, such as SS Cyg, do not have outbursts that all look the same; they can usually be divided into narrow and wide outbursts. This is also true of the SU UMa stars, but their wide outbursts, of duration several hundred orbital cycles, are dramatically wider than normal and are called *superoutbursts* (see figure 10.8). First catalogued by Brian Warner (Cape Town), the SU UMa stars display some extraordinary properties, the key ones being:

(1) during superoutburst the light curve shows broad maxima called *super-humps* which are periodic, but at a period which is a few per cent (usually 3~7%) longer than the orbital period. This is no trivial phenomenon since it represents 30% of the total light output of the system. They are called *superhumps* to distinguish them from the *orbital hump* (or bright spot) which is seen in dwarf novae during *quiescence*.

Fig. 10.8 Superoutburst optical light curves for the eclipsing SU UMa system, Z Cha, obtained at SAAO in May 1984 and December 1985 by Paul Barrett and Darragh O'Donoghue (University of Cape Town). The eclipse of the white dwarf by the companion star is obvious. The hump (marked by vertical lines) is the famous superhump *that is the principal defining characteristic of superoutbursts, and it has a period slightly longer than the orbital period. These light curves clearly show the resulting motion of the hump relative to the eclipse.*

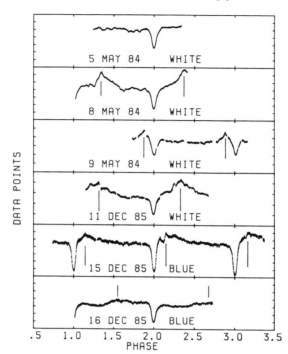

(2) the size of the superhump does not seem to depend on the angle at which we are viewing the binary system. i.e. the superhump is just as easy to see in the eclipsing systems Z Cha and OY Car, which are viewed from a direction close to the plane of the orbit, as it is in those viewed at lower inclinations such as VW and WX Hyi.

(3) the motion of the disc, as revealed by optical spectroscopy, suggests some periodic bulk motion (i.e. first towards us and then away from us) which is related to the superhump phase and not (as might be expected) the orbital motion.

A number of models to explain this extraordinary behaviour were proposed in the late 1970s and early 1980s. None of them could account for *all* the observed properties and they all suffered from serious physical difficulties. One of the major shortcomings, however, in trying to understand the behaviour of SU UMa systems was the lack of adequate data covering all wavelengths during superoutbursts. This was realised by theorists Geoff Bath and Rob Whitehurst (both then at Oxford), and together with graduate student Tim Naylor and IUE Astronomers Barbara Hassall and George Sonneborn and one of us (PAC), a campaign was planned to rectify this situation.

The basic problem is one that has always dogged studies of CVs in outburst, namely the irregularity of the event. Only it is worse in this case as superoutbursts are much rarer (they recur on timescales of ~1 year, compared to months for typical normal outbursts) and in any case start 1 or 2 days after a normal outburst has begun. Observational material was thin and random in coverage. The help of the amateur variable star observers was consequently invaluable in providing rapid information on the visual state of our target objects, the two eclipsing SU UMa systems Z Cha and OY Car. As these are both in the far south, observations were organised by Frank Bateson of the Royal Astronomical Society of New Zealand Variable Star Section, which has become renowned for its CV work over the last decade. Although both stars are too faint in quiescence (around 15th–16th magnitude) to be seen by the amateurs, in outburst they reach 11th–12th magnitude and are easily observed with modest equipment.

Once a superoutburst was discovered by the variable star observers and confirmed, then a *Target of Opportunity* programme could be implemented. Such a programme is one in which a certain amount of time on a space observatory or telescope is reserved, but the date is left open until word is received that a *confirmed* superoutburst has begun. At that time the scheduled observers are over-ridden and, at least in the case of the IUE and EXOSAT satellites, their observations are rescheduled for a later date.

The chances for OY Car *and* Z Cha both came in 1985. In late April it was confirmed by Frank Bateson that OY Car was in superoutburst. The IUE and EXOSAT over-rides were invoked in order to obtain UV and X-ray observations as early in superoutburst as possible. To extend the wavelength coverage from the ground a network of observers at ESO, SAAO and AAO obtained optical and IR photometry on several occasions during the superoutburst.

The eclipsing systems had been chosen so that the red star could be used as a probe of the superhump and the distribution of light in the disc (see figure 10.3). From the over-ride it was found that:

(1)   the X-ray emitting region is *not* eclipsed, whereas the geometry of the binary is such that we know the white dwarf *is* eclipsed. The X-rays are therefore not coming directly from the region of accretion onto the compact object, but instead must originate in an extended, hot corona above the disc.

(2)   the UV continuum and lines are *partially* eclipsed indicating that some of the UV emission is also extended. The remainder is close to the white dwarf or inner part of the accretion disc. In fact, the hotter (higher ionisation and shorter wavelength continuum) emission is *more* eclipsed and therefore nearer to the white dwarf than the cooler material, which is very likely a wind outflowing from the disc.

(3)   there is no sign of the superhump in the UV or X-ray light curves. The superhump must therefore be cool and, from the ratio of the optical to infrared fluxes, it is found to have a temperature of about 9000 K.

(4)   the superhump itself is not totally eclipsed and must therefore be large, probably comparable in size to the projected area of the disc.

(5)   dips are seen in the UV light curve just prior to eclipse which imply that part of the disc is *thickened*. This is supported by observations of independent variations in the optical and IR.

10.9 Precessing accretion discs

In parallel with these observations, Rob Whitehurst had been investigating the theoretical problem of how a disc, such as those present in Z Cha and OY Car, would behave during the conditions prevailing in both normal and superoutbursts (see figure 10.9). His numerical simulations of the hydrodynamics of the flow were the first to include the gravitational effects of both stars in a typical CV binary. Usually the secondary is ignored because it can be as much as 10 times less massive than the white dwarf primary. This is a reasonable assumption when calculating the behaviour of the inner part of the disc near the white dwarf, but can lead to severe inaccuracies in the outer part of the disc. And it is just this section that is important during outbursts, particularly during extended periods of additional mass transfer from the secondary.

The computer models showed that during a normal outburst, the disc undergoes an expansion as a result of a change in its viscosity, and it brightens substantially. After a day or two the viscosity drops causing the disc to contract and become dimmer. Occasionally the outburst of the disc somehow triggers an instability in the companion star which results in it dumping a lot more material into the disc. This causes the disc to grow even larger and it expands beyond its own *stability radius*. At this point the disc becomes elliptical in shape and starts to precess.

Fig. 10.9 Numerical simulations of how the accretion disc in Z Cha will evolve with time were performed by Rob Whitehurst as part of his Oxford D.Phil. thesis. After a large injection of matter, the disc switches from (normal) corotation in (a) to precession in panels (d) to (f). The maximum tidal stress (due to the companion star) occurs in panel (c), giving rise to the 25% increase in brightness that we see as the superhump. *It is therefore a phenomenon arising in the disc as a whole and not localised (which is why the superhump shows no appreciable eclipse). The disc pictures are taken three orbital cycles apart (about 5 hours) so that the disc precession is more clearly visible. The solid curve represents the disc's* tidal lobe. *Any material that ventures outside this curve is likely to be lost from the system.*

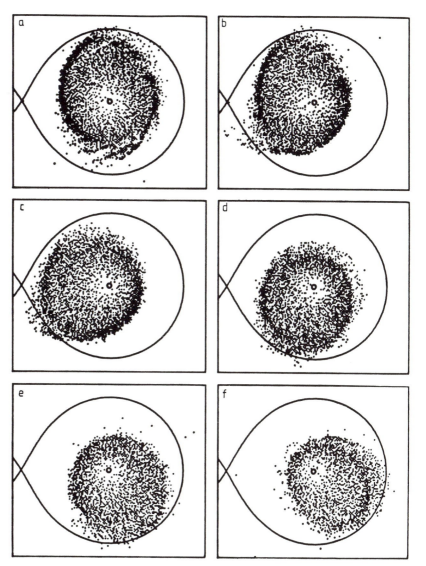

But why should a precessing disc vary in brightness? It is here that the red (secondary) star influences things. When the elliptical disc precesses around so that it is pointing towards the red star, the tidal effects on the disc are at a maximum. This stress releases extra energy from the *whole* disc which we see as a superhump.

One of the remarkable features of these simulations was that there was only one free parameter describing the system and that was the ratio of the mass of the white dwarf to that of the red star. By fixing this at the *observed* ratio for Z Cha, the simulations predict a certain precession period for the disc. The agreement with the observed superhump period is excellent. The reason that superhumps are not seen in the *wide* outbursts of longer period CVs (such as SS Cyg) is that the mass ratio is too low. In those cases the stability radius is beyond the disc's tidal radius and so *cannot* be reached, as shown in figure 10.10. If the disc were to expand beyond its tidal radius the expanding matter would be lost from the system. Hence the disc is unable to precess.

Fig. 10.10 (a) Mass ratio (white dwarf divided by companion star) as a function of period for all CVs where it has been measured. The two low mass-ratio SU UMa estimates use a method that may be inaccurate. Otherwise, the SU UMa systems all appear to be high mass-ratio systems, implying a very low-mass secondary (mass-losing) star. (Based on data compiled by Rob Whitehurst, Leicester University.) (b) Stability radius and tidal radius of an accretion disc as a function of mass ratio. Low mass-ratio CVs have a stability radius that is larger than their tidal radius. This means that the disc is always stable and cannot precess. However, at high mass-ratios (as are prevalent in SU UMa systems) the stability radius is less than the tidal radius. Hence, if the disc expands outside this stability radius (as probably happens during a superoutburst due to the extra influx of material) it can remain within its tidal radius (beyond which matter would be lost from the system) and still become unstable, i.e. it will start precessing.

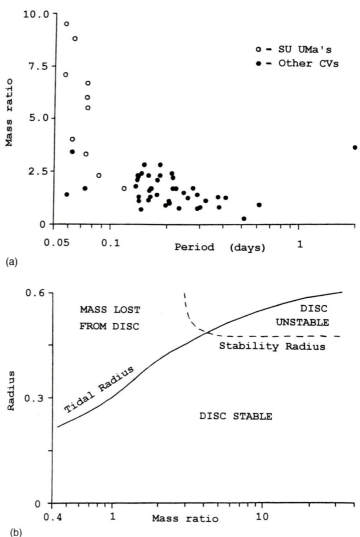

This also explains why such bizarre behaviour is confined to the short-period systems. A large mass ratio is required of a short period binary if the mass-losing star is to be filling its Roche lobe and transferring mass onto a typically 1 $M_\odot$ white dwarf. The other short-period binaries, the AM Her systems, do not have the opportunity to demonstrate the effect since the magnetic field has prevented the formation of a disc.

This work has demonstrated the power of both Target of Opportunity programmes *and* multi-wavelength observations. They are difficult and time-consuming to organise and implement but the results show that they are well worth the effort. Combined with theoretical developments there has been considerable gain in the knowledge of how accretion discs behave. It is clear that discs do not behave as *rigid* bodies, but instead they are *fluids* whose shape can change and evolve on short timescales. The discovery of the capability of the disc to precess shows that, to properly follow its behaviour with time, we need to take into account the effects of the secondary star.

There is now a model which gives a self-consistent explanation of the SU UMa phenomena and may be applicable to other high mass-ratio systems (such as those involving neutron stars and black holes). Future developments in this field present an exciting prospect.

11 Are there black holes in our galaxy?

11.1 Introduction

All of our discussion of interacting binaries so far has usually been in terms of a *compact object* which is accreting matter from its companion. In most cases this has been either a neutron star or a white dwarf. But the basic model of the binary is almost independent of what type of compact object it is. As long as we have some idea of the distance of the object then we can usually infer its nature on the basis of its X-ray luminosity. This is because the gravitational potential of a neutron star is so much deeper than that of a white dwarf (by a factor 1000) that it can produce that much more luminosity. However, this is not always true (e.g. when an X-ray binary enters a *quiescent* phase in which mass transfer onto the compact object is limited for some reason), and the discrimination between the two types can then be quite difficult.

How then can we even begin to contemplate searching for *black holes* in X-ray binaries? Especially since the difference between the gravitational potential for a neutron star and black hole is much smaller! But black holes have attracted people's imaginations perhaps more than any other kind of object in the cosmos. It is known that there must be a limit to the mass of a neutron star, and this limit is almost certainly somewhere in the region of $3\ M_\odot$. Most theorists believe it to be somewhat less than $3\ M_\odot$, but this is currently an intensely debated topic. Some have argued that it is not necessary to know the equation of state of the material inside a neutron star (about which much has been inferred from observations of radio pulsars), but simply assume that Einstein's *General Theory of Relativity* is correct and that causality is obeyed (i.e. the requirement that the speed of sound inside the star is less than the speed of light). From these two assumptions, theorists have shown that a (non-spinning) neutron star cannot be bigger than about $3\ M_\odot$. Allowing for rotation increases this value by about 20%. However, as seen in chapter 7 (figure 7.7), the *observed* masses of neutron stars, as derived from the binaries involving X-ray and radio pulsars, are all consistent with values substantially below $3\ M_\odot$, with the average being close to the canonical value of $1.4\ M_\odot$.

Massive objects are of enormous interest to scientists in many fields. The equation of state of matter at such densities and the very nature of what would be happening inside a black hole could not be inferred by observation in any other way. Indeed, our current physical laws are based on Einstein's Theories (of Relativity) and yet these have only been tested in cases where the perturbations to Newton's Laws are really quite small! A black hole represents Einstein's ultimate test.

The evidence for the very existence of black holes must then come from accurately determining the masses of as many compact objects as possible and then finding those that are clearly above the maximum mass that could be sustained as a neutron star. Such work is confined to objects in our galaxy since we are as yet unable to estimate accurately the masses of extragalactic objects. The very presence of X-rays in a galactic binary tells us that we must be dealing with a compact object of some kind. A high luminosity takes it further by limiting the choice solely to neutron stars and black holes. Can we therefore determine the mass of the compact object by direct measurement alone, and what are the limitations of this approach?

11.2 Measuring masses

The most direct measurement we can make is by optical spectroscopy which gives us the orbital radial velocity curve of either the secondary (mass-losing) star or, if there is emission present, the accretion disc around the compact object. From these measurements alone we obtain the *mass function* (as described in box 7.3). Ideally, the compact object itself can be followed if, for example, it is an X-ray pulsar whose period variations define the orbital motion directly (see figure 7.6). However, there is one fundamental difference with potential black-hole systems in that we cannot measure K_x (the velocity amplitude of the compact object) in this way because, by their very nature, the black holes cannot provide a periodic signature (such as a pulsation) to be followed around the orbit. Indeed, the presence of a pulsation tells us at once that we cannot be observing a black hole, and that the compact object must be a neutron star. For a black hole, such a signature can only be produced inside the event horizon which we cannot observe. Instead, we are measuring $K_n \sin i$ (the velocity amplitude of the normal star), and so the mass function is

$$f(M) = \frac{M_x^3 \sin^3 i}{(M_n + M_x)^2} = \frac{PK_n^3 \sin^3 i}{2\pi G}$$

which depends only on the directly observable quantities P and $K_n \sin i$ (we don't usually know i, the inclination of the orbit to our line of sight). The limitations are clear from this equation. If M_n is large (as in the MXRBs Cyg X-1 and LMC X-3, see next section), then the mass function is a small number. In order to derive a reliable value for M_x, we must know M_n and i accurately. M_n can only come from the optical spectral type of the star and our knowledge of the masses of other stars of the same type. Unfortunately, such studies, whilst being accurate, are confined to nearby, non-interacting binaries, whereas a black-hole candidate, again by definition, is a very close, interacting binary in which the components have transferred large amounts of mass between them and reached their current configuration by very unusual evolutionary processes (which are largely not understood!). This introduces the potential for large systematic uncertainties in the mass estimation.

However, if M_n is small (i.e. the system is an LMXB, see section 11.4), then the mass function itself represents a firm *lower* limit to the value of M_x (if you simply set $M_n = 0$ and $i = 90°$ in the equation, then $f(M) = M_x$). Obviously M_x must be larger than this when realistic values for M_n and i are inserted. Let us now look at the individual cases.

11.3 Cygnus X-1 and LMC X-3

These two objects are currently held to be the best black-hole candidates amongst the MXRBs. Their properties are summarised in table 11.1.

Cygnus X-1 was identified in 1971 with an already known bright supergiant, HDE226868, and has been considered a strong black-hole candidate ever since. The reason for this is quite simple. The supergiant should have a

Table 11.1 Properties of the black-hole candidates

Object	X-ray luminosity (erg s^{-1})	Distance (kpc)	Spectral type	V	Orbital period (days)	K velocity (km s^{-1})	Mass function ($M_\odot$)	Mass ($M_\odot$)
Cyg X-1	2×10^{37}	2.5	O9.7I	9	5.6	76	0.25	>3
LMC X-3	3×10^{38}	55	B3V	17	1.7	235	2.3	>6
A0620–00	1×10^{38}	1	K5V	12–18	0.32	457	2.91	>3
V404 Cyg	2×10^{39}	4	K0	13–19	6.5	210	6.3	>6

mass of around 30 $M_\odot$ (at least, it would have if it were a normal star of this spectral type), and yet optical spectroscopy of this star revealed a surprisingly large radial velocity amplitude of almost 50 km s^{-1} which, when combined with its period of 5.6 days, inferred a mass for the compact object of over 15 $M_\odot$ (using the technique described in box 7.3). This is well outside any conceivable mass for a neutron star, hence the implication that it must be a black hole. A similar approach was used for the second massive X-ray binary candidate, LMC X-3. The system sizes are shown to scale in figure 11.1.

However, the fundamental limitation in this analysis (and which is appealed to by all who maintain that there may not be black holes in these systems) is that the mass of the compact object depends on us knowing the mass of the optical primary, and, as described in the previous section, this has to be inferred from *optical* spectroscopy. This means that the masses are pinned to the mass of the mass-losing star which is determined from its observed spectral-type. The problem then is, why should a star of a given apparent spectral type in an X-ray binary be of the same mass as a similar star in a non-interacting binary? Since we know that mass loss must be taking place (onto the putative black hole), then these assumptions are indeed questionable.

Examining table 11.1 you might be inclined to think that LMC X-3 was a much better candidate for a black hole. After all, it does have quite a large mass function even though it has an early-type spectrum and thus presumably a massive secondary star. Also, being in the LMC (the Large Magellanic Cloud, one of our nearest neighbouring galaxies), its distance is quite accurately known at 55 kpc. The observations infer a massive compact object of

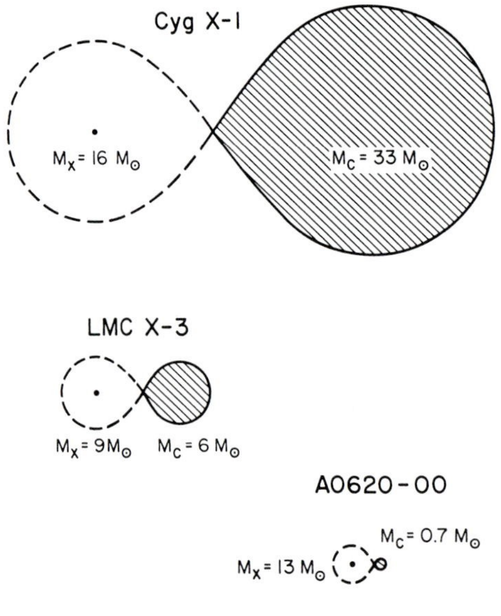

Fig. 11.1 The three best black-hole candidates, Cyg X-1, LMC X-3 and A0620–00. The curves (solid and dotted) represent the Roche lobes of each component of the binary. The secondaries (mass-losing stars) are shown hatched; they must fill their Roche lobes in order to be able to transfer matter onto their compact companion. The relative sizes of these binaries are correctly shown. (Based on a diagram produced by Jeff McClintock, CfA.)

Cyg X-1

$M_x = 16\ M_\odot$ $M_c = 33\ M_\odot$

LMC X-3

$M_x = 9 M_\odot$ $M_c = 6\ M_\odot$

A0620–00

$M_c = 0.7\ M_\odot$ $M_x = 13\ M_\odot$

at least 6 $M_\odot$, well in excess of that considered plausible for a neutron star. However, this interpretation has also been challenged. The mass estimates are based on the assumption that all the optical light we see is coming from the secondary star. Whilst reasonable in the case of Cyg X-1 (which is an extremely bright supergiant star), it is clear from more detailed observations that there is evidence for an accretion disc contributing to the light we see from LMC X-3. Allowing for this the mass of the secondary star could be as little as 0.7 $M_\odot$, which then gives the compact object mass as only 2.5 $M_\odot$. (The reason the optical star appears like the much more massive B3V star in this explanation is because we are seeing the hot core of an evolved star, sometimes called an OB-subdwarf, and not a large, early-type main-sequence star.) Clearly, interpreting the observations of compact objects with large or apparently large companion stars is not at all straightforward.

11.4 Soft X-ray transients

X-ray transients were introduced in chapter 7 and have been known from the early days of X-ray astronomy. Those associated with the massive *Be* stars (section 7.7) have hard (hot) X-ray spectra and are quite different from the *soft X-ray transients*, in which the mass-losing star is of low mass. In most other respects the soft X-ray transients are indistinguishable from the low-mass X-ray binaries described in detail in chapter 8, but for an as yet unknown reason the mass transfer onto the compact object is highly erratic. For years at a time the system is quiescent (no X-rays detectable by our current technology) and optically faint (typically around 20th magnitude). But then the mass transfer onto the compact object turns on, a powerful X-ray outburst occurs and the accretion disc brightens dramatically, sometimes by as much as 7 or 8 magnitudes. Indeed, at this brightness they can be observed by amateur variable star observers, and one such object (V404 Cyg) was first noted more than 50 years ago. The peak brightness is reached in a matter of days, and then a gradual decline sets in that can take months (see figure 11.2).

The third source in our list of black-hole candidates is A0620–00, which has become the archetypal soft X-ray transient. This source was, for a time, three times as bright as Sco X-1, normally the brightest source in the sky. A0620–00 was discovered by Martin Elvis and colleagues (at Leicester) when it flared on August 3, 1975 and was immediately detected by both Ariel V and SAS-3, two spacecraft which provided a long-term monitoring capability. The outburst lasted for 8 months, and its overall X-ray light curve is shown in figure 11.2. (This figure also contains the Ginga observations of GS2000+25, an April 1988 transient that is remarkably similar to A0620–00. Only future optical work will tell us if this too might contain a black hole.)

The optical counterpart of A0620-00, designated V616 Mon, was rapidly identified by F. Boley and co-workers at Dartmouth and MIT through its nova-like behaviour (it had brightened by over 7 magnitudes from its quiescent 18th magnitude). In quiescence it was found to consist of a K5V star bound to its (optically) unseen companion in a 7.8 hour binary. Although faint for accurate spectroscopy, this was achieved by Jeff McClintock and Ron Remillard (MIT) in a heroic battle against technical problems using the KPNO 4 m telescope. Their radial velocity curve is shown in figure 11.3.

Since the source does not eclipse, the very high velocity amplitude was again surprising and inferred a high compact object mass of at least 3.2 $M_\odot$ (for an inclination of 90°). Reasonable assumptions for the inclination and mass of the normal star will lead to masses for the compact object greater than this. As explained in section 11.2, this limit is much firmer than the mass constraints for the MXRBs. There is a further constraint imposed by the observation of geometrical distortions in the secondary star caused by the gravitational field of the compact object. A Roche geometry star presents a

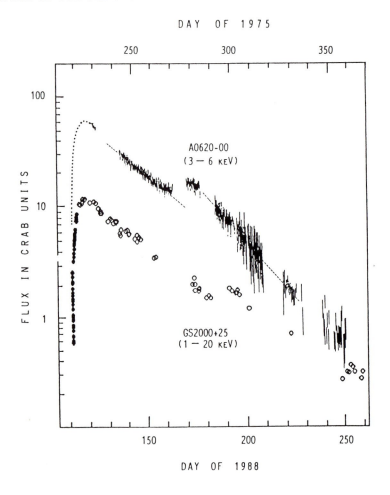

Fig. 11.2 The outburst light curves of the soft X-ray transients A0620–00 and GS2000+25. A0620-00 erupted in 1975 and was observed by the Ariel V X-ray satellite for more than 8 months. At its peak, A0620–00 was the brightest ever source in X-rays. More recently (April 1988) the Ginga satellite discovered GS2000+25, another soft X-ray transient whose properties are remarkably similar to A0620–00. It may contain a black hole too! (Diagram by Yasuo Tanaka, ISAS.)

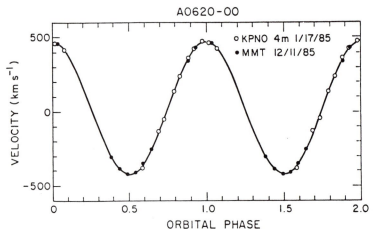

Fig. 11.3 Radial velocity curve showing the motion of the companion star in the A0620–00 binary system. These measurements were made from optical spectroscopy at KPNO and MMT; the solid curve is that expected for a circular orbit. The amplitude of this motion is very high (almost 800 km s⁻¹ peak to peak) and gives the strongest evidence so far for the existence of galactic black holes. (Diagram by Jeff McClintock, CfA and Ron Remillard, MIT.)

varying size to the observer and so the star varies slightly in brightness (as shown schematically in figure 11.4). Modelling this effect confirms the high mass, indeed the limit is then raised to at least $7\ M_\odot$, but this value is more model-dependent than the firm limit given earlier. Nevertheless, A0620–00 is a much stronger black-hole candidate than Cyg X-1 and LMC X-3 because it requires fewer questionable assumptions.

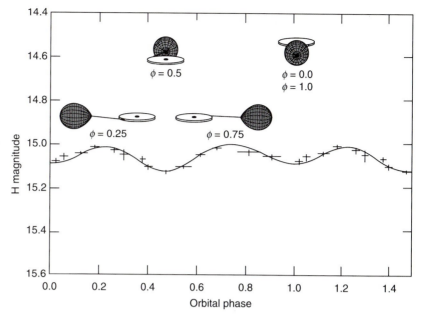

Fig. 11.4 Ellipsoidal variations in a close binary. The powerful gravitational influence of the compact object distorts the shape of its mass-losing companion. It adopts the characteristic pear shape of the Roche lobe. The net result of this is that the light output of the star varies with binary phase because of the variation in radiating area around the orbit. This is shown schematically together with the observed infra-red light curve of Cen X-4. These effects are easiest to detect in the red where the companion star's light usually dominates in this class of X-ray source. (Diagram by Tariq Shahbaz, Tim Naylor (Keele University) and PAC.)

However, those working in this field have longed to find a stellar-size object whose *minimum* mass, defined from the most straight-forward and unambiguous observations, clearly excludes the possibility of its being a neutron star. This *Holy Grail* of an object must have a mass function of at least 5 $M_\odot$, putting it clearly outside the maximum mass possible for a neutron star. Whilst writing this book, one of us (PAC) found such an object. Again it was an X-ray transient, discovered when it outburst in May 1989 by the Ginga X-ray satellite and designated GS2023+338. This was quickly identified by Brian Marsden (the renowned comet-watcher of CfA who also operates the IAU Circulars) with a previously known recurrent nova, V404 Cyg. First recorded in 1938, it has outburst several times since (optically), but its unusual nature was not recognised until the enormous X-ray output was detected.

Such an X-ray flux (almost as bright as Sco X-1) showed that it could not be a typical recurrent nova (which is actually a type of cataclysmic variable, see chapter 10), but must at least involve a neutron star. And its change in brightness of more than 7 magnitudes from quiescence to outburst showed that it must be similar to A0620-00. Consequently, when it returned to quiescence in early 1990, attempts were made to find its orbital period by photometry and spectroscopy, and to find evidence for the presence of the secondary star. These were not completely successful, because it rapidly became clear that there was still an accretion disc present (strong Hα emission) which would mask the companion star. However, the Hα emission itself showed a 6 hour modulation in velocity which at first seemed to be the orbital period, but the result was not one that was clear-cut.

Hence, in summer 1991 we (PAC, together with Spanish graduate student Jorge Casares, and Tim Naylor) returned to V404 Cyg with the power of the 4.2 m William Herschel Telescope (WHT) on La Palma, and were rewarded with superb observing conditions throughout July and August. From these we found immediate evidence for the secondary star (in the form of absorption lines in the spectrum due to calcium and iron) which indicated an early K star. And, as expected, these lines showed large radial velocity variations (amplitude 210 km s^{-1}) very similar to those of A0620–00 (shown in figure 11.3). But to our amazement the period was not 6 hours, but 6.5 *days*! From these two directly observed parameters, we immediately obtain the mass function of 6.3 $M_\odot$. This is the absolute minimum mass for the compact object, and hence V404 Cyg is the best black-hole candidate yet found.

This object is even more intriguing when we consider the presence of the 6 hour periodicity as well. Could it be a triple system? If it were, the compact object would still have to be a black-hole because the companion star in a 6 hour binary would have to be of very low mass ($<0.5\ M_\odot$) in order not to have been seen by our WHT observations. Not surprisingly we are planning much more extensive observing campaigns for V404 Cyg.

11.5 Other properties of black holes?

A natural approach to take in identifying other black-hole candidates was to look for X-ray sources showing some of the particular properties of Cyg X-1. One of the best studied of these was short timescale (millisecond) flickering (or shot noise). Subsequently, the (at times) soft X-ray spectrum of Cyg X-1 was considered another potential black-hole property. Unfortunately, other sources which showed the flickering and soft spectra have now been ruled out as black holes on the basis of regular pulsations, QPOs or bursts being subsequently discovered from them (see chapter 8). The pulsations and QPOs both arise in a magnetosphere, which cannot be observed in a black hole; bursts require a neutron star surface. Such properties are therefore not definitive on their own, but instead suggest that the object may be worthy of further study. It is at least a neutron star!

The study of the black hole candidates has shown how important it is to marry X-ray astronomy with other branches of astronomy. The prodigious X-ray output, combined with certain peculiarities of behaviour, gives us clues, but the accurate masses have all come from optical spectroscopy of the mass-losing companions. However, it must be remembered that these masses do *not* tell us that a black hole is present. The argument is indirect. If the compact object has a mass of, say, 8 $M_\odot$ then we infer the presence of a black hole because (i) we do not believe that neutron stars this massive can exist, (ii) relativity tells us that a black hole is the only alternative. The search for black holes will therefore always be a difficult one, but it offers the exciting prospect of observationally approaching the frontiers of the laws of physics.

12 SS433 – the link with AGN

12.1 The strangest star of all?

After the extraordinary variety of behaviour that is possible from X-ray binaries described in the previous chapters why should we devote an entire chapter to one particular star? The answer is that this object, SS433, exhibits properties of both *normal* galactic X-ray binaries and the exotic radio galaxies and quasars that will be the subject of the next chapter. No, the *SS* does not stand for *strange star*, but it would have been appropriate if it did. In fact, *SS* stands for the initials of two astronomers, C. B. Stephenson and N. Sanduleak, who compiled a list of emission-line objects that was published in 1977. SS433 is the 433rd entry in that list.

How and why are such lists compiled? The great majority of stars in our neighbourhood of the galaxy are normal stars like the Sun, whose optical spectra show a continuum with weak absorption lines that are characteristic of the elements in their atmospheres. However, the unusual objects such as X-ray binaries, cataclysmic variables, Wolf–Rayet stars and others have totally different optical spectra (see box 12.1). They tend to be dominated by powerful emission lines usually of hydrogen and helium, often indicating very high velocities due to mass infall and rotation. But how can we find such objects? It would be exceptionally tedious and a waste of large telescope time if we were to simply try to obtain optical spectra of all the stars we could see in our galaxy. Instead, a well-developed technique is used called the objective prism survey which allows for crude optical spectra to be obtained of all the stars in a given region down to quite faint limiting magnitudes. This is shown in figure 12.1 for the region around SS433.

However, this survey, including SS433, was published over a year before SS433's truly remarkable properties were discovered. Furthermore, a paper discussing several of the stars in the SS survey, again including SS433, was published two years earlier. Unfortunately, there was an error in the coordinates given for SS433 which would have prevented any observations of the object! Around the same time, the Cambridge 5 km radio telescope was completing the 4C survey for radio sources, which happened to include the region of SS433. SS433 was in fact detected in this survey as a strong radio emitter but, remarkably, the published coordinates for it too were in error! The reason for the inaccuracy this time is illustrated in figure 12.2, a radio map of the region of SS433 prepared by the 100 m dish at Effelsberg. It shows an extended source of radio emission, rather like a spherical shell with *ears* attached on either side. More importantly, it shows the strong point

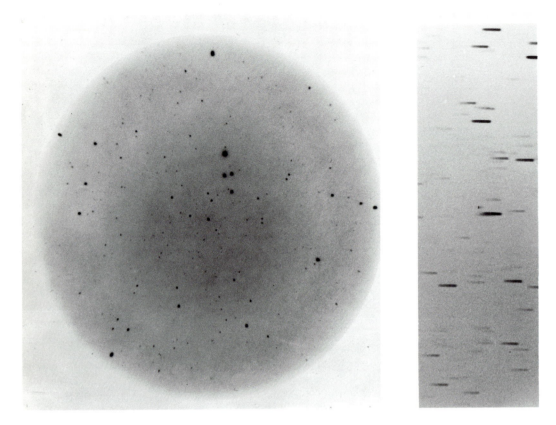

Fig. 12.1 Optical photographs of the region of sky containing SS433 (which is the star exactly at the centre). The normal picture (left) indicates nothing unusual (indeed, it is remarkable for its blandness), whereas, when an objective prism is inserted into the telescope (right), a small spectrum is obtained of the same stars, thereby enabling the strong emission lines of SS433 to stand out. SS433 appears essentially as a line image, whereas normal stars are simply elongated. (Photographs by Gene Harlan of Lick Observatory (left) and Bruce Stephenson from the Case Western Reserve survey, kindly provided by Bruce Margon, Univeristy of Washington.)

Box 12.1 Emission and absorption spectra

The energy generated by our Sun originates deep within its interior (actually within the central 1.5% of its volume) where the density and temperature are highest. By a process of absorption and re-emission this energy eventually finds its way to the Sun's *visible* surface. This is the point at which photons now have a reasonable chance of escaping from the Sun without further absorption. The temperature of this surface can be estimated from the Stefan–Boltzmann Law:

$$L=4\pi R^2 \sigma T^4$$

where R is the radius of the Sun, and σ is Stefan's constant. L is the luminosity released at the centre of the Sun, but we see it finally radiated from its large surface area. This is why the Sun appears as a cool body.

As a result of this, most of the atoms in the Sun's atmosphere are in low ionisation stages (e.g. FeI, FeII, CaII) and they absorb radiation at characteristic wavelengths from the continuum which rises from the Sun's centre.

Because they are cooler than the inner regions, these lines appear as dark absorption features in the spectrum. First found and catalogued by Fraunhofer they are still called Fraunhofer lines today and can easily be seen in the solar spectrum with a simple prism.

However, what happens if the cool material is not located directly between us and the hot continuum source? As shown in figure 12.B1, the result of this is that we then see bright *emission* lines. This is because the energetic photons from the central source (say, a hot, early-type star) excite the atoms in the surrounding interstellar material to higher energy levels (in fact, the photon is absorbed). Within a very short space of time (typically $\sim 10^{-8}$ seconds) such an atom returns to its original energy level, in the process emitting a photon of a characteristic wavelength. Because this material is spatially removed from the exciting star (as seen from Earth) we see these photons as emission lines. HII regions or planetary nebulae are excellent examples of this, with the exciting star being a hot, massive star and a hot white dwarf respectively.

Fig. 12.B1 Schematic diagram of how diffuse material surrounding a hot central object produces an emission nebula. The central object is usually a hot star with an essentially continuous spectrum. This radiation excites the diffuse gas (usually hydrogen) to higher levels from which it immediately emits a characteristic emission line photon. This radiation is essentially uniform in all directions. There is a sphere (called a Strömgren sphere) of line emitting gas surrounding and excited by the hot central star.

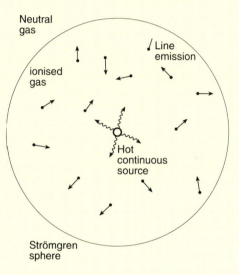

source of radio waves at the centre of the shell. Unfortunately, the confusion in the output from the Cambridge interferometer caused by the surrounding diffuse shell of radio emission resulted in the inaccuracy in the 4C survey position for the point source.

This shell is in fact W50, a supernova remnant first catalogued in the 1950s (!) but then largely ignored (see chapter 3). It was mapped with greater accuracy by David Clark and his Australian colleagues in the mid-1970s, but the map showed only the northern half of the remnant and thereby missed the fact that the point radio source associated with SS433 is in fact at the centre of the remnant's shell. Had this been noted, the radio *star* would have been immediately thoroughly investigated as a candidate for the remnant of the core of the exploded star.

One of us (FDS) was also involved in early observations of the SS433 region, this time at X-ray wavelengths in 1976 using the British Ariel V satellite. It had been thought that the X-ray source A1909+04 might be another extended source associated with the supernova remnant, W50. However, a detailed analysis showed that the X-ray source was in fact variable and hence could not be associated with an extended object because of the very large

Fig. 12.2 A radio map produced
by the 100 m radio telescope at
Effelsberg (near Bonn) of the Max
Planck Institute for Radio
Astronomy, showing SS433 itself
as a central radio star surrounded
by the shell of the supernova
remnant W50. (Radio photograph
by Barry Geldzahler, George
Mason University.)

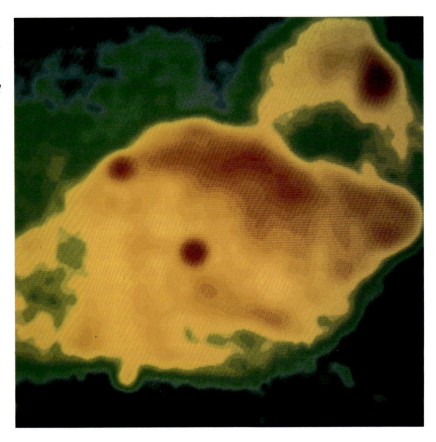

light travel time across the remnant. FDS and his colleagues also noted that the X-rays might originate from some compact remnant of the supernova explosion.

Finally, during a deliberate search for stellar remnants of supernova explosions by looking for optical stars coincident with point radio sources (which mostly turn out to be quasars and not galactic objects), it was realised by Martin Ryle and his colleagues that the strong radio source at the centre of W50 was at the same position as a 14th magnitude star, which they did not even realise had already been catalogued by Stephenson and Sanduleak. Due to a variety of errors and coincidences, the peculiar properties of SS433 had been overlooked for several years after its initial discovery by the survey, but, at last, in mid-1978 the first optical spectra of SS433 were obtained by Clark and Paul Murdin using the Anglo-Australian Telescope.

12.2 The *moving* lines

These spectra of SS433 displayed strong emission lines, which was not surprising considering its discovery in the SS survey. Clark and Murdin also found other emission lines at odd wavelengths that they could not identify, but they were the first to point out that the radio star, the X-ray source and SS433 were in fact the same object, and the location was at the centre of W50! This work came to the attention of University of California astronomer, Bruce Margon, who began taking a series of optical spectra at Lick Observatory in order to study their temporal variations (see figure 12.3). The principal Balmer line, Hα, appears to stay fixed throughout. However, there are two other emission lines, at atypical wavelengths, which,

Fig. 12.3 A compilation of optical spectra of SS433 obtained over a 5 month period from Lick Observatory with the 3 m Shane Telescope. The strongest features are those that made SS433 stand out in figure 12.1, which is the spectral region around Hα. The periodic motion, in opposite directions, of the blue and red moving lines is clear. The amplitude of the motion seen here is a staggering 40 000 km s⁻¹. (Diagram by Bruce Margon.)

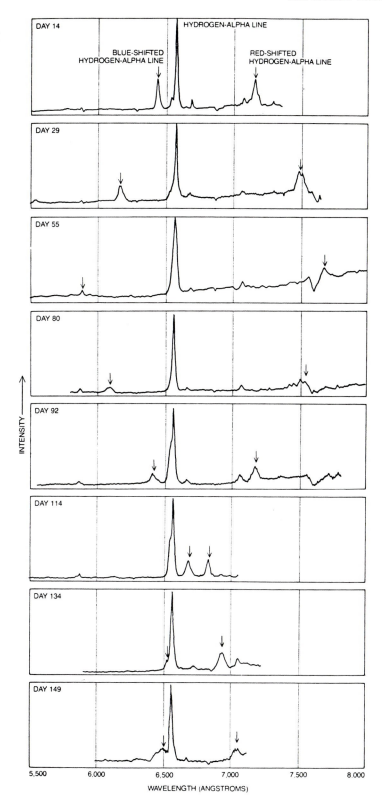

astonishingly, move from one night to the next. The velocity *changes* implied by our being able to actually see the movement on such a simple plot are staggering.

Worse still, it is clear from examining these spectra over many months, that these extra lines, one to the *blue* of Hα, the other to the *red*, move smoothly back and forth through the spectrum. Margon also discovered that the other emission lines have moving components too. At certain times, this means that the spectrum can become extremely complicated to interpret. Overall though, each line's set of *moving* components appears to follow the same pattern (figure 12.4), repeating their spectral motions smoothly about every 160 days. There are two obvious and quite remarkable properties associated with this motion:

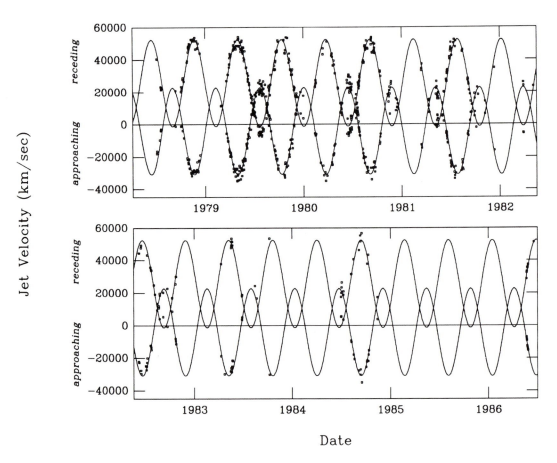

Jet Velocity (km/sec)

Fig. 12.4 Complete plot of all radial velocity measurements of SS433 against phase of the 164 day period, showing the excellent stability of the motion of the jets. (Diagram by Bruce Margon.)

(1)   the amplitude of the movement is enormous. It implies a velocity of about 40 000 km s^{-1}, far larger than anything ever seen in a galactic object. Indeed, it is quite easy to show (box 12.2) from Kepler's Laws that such a motion, if due to two bodies orbiting each other with the 160 day period, implies that their total mass must be about 2 *billion* times that of the Sun! It is, of course, inconceivable that such a massive object within our galaxy could have escaped detection until now;

(2)   the average, or mean, velocity of the motion is itself about 12 000 km s^{-1}. Taken at face value it would imply, on the basis of the observed recession of the galaxies (i.e. Hubble's expanding universe) that SS433 is extremely distant and certainly extragalactic. However, the

Box 12.2 Why can the moving lines not represent orbital motion?

The radial velocity curve shown as figure 12.4 is, at first sight, remarkably similar to that expected as a result of binary motion i.e. two stars orbiting each other under the influence of gravity. If this is true then we can use Kepler's Laws to calculate the mass of the stars involved. The masses and separation of the two stars are shown in figure 12.B2.

Fig. 12.B2 A simplistic binary interpretation of the velocities observed in SS433.

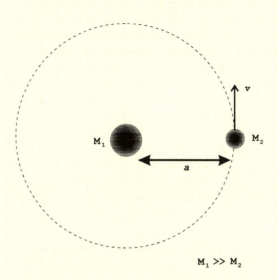

$M_1 \gg M_2$

The observed velocity amplitude is 42 000 km s^{-1} or about 10 000 AU yr^{-1}. The reason for using such apparently odd units is that Kepler's Third Law is very straightforward if we measure masses in solar masses, periods in years and distances in AU:

$$M_1 + M_2 = \frac{a^3}{p^2}$$

(it is obvious that such must be the case if you just consider the Sun–Earth system which has $M_1=1$, $M_2=0$, $a=1$ AU and $P=1$ yr). In this equation, a is the separation of the 2 stars and, strictly speaking we should also consider the inclination of the plane of the orbit to ourselves. However, the minimum masses involved obviously occur for an assumed inclination of 90° (i.e. we are observing in the plane of the orbit) and we shall also assume for simplicity that M_1 dominates so that we can estimate a from the circular velocity:

$$v = \frac{2\pi a}{P}.$$

If we substitute a from this equation into Kepler's Third Law above we obtain

$$M_1 + M_2 = \frac{Pv^3}{8\pi^3}.$$

We know that $P = 0.5$ yr and so $M_1 + M_2 = 0.5 \times 10\,000^3 / (8\pi^3)$ or *2 billion* solar masses!

stationary emission lines are *not* at such a velocity, but are consistent with a very much lower velocity, as one would expect from an object in our galaxy.

12.3 Relativistic jets

The physical origin of such fantastically high velocities emanating from a star that appeared to be in our galaxy was a problem that provoked tremendous interest in the year or so after the emergence of SS433. The solution to this problem and now widely accepted as the basis for understanding the nature of SS433 is the *kinematic model* developed independently by Andy Fabian and Martin Rees at Cambridge and Mordy Milgrom at the Weizmann Institute, and fit to the optical characteristics of SS433 by Margon and the late George Abell. This model consists of a pair of oppositely directed jets of gas travelling at relativistic speed (figure 12.5). These jets precess about a common axis with the 164 day period, thus varying their orientation with respect to us and hence producing the observed Doppler velocity modulation.

Fig. 12.5 Schematic of the kinematic model of SS433. This shows the almost pencil-thin beams emanating along the rotation axis of an accretion disc which precesses with the 164 day period. Each beam or jet produces emission lines which we see, due to the Doppler shift, as a moving line. The angular extent of the precession is such that there is a time when the beams are travelling perpendicularly to our line of sight, and thereby the moving lines appear to cross over in the spectrum. (Diagram by Bruce Margon.)

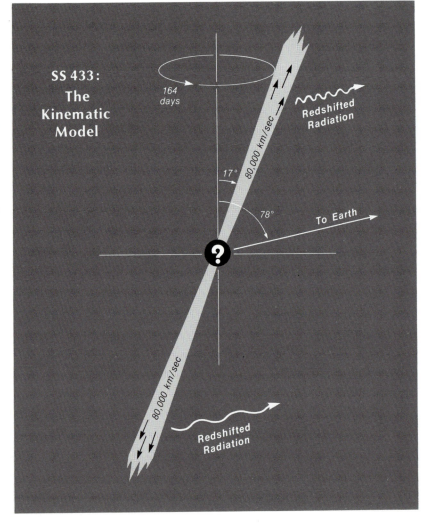

It is not immediately obvious from this, however, why the average speed of the jets at any instant and, in particular, their speed when travelling in a direction that is perpendicular to the observer, is not zero. To understand this fully requires some detailed knowledge of special relativity (see box 12.3), but there is an excellent analogue with a well-known feature of relativity that illustrates the effect. Most people have seen or heard of the apparent paradoxes that can arise when observers and clocks are travelling at relativistic speed. If two clocks are synchronised and then one is mounted on a spacecraft capable of very high speeds some remarkable features emerge. The Earth-bound scientist will observe the spacecraft's clock to be running slow compared to his own as it passes by him at high speed. (It is also true that, to an observer actually on the rapidly moving spacecraft, the Earth-bound clock will appear to be running slow also! This is a well-known *paradox* of special relativity which is not relevant here. An interested reader can find a full account of this phenomenon in the book *Spacetime Physics* by Taylor and Wheeler). This result is called *time dilation*. What is important about it is that it is independent of the direction in which the moving spacecraft is actually travelling. The clock appears to be running slow whether the spacecraft is coming towards you, away from you or travelling from side to side.

What is more, the notion of time dilation can be extended from these simple ideas of moving clocks to the processes of atomic physics. The light emit-

Box 12.3 The transverse Doppler shift

Consider an object travelling with a velocity v at an angle θ to our line of sight. The component of this velocity towards us is simply $v \cos \theta$. If v is small then any radiation emitted by the object at wavelength λ_0 will be observed by us at a wavelength λ given by the classical Doppler shift:

$$\lambda = \lambda_0 \left(1 + \frac{v \cos \theta}{c} \right).$$

Note that this gives us the result we expect when $\theta = 90°$, i.e. there is no Doppler shift when the object is travelling perpendicularly to our line of sight. Also, the observed wavelength shift in a spectrum, defined by $z = (\lambda - \lambda_0) / \lambda_0$ is, of course, simply equal to $v \cos \theta / c$.

But when v is a large fraction of c we must use a result from Special Relativity which gives the observed wavelength as:

$$\lambda = \lambda_0 \frac{\left(1 + \dfrac{v \cos \theta}{c} \right)}{\left(1 - \dfrac{v^2}{c^2} \right)^{1/2}}$$

and this is sometimes called the *relativistic Doppler shift*. Note that the emission of spectral lines can be considered as an atomic clock which, at relativistic speeds, will appear to us to be running slow. The factor by which it does run slow is the $(1 - v^2/c^2)^{-1/2}$ term in the above equation, and is known as the *Lorentz gamma factor*, and explains the effect we call *time dilation*.

There are two important things to note about this equation. Even if $\theta = 90°$, the observed wavelength is not the same as that emitted by the moving object; and this observed wavelength is *always* greater than that emitted (i.e. a REDSHIFT) *regardless* of the actual direction in which the object is travelling. It is this constant term in the relativistic equation that we observe in SS433 as the 12 000 km s^{-1} average of the moving lines. From this we can simply calculate the true space velocity of the moving material as $v = 0.26c$.

ted by a moving atom will, because of this effect, be seen by a stationary observer (i.e. us) to have less energy than expected for that particular atomic transition. Less energy means that it will have a longer, or more red, wavelength. The independence of direction is most important as this is what distinguishes this effect from the *classical* Doppler shift of wave motions which depend in a very straightforward way on direction. A train whistle or police car siren will audibly change their frequency when travelling towards or away from the listener, whereas no change is heard if the vehicle is traversing across our field of view at some distance. This is simply because the vehicle has no component of velocity in our direction in the latter case.

We can now see how to interpret the velocity curves of the two jets in SS433. The time-varying components are due to the normal Doppler shift as the direction of motion of the jets changes with respect to us (see figure 12.5). The average of the jets' velocities, or cross-over velocity, is the time dilation effect, often called the *transverse Doppler shift*. Because it depends *only* on the absolute velocity of the jets and not their direction of motion, we can use this result to determine the jet velocity directly (box 12.3). It is $0.26c$, or $78\,000$ km s^{-1}. The relativistic nature of these jets had been established beyond doubt.

12.4 Direct observation of the moving jets with the VLA

The discovery of the remarkable properties of SS433 coincided with the commencement of operations at the Very Large Array (VLA) in New Mexico. The VLA consists of 27 separate radio telescope dishes (each 26 m in diameter) spread out over a baseline of 27 km on a plateau surrounded by mountains. This configuration enables the ensemble of dishes to achieve the same angular resolution on the sky as a single radio telescope whose diameter was equal to this huge baseline (and is called an *interferometer*). The early radio telescopes, and even single large dishes now, had angular resolution substantially worse than that achievable optically and this greatly limited the extent to which optical and radio maps could be correlated. The VLA on the other hand now regularly produces radio maps at a resolution that is over 10 times better than that achievable optically from the ground.

This extremely high resolution was utilised to good effect when applied to SS433. The central *point* radio star was found by the VLA to consist of a highly variable point source surrounded by extended components which moved with time. These extended components were found to move outwards in opposite directions at a rate of 3 arcseconds per year as is clearly visible in figure 12.6. If this is equated to the optically determined velocity of $0.26c$ and the geometry of figure 12.5 is used, then it is possible to determine the distance of SS433 accurately and unambiguously. It is 5.5 kpc distant. What is more, the detail in this motion on the sky, when followed over several years gives excellent support for the kinematic model.

12.5 Interaction of the jets with the supernova remnant W50

The large amount of energy continuously pumped out in these jets must surely have an observable effect on the large scale structure of the medium surrounding it. With hindsight this can now be seen in the radio maps of W50 (figure 12.2) which clearly show an essentially spherical shell with two *ears* or lobes that are aligned with the axis of the jets as defined above. Furthermore, the relativistically moving jet material must produce strong shock heating when colliding with the slower moving gas in the shell, thereby heating it to X-ray temperatures. The Einstein IPC has been used to map SS433 and the result is shown in figure 12.7. A superposition of this X-ray image on the latest VLA map of SS433 (figure 12.8), includes the jet axis and the precession cone.

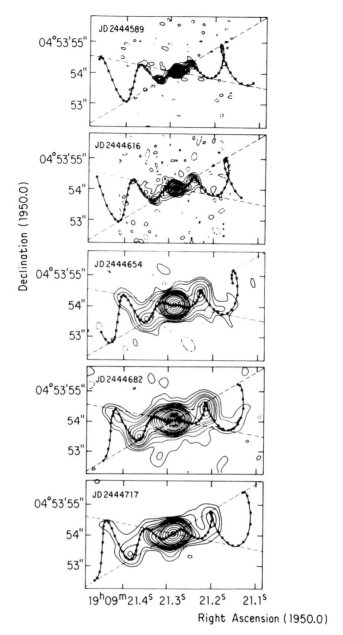

Fig. 12.6 VLA maps of SS433 produced over a 4 month period showing the growth of the radio lobes as predicted by the corkscrew motion (solid curve) of the precessing relativistic jets in the kinematic model. (The dashed lines define the cone within which the jets precess.) (Produced by Bob Hjellming and K. Johnston)

12.6 SS433 as an X-ray binary

So far we have concentrated on the dramatic properties of the *moving* lines in the spectrum of SS433 and the directly observable radio jets. However, the spectra in figure 12.3 also show apparently normal, non-moving lines. These features were studied in detail by the Dominion Astrophysical Observatory (DAO) group of John Hutchings, Anne Cowley and David Crampton who found that these lines were not quite stationary, but were in fact slowly moving with a period of 13.1 days. The velocity curves are not at all unusual and are typical of a normal binary system, in which the mass-losing star is an early-type B star.

The picture that is now emerging of SS433 is that of an interacting binary system involving substantial mass loss from the B star into an accretion disc surrounding a compact object, which must be either a neutron star or black

Fig. 12.7 X-ray picture of SS433 obtained with the Einstein Observatory IPC. The strong point source is the accreting compact object of SS433 itself, whereas the X-ray lobes almost certainly delineate the precession axis of the jets. The X-ray emission mechanism remains unknown.

hole. The disc is precessing with respect to the orbital plane, thereby giving rise to the 164 day precession period of the jets (figure 12.9).

But why is SS433 the only X-ray binary to clearly exhibit the property of possessing relativistic jets? The answer to this is unclear. It may be that the system is at a stage of its evolution where the mass-losing star is transferring material at an unusually high rate and the *normal* flow of matter through the disc onto the compact object is disrupted. This is sketched in figure 12.10, where the inner part of the accretion disc has so much material entering it that a substantial fraction is not accreted onto the compact object, but is ejected along the rotation axis of the disc. This will bring us, in the next chapter, directly into the processes occurring in the most powerful galactic nuclei.

12.7 The engine driving the jets

But we have still not addressed the problem of how this material, ejected along the disc's rotation axis, is then accelerated to such a high velocity. As X-ray binaries go, SS433 is not that bright, it is only radiating about 3×10^{35} erg s^{-1} as X-rays. Indeed, there are many binaries that are 100–1000 times more luminous! Surely if the radiation output of the matter falling onto the compact object were somehow to power the jets then we would expect to see such behaviour in many other sources? It is here that recent observations by EXOSAT have given valuable new insight.

Moving X-ray lines

Although concentrating on long, uninterrupted observations of X-ray sources with the large proportional counter arrays, EXOSAT also had a novel X-ray detector on board called the *gas scintillation proportional counter* (GSPC). As mentioned in chapter 2, this device is essentially a normal X-ray counter, in which

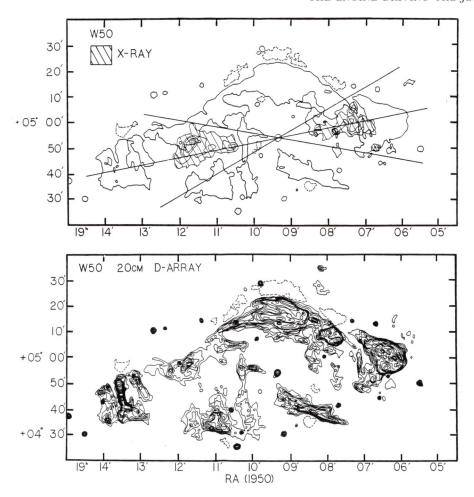

Fig. 12.8 Overlay of the X-ray picture of figure 12.7 (the hatched contours) onto an outline of a new complete radio map of SS433 produced by the VLA. In addition, the straight lines show the cone of the precessing jets as taken from figure 12.6. It is this part of the large radio shell of W50 which appears to have been blown up, rather like giant ears, almost certainly by the continued blasting by the relativistic jets. (Courtesy S. Baum, STScI.)

the X-rays are detected by the optical *scintillations* produced when the ionised gas atoms recombine. The optical light can be detected without the need for substantial amplification in the gas itself, and this means that the spectral resolution is better than that of a normal counter. However, the EXOSAT GSPC was basically a test-bed device; it was rather small and not as sensitive as the large array. Nevertheless, for sources that were not too faint, the GSPC was able to produce very good quality X-ray spectra and it was employed to observe SS433 at different times in the 164 day precession cycle.

The Leicester and Max-Planck X-ray groups (led by Mike Watson) reduced these observations and found that the X-ray spectrum could be represented by a continuum and a *moving* emission line (figure 12.11). The emission line is that due to highly ionised iron which has lost all but one of its electrons, and is thus at a high temperature (50–100 million degrees). When the position of the line is shown as a function of the phase of the 164 day cycle (figure 12.12) another very interesting fact emerges. Only *one* X-ray line is detectable, and its behaviour is consistent with the *blueshifted* line in the kinematic model. For some reason in the geometry of SS433 we cannot see the redshifted X-ray line. An explanation has been offered by Watson and his colleagues which is demonstrated in figure 12.13.

This has important ramifications for the understanding of SS433. Assuming that the X-ray line is coming from hot gas, it is possible to estimate the amount of matter flowing out of SS433 in the jet, and this is about 10^{-6} $M_\odot$ per year, a very high rate. Given that we know the speed of the jet we

Fig. 12.9 Artist's impression of the SS433 binary system showing all the important components. (Painting by William K. Hartmann for Cycles of Fire, *Workman Publishing Co.)*

can calculate the kinetic energy being given to the jet by the acceleration mechanism every second, and it is about 2×10^{39} erg s^{-1}. And yet we only see an X-ray radiation luminosity 10 000 times smaller than this! Perhaps we never see the X-ray source directly, as already mentioned in chapter 8 for the *accretion disc corona* sources. We already know that we are observing SS433 at a fairly high inclination and so, as is occasionally true in Her X-1, the rather thick disc never allows us to see the true or intrinsic X-ray luminosity. This would imply that the accretion luminosity is somehow related to the mechanism which drives the jets, the so-called *central engine*.

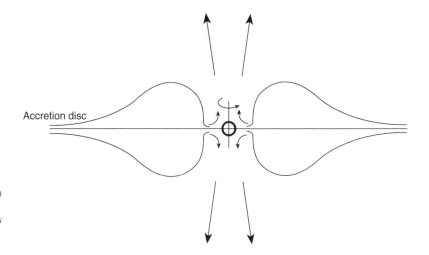

Accretion disc

Fig. 12.10 Schematic of how material flowing in the thickened inner part of the accretion disc can be constrained to be ejected along the disc's rotation axis. Such ideas recur in chapter 13.

Fig. 12.11 X-ray spectra of SS433 obtained by the EXOSAT GSPC at two different phases of the 164 day period. The dominant feature is due to highly ionised iron, which moves from 6.5 to 7.8 keV in phase with only one of the optical components. (EXOSAT Observatory, ESA.)

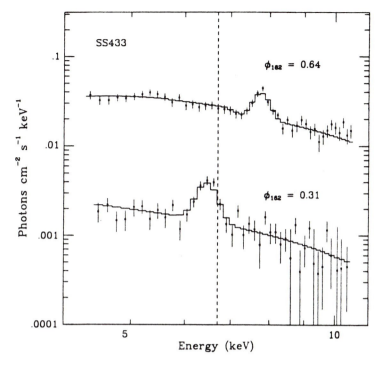

Fig. 12.12 X-ray radial velocity curve as a function of the 164 day precession period for the iron line of figure 12.11 shows that the line moves in agreement with the kinematic model (solid line). However, only the blue-shifted component is seen, never the red-shifted one. The solid points are from the EXOSAT GSPC detector, the open circle is from the Japanese Tenma satellite. (Based on an original diagram by Hakki Ögelman.)

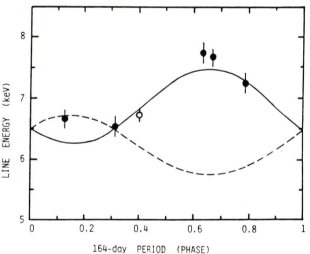

An X-ray eclipse

Further EXOSAT observations by the Leicester group (under Gordon Stewart) have dramatic consequences for the nature and mass of the compact object. The X-ray light curve produced from EXOSAT's capability for long, uninterrupted observations has revealed that it is eclipsing (confirming the high inclination hypothesis). The eclipse lasts for longer than 1.2 days which sets a lower limit on the size of the mass-losing star. This can now be combined with the information about that star obtained optically by the DAO group and requires that the compact object mass be in the range 30–60 $M_\odot$ with a firm lower limit of 10 $M_\odot$. It therefore has to be a black hole! The X-ray eclipse has been confirmed by Ginga, and the motion of the X-ray iron line has also been found to fit the ephemeris of figure 12.12.

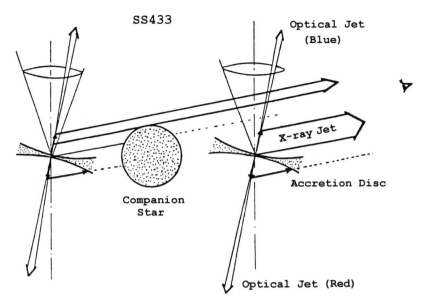

Fig. 12.13 The presence of only the blue-shifted X-ray component means that the X-rays are produced close to the compact object and that the red-shifted component is obscured from us by the accretion disc itself. The geometry needed to produce this effect is demonstrated here, with the accretion disc shown at opposite sides of the 13 day orbit. The upper X-ray jet will suffer a partial X-ray eclipse when on the far side of the companion at the same time that the optical emission from the disc is also eclipsed. The lower (receding) X-ray jet is never visible to us, because it is always obscured by the accretion disc itself. (Based on an original by Masaru Matsuoka, ISAS.)

12.8 The jet acceleration mechanism

Interestingly enough, these observations of moving X-ray lines have shot down one of the most dramatic models that accounted for the high velocity of the jets. Milgrom noted that the relativistic speed, $0.26c$, corresponded to a redshift $1 + z$ of 1.33 which was almost exactly equal to the ratio 1215/912. These two numbers represent the wavelengths of Lyman α and the Lyman limit of hydrogen (the Lyman series are the characteristic absorption or emission lines of hydrogen that are associated with the ground state of the hydrogen atom). Although more fully described in box 12.4, this coincidence suggested that some kind of *line-locking* was taking place. In this the intense radiation pressure due to the intrinsically high luminosity was actually accelerating the material which was channelled by the disc into the narrow exit beam along the disc's rotation axis. This was physically simple and highly appealing. It also predicted that, if we ever found one, the jet velocity in *another* SS433 would be the same.

Unfortunately, one basic ingredient of the model is the requirement that the hydrogen in the jets remain neutral so that it can absorb radiation and thereby be accelerated. The observation of highly ionised iron lines obviously moving close to the source and already accelerated eliminates this model since, if iron is ionised, all the hydrogen certainly will be, and hence unable to absorb the radiation. Another accelerating mechanism is needed.

We are left then with this one glorious example of a *star* with relativistic jets. In spite of diligent observations of other stars in the SS catalogue and of X-ray binaries, no other evidence for such rapidly moving material has been uncovered. It is difficult to believe that SS433 is the only star in our galaxy to have these properties. It may be the brightest, or one of the brightest examples, but there must be others, ripe for discovery, if only we knew where to search for them. At this point other examples are needed in order to make progress on the definition of the acceleration mechanism and the nature of the compact object in SS433 itself.

Box 12.4 Milgrom's line-locking model

The energy levels of the hydrogen atom are shown in figure 12.B3(a) together with the transitions corresponding to the Lyman α line and that needed to ionise the atom. The wavelengths of these two transitions are 1215 Å and 912 Å respectively, and it was noted by Milgrom that the ratio of these two numbers (1215/912) is 1.33 which is exactly the observed velocity of the jets in SS433 ($v = 0.26c$ corresponds to $1 + z = 1.33$ when the relativistic Doppler equation is used, see box 12.3). Is this a coincidence or is it telling us something about the physics of the motion of the jets?

Fig. 12.B3 (a) Energy level diagram for the hydrogen atom. (b) Required intrinsic spectrum for the SS433 central object. (c) Spectrum of central object as seen by moving gas in the jets at different speeds. If v were greater than 0.26c the photons would have insufficient energy to excite to the n=2 level and hence could not accelerate the gas by radiation pressure.

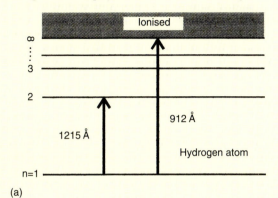

(a)

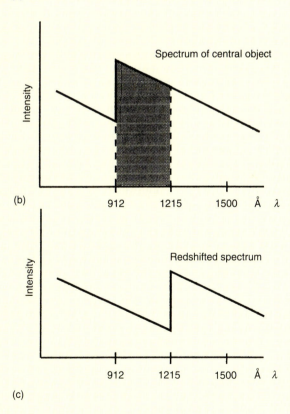

(b)

(c)

Milgrom suggested that if the spectrum of the central object was as depicted in figure 12.B3(b) then the bright continuum beyond 912 Å would be able to accelerate the material in the jet by radiation pressure. When an atom absorbs a photon and makes a transition to a higher energy level it also absorbs the momentum of the photon. Although small, if the luminosity is high enough this momentum can accelerate the absorbing material. But why should it be accelerated to a particular velocity and not just go on to some higher velocity?

The trick is the break in the spectrum at 912 Å (figure 12.B3(b)). As seen by the moving gas in the jet, the central object's spectrum will appear to be red-shifted as shown in figure 12.B3(c). When the gas is travelling at $0.26c$, the edge in the bright continuum will appear to be at a wavelength of 1215 Å and not 912 Å. Since the light cannot be absorbed by the atom unless it has enough energy to excite it from the lowest energy level ($n = 1$) to at least the next level ($n = 2$) then the acceleration will only work if there are photons with wavelength less than 1215 Å. Once the break is red-shifted to 1215 Å it cannot accelerate the jets anymore. Hence, in this model the jets *must* have the specific velocity of $0.26c$.

But what causes this break in the spectrum in the first place? Milgrom suggests that this is due to hydrogen much nearer the central object. It readily absorbs this region of the spectrum and becomes ionised. However, the problem of course is how to have enough neutral hydrogen to do this absorbing so close to the X-ray source. There are many other difficulties with this model, but it also has an appealing simplicity and a fundamental prediction for the *next SS433*.

13 Active galactic nuclei

13.1 Introduction

Normal galaxies like our own, when viewed from great distances, appear to be peaceful and unchanging aggregations of stars, whose well-being is only slightly disturbed by the occasional supernova explosion. However, violent processes far more powerful than supernovae have been known since early in this century. The optical jet emanating from the giant elliptical galaxy M87 (in the Virgo cluster of galaxies) was found in 1917. After the Second World War the founding of radio astronomy led to the discovery of luminous extragalactic radio sources such as Cygnus A. And short exposure optical photographs showed that some apparently normal spiral galaxies actually had very bright, almost star-like, nuclei, the prime example of which is NGC 4151 (figure 13.1). Hence the term *active galactic nuclei*, or AGN.

These are referred to as *Seyfert galaxies* after their discoverer, Carl Seyfert. But even these exotic objects paled in comparison with the enormous energy output of *quasi-stellar objects* (known as quasars or QSOs), discovered through their radio emission in the early 1960s. Their very high redshifts (see box 13.1) implied that these were immensely distant, and hence the most luminous objects known. [1]

The nearest quasars have subsequently been found to be embedded in apparently normal spiral galaxies and hence can be considered as extreme examples of Seyfert galaxies. Indeed, at the other end of the scale, even local normal galaxies like our own appear to have a low level of nuclear activity, the nature of which is as yet unexplained.

13.2 The range of activity in active galaxies

The similarity of the optical spectrum of the nucleus of a Seyfert galaxy to that of a quasar infers that both are the product of the same mechanism. Indeed, if a Seyfert galaxy were to be moved sufficiently far away for the

---

[1]The debate over whether or not to interpret the huge redshifts (and hence velocities) of quasars as being of a cosmological origin was one of the most heated in the astronomy of our time. We will not repeat it here, but point out that deep exposure photographs of the relatively nearby quasars 3C273 and 3C48 have revealed the presence of a companion galaxy to 3C273 and a host galaxy in which 3C48 is embedded. Spectroscopy of these very faint galaxies (which appear completely normal) shows that they are at exactly the same redshift as the quasars. In addition, the quasar 3C206 has now been found to be associated with a cluster of galaxies at the same redshift. We believe that the evidence in favour of the cosmological interpretation of quasar redshifts is now overwhelming.

Fig.13.1 Short, medium and long exposures of NGC 4151 showing the bright star-like nucleus and the otherwise normal, surrounding spiral galaxy.

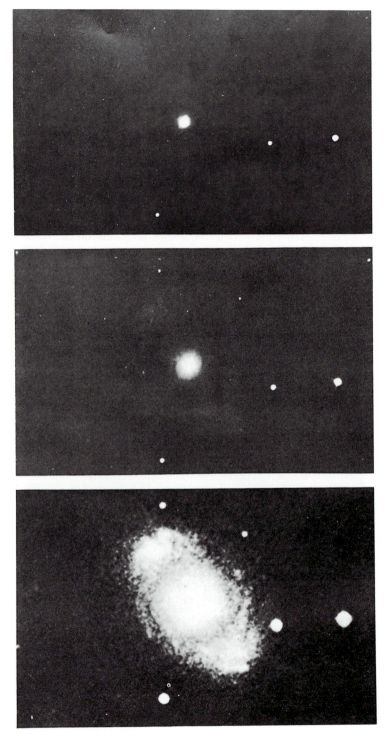

spiral galaxy to be undetectable, we would classify it quite simply as a quasar. The discovery of fuzz (host galaxies) around quasars reinforces this, and it is not surprising that the highest luminosity (most distant) Seyferts overlap with the lowest luminosity (closest) quasars. This range of activity is well demonstrated by the series of optical spectra shown in figure 13.2. Table 13.1 sum-

Box 13.1 H_0 **and the expansion of the universe**

The distances of most of the extragalactic objects referred to in this book are sufficently small that we can estimate their distance, and hence their intrinsic luminosity, by the methods of classical astronomy. In addition, Hubble had shown in the 1920s that there was a linear relationship between the distance of a galaxy and its velocity of recession from us. This is Hubble's Law and led to the concept of an expanding universe. The constant of proportionality in this relationship is now believed to lie in the range 50–100 km s^{-1} Mpc^{-1} and is known as Hubble's constant, or H_0. In this book we have assumed that $H_0 = 50$ km s^{-1} Mpc^{-1}.

The observed redshift of an object is defined quite straightforwardly as $(\lambda_{obs} - \lambda) / \lambda$, the ratio of the wavelength shift of a particular feature (e.g. emission lines of hydrogen) to the rest wavelength of that feature, and is called z. However, to relate z to the line of sight velocity of the object is not straightforward. At low velocities ($z \leq 0.1$), it is quite simply equal to v/c, but at larger values of z this breaks down (especially now that we are approaching z values of 5 observationally). The full relativistic expression to use is:

$$\frac{v}{c} = \frac{(1+z)^2 - 1}{(1+z)^2 + 1}$$

from which the following values can be obtained :

z	0	0.05	0.10	0.50	1.00	2.00	5.00	10.00
v/c	0	0.05	0.10	0.38	0.60	0.80	0.95	0.98

marises the X-ray properties of the complete range of active galactic nuclei and gives them with respect to normal galaxies.

13.3 The origin of the radio jets

Over the last 20 years various models have been proposed to account for the remarkable activity of quasars. An extraordinarily powerful *central engine* is required that is capable of generating enormous amounts of power in only a small volume of space. However, the similarity of certain of the radio galaxy jets to the much smaller scale operation in SS433 is quite striking and implies that *accretion power* must be the key. 3C129 is an excellent example of this in which a precessing beam has been *blown* backwards by passage through the cluster medium (figure 13.3). It is simply a larger scale version of the model that we envisage for SS433. Just compare this radio image of 3C129 with figure 12.6 showing SS433's precessing radio jets.

In fact, this observation provides direct evidence for the existence of the medium, which is, despite its low density, very hot and produces thermal X-rays which we see as diffuse emission from the cluster as a whole (as discussed in chapter 14). It would seem then that we should look to SS433 for clues that can be applied to active galaxies.

As described in chapter 12 the basic energy source in SS433 is accretion. Gas from an evolving *normal* star is transferred into an accretion disc about the stellar mass sized compact object (either a neutron star or black hole). For reasons that we do not fully understand in SS433 a significant fraction of the material does not accrete onto the compact object, but is instead ejected along the spin axis of the object at a velocity of $0.26c$. The jets precess with a 164 day period about the (presumed) spin axis of the disc. But SS433 is a stellar-sized object producing *only* about 10^{36} erg s^{-1} at X-ray wavelengths, although its total power in the jets must be 10 or 100 times greater. How

Fig.13.2 Montage of optical spectra showing the range of activity in AGN. For comparison we start with the spectrum of a normal galaxy (which is dominated by normal stellar absorption lines of sodium, magnesium and calcium), then progress to a narrow emission-line galaxy, a Seyfert galaxy and finally a quasar (PKS 1217+02). Note the increasing width and power of the emission lines in this sequence, although the spectrum of a powerful Seyfert is harder to distinguish from that of a quasar.

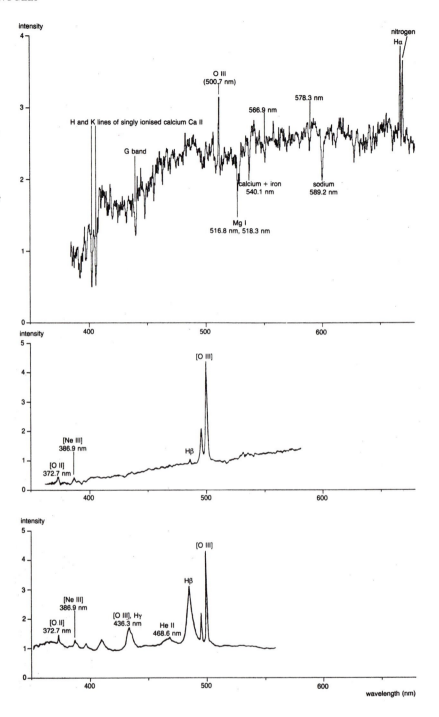

can this be related to the enormous energy output of quasars and radio galaxies?

The answer is simply one of scale. The classical binary nature of SS433 tells us that the compact object must be either a neutron star or relatively small black hole. To get out of the system a factor of a million or more times more energy requires the mass of the compact object to be greater by exactly the same factor (see box 7.2). Such an object must be a black hole. However, the fuel for the accretion mechanism must also be found. We can no longer

Fig. 13.2 For legend see opposite.

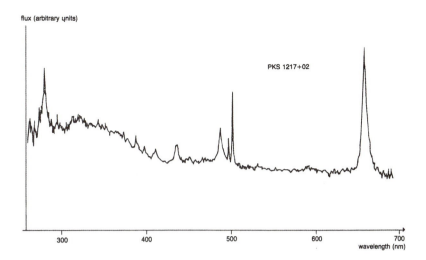

Table 13.1 Properties of active galactic nuclei

Type	L_x (erg s^{-1})	Host galaxy	Example
Normal galaxy	$10^{37} - 10^{39}$	–	M31
Narrow emission line galaxy	$10^{40} - 10^{43}$	–	Mkn 176
Seyfert galaxy	$10^{43} - 10^{45}$	Spiral	NGC 4151
BL Lac objects	$10^{44} - 10^{46}$	Elliptical	OJ 287
QSOs	$10^{45} - 10^{47}$	?	3C273

consider a single mass-losing star as in the X-ray binaries, but must search for sources of gas on much larger scales. Powerful quasars require accretion rates of several $M_\odot$ per year in order to explain their prodigious energy output. Although the higher density of normal interstellar material (from supernovae, novae and stellar winds) near the galaxy's centre may account for some of the fuel, it is likely that the higher stellar density combined with the black hole's intense gravitational field could tidally disrupt entire stars and generate a massive disc around the hole. Such a system is shown schematically in figure 13.4.

It is conjectured that the combination of gas pressure from the disc and intense radiation pressure near the hole leads to a naturally collimated jet ejected along the spin axis of the hole. The process is complicated by the presence of the associated magnetic fields, but the advent of *supercomputers* has opened up entirely new fields of research into the properties of jets and the interpretation of the very detailed radio images. But how can we find out what is going on in the *central engine* and what is powering it?

13.4 The supermassive object in NGC 4151

As is often the case in astronomy, exotic ideas, such as supermassive black holes ejecting relativistic jets of material over vast reaches of space, are often best tested and examined by studying related, but less extreme, objects on our own *doorstep*. The prototype (and brightest) Seyfert galaxy, NGC 4151, which was mentioned at the beginning of this chapter is certainly an interesting and striking object, and its brightness makes it possible to obtain high

Fig.13.3 The radio galaxy 3C129 exhibits precessing jets that are observed in the radio (VLA) and compared (below) with a model of a precessing jet from a nucleus which is travelling through a cluster medium. This medium causes the jet material to be swept back *by ram pressure. The model can generate the observed structure extraordinarily well. (Diagrams by Vincent Icke, Leiden.)*

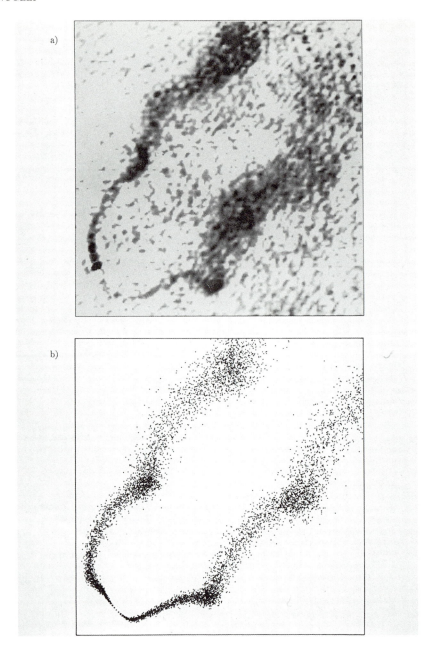

quality data in reasonable observing times. The late Mike Penston (RGO) and his group of *old LAGs* (*lovers of active galaxies*) have observed NGC 4151 from the ground and in space for the better part of two decades now. One of the startling results of their monitoring is shown in figure 13.5 which contains optical spectra of the galaxy's nucleus taken with the 2.5 m Isaac Newton Telescope (INT) in 1974 (when it was based in Sussex at Herstmonceux Castle) and 1984 (from its new La Palma site). The enormously broad emission line component that signifies its classification as a Seyfert 1 galaxy (the most active variety) had vanished. If NGC 4151 had been observed then for the first time it would have been classified as the much less active Seyfert 2. Clearly these changes were taking place close to the core of the galaxy.

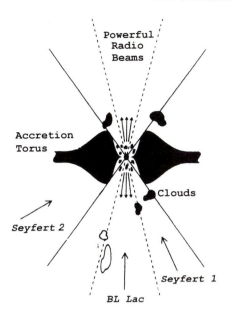

Fig.13.4 Schematic model of an AGN, consisting of a black hole, accretion disc and jets. This is very similar to the model of SS433 described in chapter 12 (see figure 12.10) as the jet collimation mechanism is the bloated inner accretion disc surrounding the compact object. Then the type of AGN that it is observed to be depends on the orientation of the disc and jets relative to our line of sight. A BL Lac designation is believed to indicate we are observing along the axis of the jets themselves. Otherwise it will be a Seyfert type 1 or 2 depending on whether the central region is obscured from our view or not.

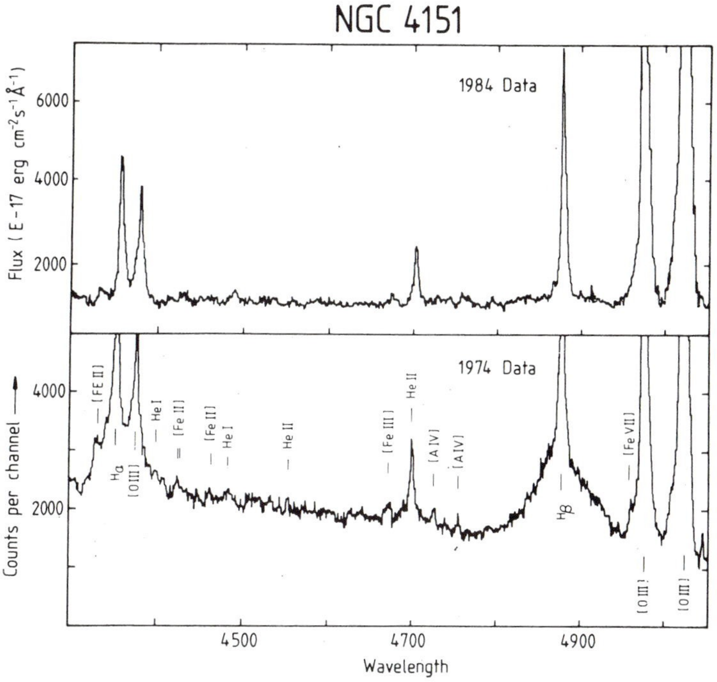

Fig.13.5 Optical spectra of NGC 4151 obtained with the INT in 1974 and 1984. The activity of the nucleus has changed dramatically in 10 years from that of a classic Seyfert 1 to the much milder Seyfert 2 (in which the broad line emission corresponding to the highest velocity material is greatly reduced). (Diagrams by Mike Penston, RGO.)

Observations in the far ultraviolet from space using IUE gave more detail on the nature of the *clouds* of rapidly moving gas in the core region. In 1979 Penston and his team were fortunate to observe (figure 13.6) a brightening of the core (the unresolved stellar component) but noted that the 'clouds' did not increase in brightness until approximately 13 days later. There is clearly a time-lag between the variation of the continuum (produced in the core) and the emission lines (that originate in the surrounding *clouds*).

This presented a remarkable opportunity to determine some of the physi-

Fig.13.6 (a) Ultraviolet spectra of NGC 4151 obtained with IUE at different times. By noting the delay in variations between the continuum and line emission it is possible to estimate the mass of the central compact object. Note the two faint emission lines, labelled L_1 and L_2. They are both CIV emission (red and blue shifted) from a jet in the centre of NGC 4151. (b) shows schematically the location of the different emitting regions. (Diagrams by Mike Penston, RGO and Sky & Telescope*)*

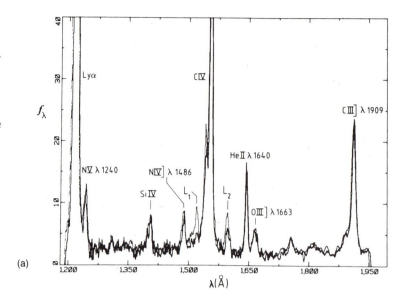

(a)

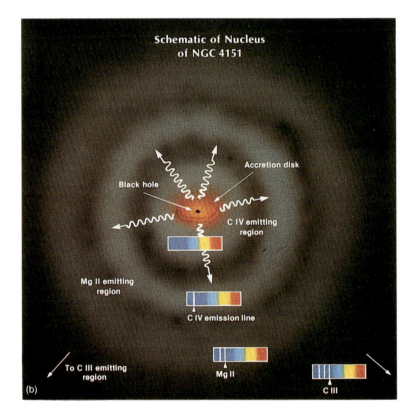

(b)

cal parameters of the nucleus of NGC 4151 if certain simplifying assumptions were made. The detailed calculation is laid out in box 13.2, but essentially we are assuming that the *clouds* are orbiting the central core in exactly the same way that the planets orbit the Sun. The observed time delay directly yields the distance of the *clouds* from the core and hence the mass of the core can be estimated. It is about 500 million $M_\odot$, a truly supermassive object, which must be a black hole. NGC 4151's activity is noticeable to us simply because

Box 13.2   Weighing the black hole in NGC 4151

The basis of this calculation is shown schematically in figure 13.B1. A super-massive central object (assumed to be a black hole) of mass M is the *central engine* that generates the blue continuum radiation. This radiation excites the material in the surrounding clouds, thereby producing the strong emission lines. The clouds are at a distance r from the central object, which they orbit with a velocity v.

Fig.13.B1 How to weigh the presumed black hole at the centre of an active galaxy.

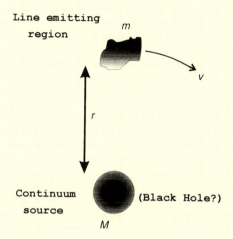

When the UV continuum radiation from NGC 4151 was observed to brighten, there was a 13 day delay before the emission lines brightened too. This means that r is 13 light days, or 3.4×10^{16} cm. The observed widths of the emission lines (e.g. CIV) give $v = 14\,000$ km s^{-1}, or 1.4×10^{9} cm s^{-1}. We now assume that this velocity is the circular velocity required to maintain the cloud in orbit about the black hole. In other words,

$$\frac{mv^2}{r} = \frac{GMm}{r^2}$$

which simplifies to give

$$M = \frac{v^2 r}{G}$$

and inserting the numerical values above:

$$M = 10^{42} \text{ g} = 5 \times 10^{8} \, M_\odot$$

of its proximity, and there is no evidence at present for radio jets as in the more powerful quasars.

13.5 Our nearest active galaxy: Centaurus A

This famous and beautiful galaxy of the southern skies is a remarkable object to observe at any wavelength. As is evident from the classical optical photograph (figure 13.7), Cen A is a giant elliptical galaxy (optically known as NGC 5128) split into two by a dark dust lane which has properties normally associated with spiral galaxies. The lower photo of Cen A shows the appearance of the supernova SN 1986G in May of that year which caused considerable excitement amongst astronomers.

Cen A's very unusual structure (with properties of both spiral and elliptical galaxies) makes it difficult to estimate its distance accurately. But the supernova,

Fig.13.7 Normal optical photograph of our nearest active galaxy, Cen A, with its famous dust lane, on which is superposed a radio map that shows the giant radio lobes. The lower photo shows a picture obtained at ESO in 1986 when a supernova within the dust-obscured area became visible, allowing an accurate distance estimate to be made. (Supernova photo by ESO.)

whose peak brightness is assumed to be similar to other such supernovae, has changed this and gives a distance of 2–3 Mpc. This means that Cen A is only 3–4 times further away than M31! Superposed on figure 13.7 are the contours of a radio map showing the extended radio lobes discovered in the 1950s. The dust

Fig.13.8 (Upper) Combined high resolution radio (VLA) and X-ray (Einstein HRI) images of the inner jet of Cen A. The X-ray emission is coloured whereas the radio is white. (Lower) The long-term variability of the X-ray source is shown by the Ariel V light curve (inset) obtained in the 1970s. (Figures by Eric Feigelson and Ethan Schreier.)

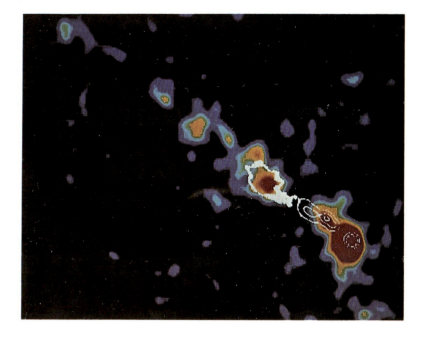

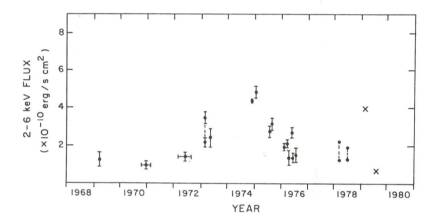

lane defines a plane which is approximately perpendicular to that of the radio jets. Using the Einstein Observatory and the VLA it has been possible to map out both the X-ray and radio emission from Cen A with unprecedented accuracy. As an outlying member of our Local Group of galaxies, Cen A is the nearest bright X-ray and radio emitting active galaxy, which means that we can observe detailed structure in the nuclear regions at the highest intrinsic spatial resolution.

These X-ray and radio images (figure 13.8) reveal similar structure, with close correlation of details such as knots. In fact, Cen A is the only object close enough for such detail to be revealed in the Einstein data. We see a one-sided jet, highly collimated in the direction of the giant radio lobe. No jet has been found in the opposite direction. Unfortunately, the relativistic model does not appear to work in this case and other explanations must be sought. One is that the jet oscillates between the two directions and, over a long period of time, deposits roughly equal amounts of energy in both lobes, with the oscillation period rather long. The enormous range of scales operating in Cen A is shown schematically in figure 13.9. It is based on the accre-

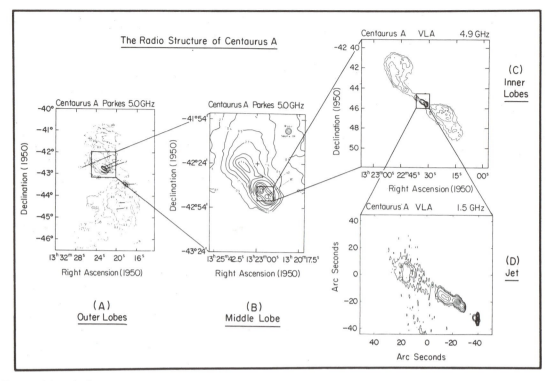

Fig.13.9 Schematic diagram showing the enormous range of scales of activity within Cen A, from the inner jets out to the ~9 degree radio lobes. (Courtesy J. O. Burns, New Mexico State.)

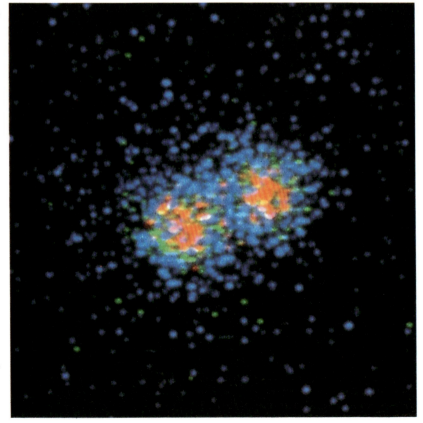

Fig.13.10 Einstein HRI image of M87 showing the X-ray jet. The bright centre of M87 is the left-hand source, the jet to the right. (Courtesy Smithsonian Institution.)

Fig.13.11 General schematic of the features of an AGN on both large (upper) and small (lower) scales. The artist's conception of a low-luminosity radio galaxy shows how the active nucleus drives two jets of oppositely directed beams, thereby creating radio lobes on scales of thousands to millions of light years.

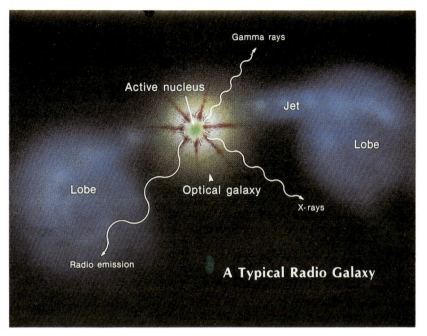

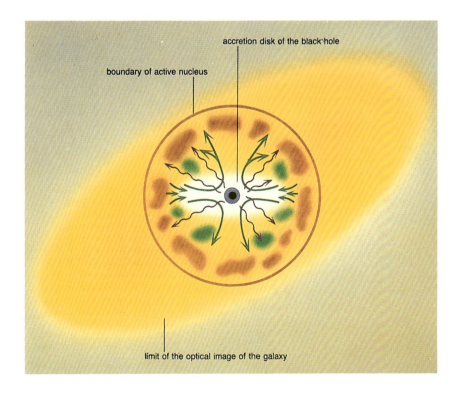

tion onto a massive black hole model developed earlier and illustrates the importance of the collimation mechanism.

The only other active galaxies in which Einstein was able to resolve X-ray jets, and which are undoubtedly similar to Cen A, are 3C273 and M87, the latter of which is shown in figure 13.10. All these elements can now be assembled into a *picture* of an active galaxy, which is drawn in figure 13.11.

13.6 Evidence for an accretion disc in AGN

This *picture* clearly represents an exotic and violent image of activity in quasars and related objects. The accretion disc is needed to provide a reservoir of material for the compact object and, perhaps, to aid in the jet

Fig.13.12 Overall (ultraviolet to infrared) spectrum of a QSO, PKS 0405–123, showing the complexity of the continuum shape. Below the observed spectrum are the individual components that are believed to exist in AGN spectra: the power law from the central object, and thermal contributions from different parts of an accretion disc (lower diagram). The sum of these components fits the observations very well and provides a natural explanation for the origin of the big blue bump as the accretion disc. (Diagram based on an original by Matt Malkan, UCLA.)

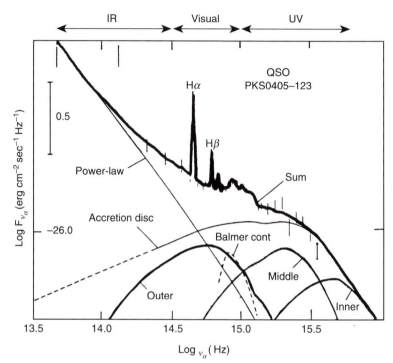

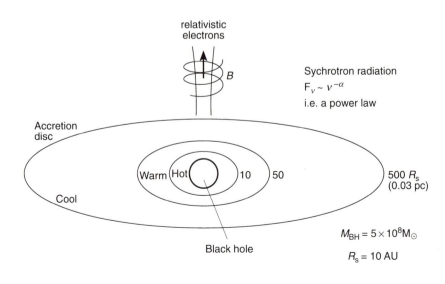

collimation procedure initially and to maintain its orientation over a long period of time. Such a disc will be much more massive than those found in galactic X-ray binaries as discussed in earlier chapters. Is there any direct evidence for the existence of massive discs in AGN? There is, and it comes from examining the overall shape of the spectrum of a quasar from radio through to X-ray wavelengths; information that again has only become available very recently. A typical quasar spectrum, that of PKS 0405-123, has been assembled by Matt Malkan (UCLA) and is shown in figure 13.12.

The radio region of the spectrum extends into the infrared and is plainly well represented by a power-law formula. As we have learnt earlier (see chapter 3) this is the signature of the synchrotron emission process and indicates the presence of relativistic electrons spiralling in magnetic fields. It is primarily associated with the jets and the extended radio lobes. However, the visual and ultraviolet part of the spectrum clearly deviate from the extension of this power law and represent a separate emitting component. This section of the spectrum peaks at around 3000 Å and is often referred to as the *big blue bump*. To account for this component Malkan developed a model of the light output from an accretion disc based on theoretical ideas laid down in the 1970s by Jim Pringle (Cambridge), Kip Thorne (Cal-Tech) and others. As shown schematically in figure 13.12, a disc surrounding a 500 million $M_\odot$ black hole can be divided into three main regions: an outer, cool edge, a warm region in the middle and a hot inner component. When these three components are added together and superposed on the power-law extrapolation, Malkan was able to fit the overall spectrum remarkably well. To account for the total energy output a mass accretion rate through the disc of about 10–20 $M_\odot$ per year is needed. This ultraviolet hump is perhaps the most compelling observation in favour of the accretion disc model.

13.7 BL Lac objects

The final group of active galaxies that must be introduced are the enigmatic and oddly-titled BL Lac objects. The prototype, BL Lac, is a bright (~13 mag), variable *star* in the constellation of Lacertae that has been known for more than 50 years! The large amplitude (more than 2 mags), irregularity and short timescale of the variations (figure 13.13) are classical properties of this group.

However, the physical nature of BL Lac has remained a mystery until very recently, because the optical spectrum of the star is completely *featureless*. A pure continuum spectrum such as that shown in figure 13.13 is extremely unusual and, of course, gives no direct clues to the nature of the object. Additional properties of BL Lac and similar objects include the polarization of the visible light, and their association with strong radio sources. The discovery of polarization has led to their sometimes being referred to as *blazars*.

With no spectral features of any kind to measure, until the mid-1970s even the distance and hence luminosity of any member of this class were unknown. However, deep photographs of BL Lac revealed a faint *fuzz* around the star which was observed spectroscopically by Joe Miller at Lick in an extremely difficult piece of work (due to the bright nucleus). The spectrum was found to be that of a giant elliptical galaxy at a redshift of $z = 0.07$. Several other BL Lac objects have been associated with elliptical galaxies in the same way, thus making them analogous to Seyfert galaxies which are essentially normal spiral galaxies that have a bright nucleus. But the absence of emission lines is a puzzle that is still not understood. The most widely accepted model of a BL Lac object is that it is an AGN in which the jet is pointing directly at us! Relativistic effects will enhance the apparent brightness of the continuum radiation and thereby swamp the outer emission line regions. However, confirming this hypothesis is proving to be very difficult.

(a)

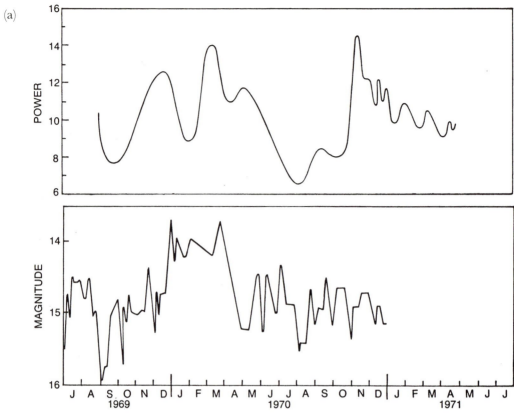

(b)

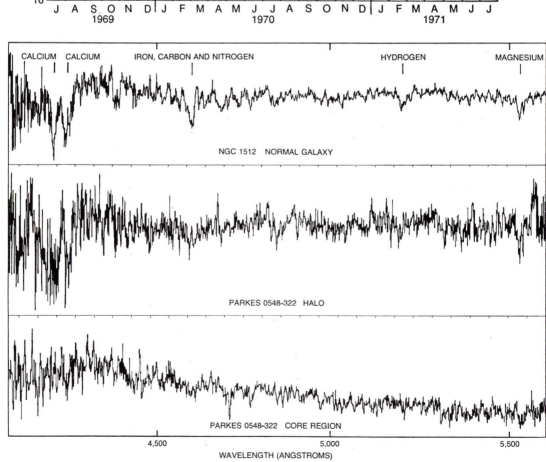

Fig. 13.13 (Facing page) The properties of BL Lac objects: erratic variability (a) and a featureless spectrum (b). The variability is demonstrated with BL Lac itself, the upper panel showing the 2.8 cm radio output, the lower panel the corresponding optical light curve. The spectrum of PKS 0548–322 shows how hard it is to unravel the nature of the beast. The features of a normal galaxy (NGC 1512) are just detectable in the halo of PKS 0584–322, but the spectrum of the core region is featureless! (From diagrams by M. J. Disney and P. Véron.)

13.8 Towards the central engine: X-ray variability

As we learnt in the introduction to this chapter, the optical variability of AGN has been known and studied for some years. Indeed, BL Lac was a well-known variable star long before its extragalactic nature was realised. Variations on timescales from years down to days clearly indicated that AGN were exotic in some way because of the small volume in which this prodigious amount of energy must be produced. Indeed, X-ray variability on such long timescales was first discovered by Ariel V in the 1970s for Cen A (see figure 13.8) and other bright objects such as NGC 4151. However, these variations represented a physical study of the outer accretion process as indicated by the long variability timescale.

To probe what was going on near the central compact object required a search for variations on much shorter timescales. But how short? Variations in a bulk region of emitting gas cannot occur faster than the light travel time across that region. And this of course must all be outside the event horizon of the black hole. From this we expect the shortest variations to be on a timescale of about $50M_6$ seconds, where M_6 is the mass of the central object expressed in millions of solar masses. Hence for a 100 million $M_\odot$ black hole we should look for 1–2 hour variations.

The discovery of intense X-ray emission from the whole gamut of active galaxies therefore opened a new window for probing even closer to the compact object itself. However, the X-ray astronomy missions of the 1970s revealed few clear indications of X-ray variability in AGN, especially on short (hundreds of seconds) timescales. One of the exceptions was NGC 6814, the HEAO-1 data from which is shown in figure 13.14.

Indeed, even as recently as 1983, it was thought that, in general, AGN did not vary rapidly! However, the nature of X-ray astronomy missions up until that time tended to make this a self-fulfilling statement because they were limited in a fundamental way for such variability studies. They were all in near-Earth orbits of period around 100 minutes which, together with the constraints of Earth occultation and charged particle regions, limited observing efficiency to around 40% with no chance of long uninterrupted observations. EXOSAT's 4 day orbit, which offered the opportunity of 90 hours *unbroken* on a single object has, in its 3 year lifetime, produced the first clear records of time variability in processes occurring near the compact object in AGN.

13.9 The EXOSAT contribution

Fig.13.14 Rapid X-ray variability in NGC 6814 on a timescale of minutes seen by HEAO-1.

These unbroken observations within one EXOSAT orbit of almost 4 days are best referred to as *short* timescales, so as to distinguish them from the *long* timescales between individual observations of particular sources (weeks to

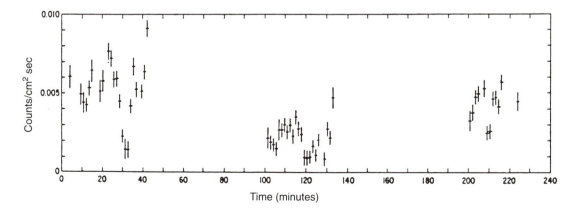

months to years). Unfortunately, the *long* timescales are naturally the hardest to investigate thoroughly with any space mission and much remains to be done in this area. However, the variations on long timescales seem to be due mainly to changes in the mass accretion rate. This is because the changes are essentially the same at all X-ray energies, implying that the spectrum remains roughly constant. The only exception to this rule is NGC 4151 which was constant in EXOSAT's LE telescope, but highly variable in the ME. This suggests that the soft X-rays are being produced in a much larger volume, such as a hot medium between the broad line clouds (see figure 13.11).

But it was on the short timescales employing EXOSAT's unbroken observations where the major contributions have come (figure 13.15). To demonstrate more clearly the problems facing other X-ray missions in conventional near-Earth orbits, we have produced alternative plots of the NGC 4051 and 3C273 observations which simulate the restricted view of the *100 minute orbit* and show how much more difficult it is to ascertain the true nature of any variability that is present. This may well account for some of the *periodicities* that have been discovered in AGN light curves, so their reality is doubtful at best.

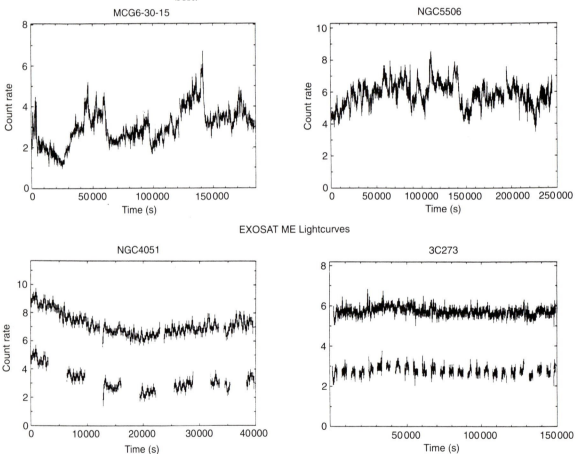

Fig.13.15 EXOSAT long looks *with the ME detectors reveal variability on minutes to hours in three of the four AGN shown here (only 3C273 remained steady on this timescale). The lower (interrupted) curves which are plotted for NGC 4051 and 3C273 demonstrate what the data would look like to a near-Earth orbit spacecraft observing for the same time period, the gaps being produced by Earth obscuration of the source (see also figure 8.5). NGC 4051 and MCG 6–30–15 were bright enough in the LE array for EXOSAT to determine their soft X-ray variability as well. (Diagrams courtesy of EXOSAT Observatory, ESA.)*

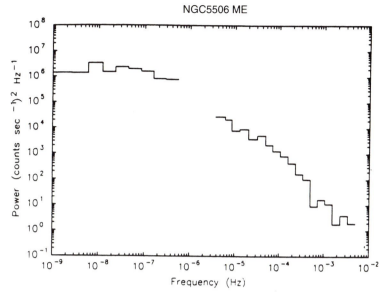

Fig. 13.16 Power spectrum (see chapter 8) of NGC 5506 performed on the data of figure 13.15. This shows that there is no characteristic frequency of variability, but on the lowest frequencies (longest timescales) there is a flattening that is similar to that seen in Cyg X-1. (Diagram by Ian McHardy, Oxford.)

Are there any indications of a *characteristic* or *fastest* timescale in these light curves? To look for this we must calculate the power spectrum of the source, exactly as was done for the study of QPOs in X-ray binaries (chapter 8), and this is shown in figure 13.16.

No such timescale appears and it is not possible to constrain the black hole mass in this way. But for NGC 5506 (figure 13.16) there appears to be a flattening at the longer timescales. If the *knee* in this plot occurs at a frequency of a few $\times 10^{-7}$ Hz (around 20 days) then we can scale the result relative to Cyg X-1 which shows the same effect. This implies a central object of less than a million $M_\odot$! In this case the timescale is due to variations in the disc itself, and is not the light-travel time across the regions closer to the black hole.

A few of the sources are bright enough at low energies to enable their variability to be examined with the LE telescopes. The light curves of MCG6–30–15 and NGC 4051 (also in figure 13.15) show that there is less variation at high frequencies (i.e. short timescales) in the LE band, compared to their ME counterparts. This is exactly what would be expected if the soft X-rays originated in a larger volume. Perhaps this is simply a less extreme analogue of the situation in NGC 4151. One of the most important observations for future X-ray observatories in this field will be to search for a time lag between the low and high energy X-rays. This would be the key to the actual emission mechanism and a probe of the geometry near the central engine.

13.10 Quasars

There has been very little if any significant variability seen in quasars on short timescales. Indeed, figure 13.15 shows the remarkable steadiness of 3C273 during an EXOSAT observation. It is on longer timescales (typically 6 months) that variation has been detected. As quasars are more distant and hence more powerful, this might seem to be what should be expected. The higher luminosities require a more massive central object, or higher accretion rate, or both! This would naturally lead to a larger length scale for variability and hence nothing on short timescales.

But the picture is not quite as bleak as this. The quasar IIIZw2 usually showed no variation, but on one occasion exhibited an X-ray flare. This suggests that in this quasar and 3C273 the normal *quiescent* X-ray emission is provided by a very large jet. Only when this is disturbed do we see flares

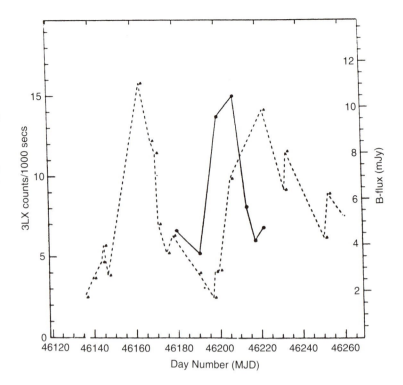

Fig. 13.17 X-ray and optical light curves of OVV1156+295, an optically violent variable quasar. The blue light (dashed curve) was monitored for almost 4 months, during which EXOSAT (solid curve) observed the source seven times in a month and a half using the soft X-ray telescope. The X-ray flare is followed one week later by an optical brightening, probably as a result of variations in the nuclear jet. (Diagram by Ian McHardy, Oxford.)

associated with the shocks so produced. As in SS433 these jets are collimated by a thick accretion disc, which then generally obscures the central source from our view. This does not happen in the lower luminosity sources. Also, the optically violent variable (OVV) quasars (believed to be very similar to BL Lac objects) have shown correlated variations in the optical and X-ray regions with delays of a week or so (figure 13.17). An optical outburst in OVV1156+295 occurred about 1 week after an X-ray flare, which suggests that shocks in relativistic jets appear first in X-rays and only later at optical wavelengths after the shock has expanded. However, there is insufficient coverage of the source's variability to be sure of this interpretation, and future multi-wavelength programmes must attempt to verify this behaviour and probe it in more detail.

13.11 X-ray spectra

Given that the Uhuru satellite was only just able to detect the brightest AGN (3C273) it might be thought that little detailed information could be obtained on the X-ray spectra of AGN because of their faint signals. However, the EXOSAT and Ginga X-ray detectors were typically 100 times larger than Uhuru, thereby greatly increasing their sensitivity, and enabling them to obtain spectra of faint sources. There are a number of important features that were revealed by these spectra, but one of the most remarkable is the uniformity of their slope. The continuum emission is a power law with an index of 0.7 and this is found to be true for objects from low luminosity Seyferts to very high luminosity quasars, covering a very wide range of luminosities (more than a factor of 10 000!). Because of this overall similarity amongst many objects, Ken Pounds (Leicester) and his collaborators were able to improve the statistical quality further by producing an *average* Seyfert X-ray spectrum using results from Ginga observations of 12 different objects. This spectrum, called Ginga-12 (see figure 13.18), shows the structure of the iron line at 6.7 keV very clearly.

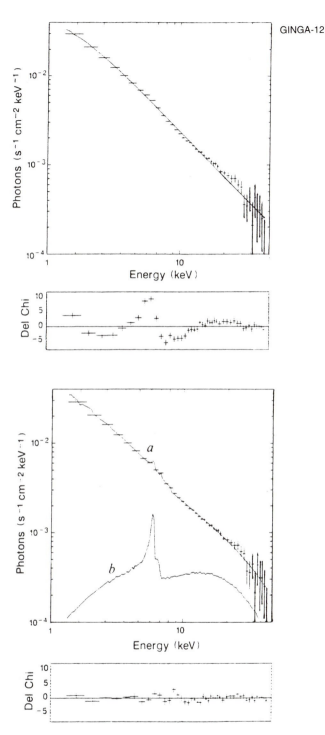

Fig. 13.18 Ginga-12, the average of the X-ray spectra of 12 AGN obtained by the Ginga satellite. A simple power law (top) leaves residuals that have considerable structure around the iron line at 6.7 keV. The lower panel shows the more sophisticated model involving partial absorption and a reflection component. (Based on a diagram by Ken Pounds, Leicester.)

This immediately tells us that we are dealing with similar emission mechanisms from the central engine in all these AGN. The presence of the jets means that shocks within them can heat up electrons to relativistic energies. Hence, synchrotron radiation results when the electrons encounter the magnetic fields which the radio observations show must be there. However, the presence of copious amounts of low energy (optical, infrared) photons, means that Compton scattering must also contribute significantly (in which the energetic electrons *upscatter* these low energy photons to X-ray energies). Unfortunately, there are few observations that enable this mechanism to be tied down any more precisely at present. Nevertheless, there are some other features in these spectra:

(i)   most have a higher absorbing column than expected on the basis of other indicators (direction of view out of our galaxy; optical reddening of the nucleus). This suggests that much of this material is intrinsic to the object, and not between the AGN and us;

(ii) between a third and half of the AGN show a soft X-ray *excess* in the spectrum. This is surprising given the amount of absorption inferred from the higher energy continuum;

(iii) the spectra are of good enough quality to detect the presence of iron emission between 6 and 7 keV. Surprisingly, the typical energy of this line was 6.4 keV, which is *below* the energy expected of neutral iron. This indicates that the emission cannot be thermal in origin, as the very hard

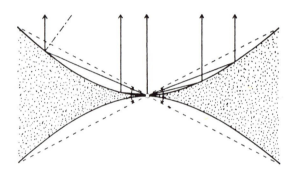

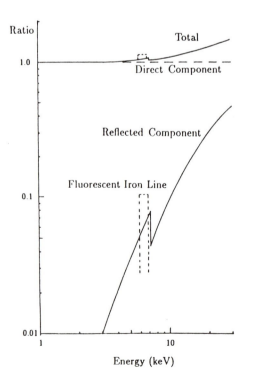

Fig. 13.19 Schematic model of the X-ray emission regions within an accretion disc surrounding a massive black hole. The central source provides the power-law continuum and ionises sufficient material to produce partial absorption of the X-rays. Additionally it invokes iron line emission further out in the disc by the fluorescence mechanism. The lower diagram shows the main spectral components relative to the direct component. This is the basic model used in fitting the Ginga-12 spectrum of figure 13.18. (Diagram by H.Inoue, ISAS.)

X-ray spectra imply very high temperatures which would lead to iron emission at energies *above* 6.7 keV, not below it.

There are also subtle features in the continuum which require more than just the canonical power law to explain them. The presence of the soft X-ray excess indicates that *normal* absorbing material cannot be used to fit the spectrum. The luminous central source will partially ionise the surrounding gas, thereby making it less effective as an absorber. Hence, it is called a *warm* absorber. This surrounding gas, which contains iron (possibly at much higher abundances than solar), will also *fluoresce* under the influence of the central source. Combined with additional absorption on its route out of the AGN it is possible to account for the wavelength of this iron fluorescence being longer (i.e. at a lower energy) than expected from neutral iron. (Another more exotic idea is to appeal to the gravitational redshift expected if the emission is close to the event horizon of the central black hole.)

The Ginga-12 spectrum can be fitted well by such a model (also shown in figure 13.18). Recent theoretical work by H. Inoue (ISAS) shows that this emission is exactly what would be expected from a thick accretion disc! Figure 13.19 contains a schematic model of the emission from the central source illuminating the inner face of the disc which produces the fluorescence, together with the calculated spectrum.

13.12 An analogy with spectra of X-ray binaries

Katsuji Koyama (Nagoya University) has taken this idea further in order to account for the variety of X-ray spectral features around the iron line seen in Seyfert 1 and 2 galaxies. More interestingly, he calls upon an analogy with the X-ray behaviour of the massive X-ray binary Vela X-1. The upper part of figure 13.20 shows the X-ray spectrum of Vela X-1 at three different orbital phases.

Out of eclipse the spectrum is normal, but as the neutron star is about to be eclipsed by its supergiant companion the X-rays suffer a grazing eclipse and an absorption edge appears. Once the neutron star is eclipsed, only scattered X-rays and a strong iron emission line are visible. In the lower panel are shown X-ray spectra of a Seyfert 1 galaxy (NGC 4051) and two Seyfert 2 galaxies (Mkn 3 and NGC 1068). The strongest iron line is in NGC 1068 in which it is inferred that we do not have a direct view of the central X-ray source, so that scattered radiation dominates, as in the accretion disc corona X-ray binaries (see chap. 8). Mkn 3 is suffering a grazing eclipse by thick absorbing material whereas NGC 4051 is not eclipsed at all. If these clouds are close to the central region of activity then we can expect variability in the spectra and hence in how they would be classified (which has already been seen optically in the spectra of NGC 4151).

Since the central engine and the thick accretion disc combine to produce the jet mechanism, these ideas can be developed further according to whether the X-ray flux is beamed towards us or not. The OVV quasars and BL Lac objects are believed to be so beamed that we are not able to see any emission lines at all. The variability that is seen is related to fluctuations within the jet itself.

13.13 High resolution X-ray spectrum of a BL Lac object

Finally we shall take a glimpse at what may be possible when truly high spectral resolution is available in the X-ray band for observing AGN. Using the Einstein Observatory's objective grating spectrometer (OGS), Claude Canizares (MIT) discovered an absorption feature at 0.6 keV (about 25 Å) in the soft X-ray spectrum of the bright (~13th mag) BL Lac object, PKS 2155–304 (figure 13.21).

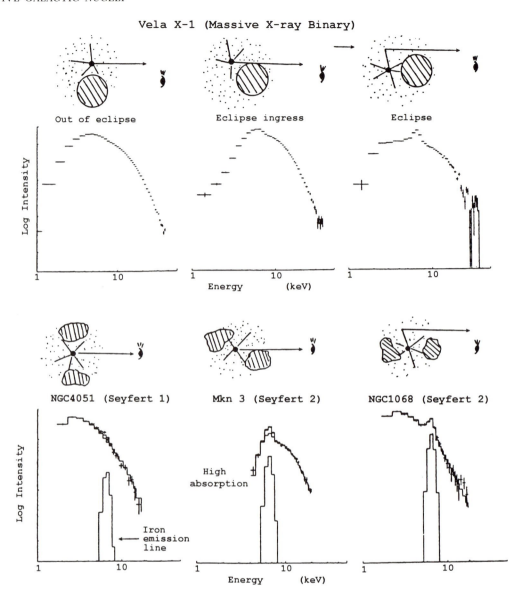

Fig. 13.20 Structure in the iron line seen in Seyfert galaxies and massive X-ray binaries suggests a geometrical interpretation of their properties. Ginga X-ray spectra of Vela X-1 (9 day period) at three different phases shows the iron absorption as the eclipse starts, to be replaced by iron emission (scattering/fluorescence) during the eclipse of the direct source of X-rays. The lower panel then shows three Seyfert galaxies. On the left in NGC 4051 (Seyfert 1) the central source is directly visible, whereas the middle spectrum of Mkn 3 (Seyfert 2) has most of the central source just obscured. Finally the right panel (the Seyfert 2, NGC 1068) exhibits the much stronger iron emission of a presumably totally obscured central source. (Based on an original by Katsuji Koyama, Nagoya.)

The energy of this feature corresponds to that of highly ionised oxygen (OVI) which, if confirmed, is very important in telling us the physical state of the material surrounding the X-ray source. If oxygen is this highly ionised then lighter elements (helium, carbon, nitrogen) will be fully ionised and therefore incapable of absorbing X-rays further. This material is then a *warm* absorber and enables far more soft X-rays to be observed (the soft X-ray

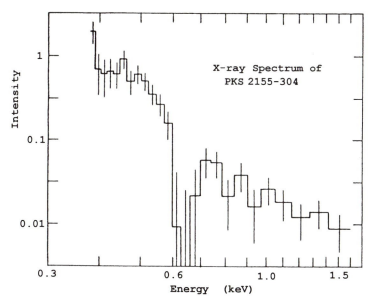

Fig. 13.21 High resolution X-ray spectrum of the BL Lac object, PKS 2155–304, obtained with the Einstein Observatory's objective grating spectrometer. The spectrum is soft but shows a sharp absorption feature at 0.6 keV which is attributed to ionised oxygen close to the X-ray source. (Based on an original by Claude Canizares, MIT.)

excess) than would be expected. However, far more detailed and better resolution observations are required before this interpretation can be taken as secure. The soft X-ray grating experiments on XMM, ESA's X-ray multimirror spacecraft, due for launch towards the end of the 1990s (see chapter 17), will revolutionise this field.

14 Clusters of galaxies

14.1 Introduction

The largest aggregations of matter that can be detected as visible entities are galaxies. The large-scale structure of the universe, as we know it, has been determined by measuring the distribution and motion of galaxies in space. Faint galaxies can be isolated with deep telescopic observations, spectra can be measured, and distances calculated by means of redshifts and the Hubble constant.

The distribution of galaxies is not random. Although on the largest scale, the arrangement of matter in the universe is uniform, on an intermediate level, galaxies are found in gravitationally bound aggregates referred to as 'groups' and 'clusters' (see figure 14.1). These exist in sizes ranging from a few galaxies to ten thousand galaxies. The gravitational potential which binds galaxies within a cluster often also binds a vast cloud of hot gas which fills the space between and around the galaxies. This gas has a temperature of tens of millions degrees. It coexists with the galaxies and, although very diffuse, is a strong source of X-ray emission.

This hot gas was discovered unexpectedly in 1971 through the analysis of X-ray observations. Now, twenty years later, the X-ray luminosities of over 200 galaxy clusters have been measured and the morphology of emission from the brighter clusters has been mapped. Using simple models, the shape of the gravitational potential of the cluster has been derived. The mass of X-ray emitting gas is equal to, perhaps greater than, the 'optical' mass derived from the observed visible luminosity of the galaxies. The X-ray data, in a straightforward fashion, have doubled the amount of observable matter in clusters of galaxies!

More importantly, however, the cluster gravitational potential required to explain both X-ray and optical measurements implies the existence of a large hidden mass. Some data require a factor of 100–300 times more mass than that observed in the form of galaxies or hot gas! (The deeper the gravitational potential well formed by a cluster, the faster the motion of the galaxies within it, and the greater the concentration of hot gas at the centre.) This is the 'missing mass', and the observation of diffuse X-rays from clusters can determine how it is distributed.

Many clusters contain an additional complication, a central dominant galaxy. This is sometimes a giant elliptical galaxy which is considerably brighter and more massive than any other galaxy in the cluster. These are the largest of any known type of galaxy. Some of these central galaxies are

Fig. 14.1 The Coma Cluster, a regular, almost spherical, diffuse source of X-rays about 1 Mpc in diameter. The Einstein IPC X-ray picture, shown in red, has been superimposed on the Palomar Sky Survey, shown in blue. Where both wavebands are bright, the colour is white. Most of the optical objects at the cluster centre are galaxies and can be seen more clearly in figure 14.4, an enlargement of this region. The region shown here is 30 arcminutes in extent. The blue cross is fiducial and marks a foreground star.

actually growing in size by accretion of the diffuse gas. This helps explain both the unusually large mass and luminosity of these galaxies and their central position within the cloud of hot cluster gas.

Because the gravitational potential can be accurately derived from the characteristics of the hot gas, X-ray observations can be used to classify clusters of galaxies in a physically meaningful way. One such classification scheme is based on the principle that clusters should evolve with time. They progress from loosely bound irregular aggregates to dense configurations with strong central potentials. These ideas and the supporting X-ray observations form the subject matter of this chapter.

14.2 Results from optical observations

Our galaxy is a member of a small cluster. The nearest neighbours of the Milky Way are the Magellanic Clouds at a distance of 55–60 kpc. Then, a bit farther away, there are about 10 other dwarf irregular and dwarf elliptical galaxies. These form a swarm of small clouds of stars around a large spiral galaxy (the Milky Way). Together with M31 and its attendant smaller systems we form the 'Local Group', all contained within a volume of radius 1 Mpc. If viewed from a great distance, our group would appear to consist of only the two large spiral galaxies – a binary group.

The nearer bright spirals that make such impressive photographs are also found in groups. Gerard de Vaucouleurs (University of Texas) has catalogued 54 groups at distances less than 10 Mpc. The nearest of these is the Sculptor Group, a loose association of seven spiral galaxies with NGC 253, an edge-on system (discussed in chapter 6 because of its starburst nucleus), being the largest galaxy. Members of this group subtend an angle of 20°, are at an average distance of 2.4 Mpc, and are only visible from the southern hemisphere. Another nearby group is centred on the galaxies M81 and M82 in Ursa Major at an average distance of 2.5 Mpc (figure 14.2). The boundary of this group may extend south toward the galactic plane to include the obscured galaxies Maffei 1 and 2 and about 30 smaller systems. And so, in this manner, groups can be listed containing familiar galaxies extending out to the Virgo Cluster, the nearest large cluster of galaxies (figure 14.3).

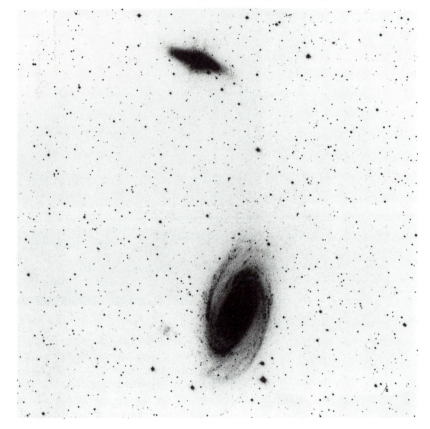

Fig. 14.2 The spiral galaxy M81 and the peculiar galaxy M82, the two principal members of the 'M81 Group' in Ursa Major. The two galaxies are separated (on the plane of the sky) by 38 arcminutes, or a linear distance of 27 kpc. The area of the field is one square degree. (Palomar Sky Survey.)

The Virgo Cluster is centered about 20 Mpc from us and subtends an angle of 12°. It contains at least 100 fairly massive galaxies and a total of perhaps 1000 galaxies. The largest galaxy is M87, the famous giant elliptical and a strong source of radio emission (Virgo A). It has an active nucleus which has been studied extensively because of the bright optical, radio, and X-ray jet discussed in chapter 13. About 40% of all clusters contain dominant galaxies such as this. The role of the dominant galaxy in the evolution of the cluster is an important one.

As one proceeds outward, listing the properties of clusters, major uncertainties arise. First, the fainter galaxies can no longer be seen, so only the brightest members of more distant clusters are catalogued. Second, the Hubble constant, and consequently the distance, is uncertain by a factor of 2. The absolute luminosities of galaxies are not well determined. Indeed, in the absence of spectral measurements, it is usually assumed that the luminosity of the brighter cluster members is the same and this is used to derive the distance. A third difficulty arises because we view a projection of a three-dimensional distribution on the plane of the sky. Membership of many clusters has been determined by only the two-dimensional grouping. Positions in the

Fig. 14.3 Galaxies in the Virgo Cluster. At a distance of 20 Mpc, this is the nearest large cluster of galaxies. This picture shows one square degree of the sky and contains galaxies of many types. The brightest are the elliptical galaxies M84 and M86. (Palomar Sky Survey.)

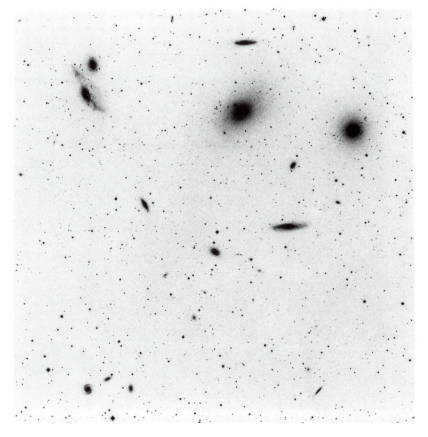

third dimension have, until recently, been lacking. To separate overlapping clusters at different distances, the redshifts of individual galaxies must be known. Because of faintness, it is increasingly difficult to measure redshifts at larger distances. Therefore the problem of overlapping-cluster confusion is most serious for the farthest clusters.

Using only optical data, clusters of galaxies have been catalogued by several authors, most notably Fritz Zwicky and George Abell. In these catalogues the membership of a cluster depends critically on how the cluster is defined, particularly in the outer regions where the density of galaxies in the plane of the sky is close to the background level. Most of the examples here are well-studied massive clusters and there is general agreement about their size and composition. We will refer to most of the clusters using their number in Abell's catalogue. The Coma Cluster, for example, is called Abell 1656 or A1656.

The 'richness' of a cluster describes the number of galaxies included. The richest clusters contain thousands of galaxies, the poorest only a few.

Some clusters are called 'regular'. These are all rich (containing at least a few hundred galaxies), show high central concentrations, and approximately spherical symmetry. They consist almost entirely of E (elliptical) and S0 (highly-flattened elliptical) galaxies, systems without concentrations of dust (which shows as absorption features in photographs of the individual galaxies). The cluster diameters are in the range 1–10 Mpc. The Coma Cluster, for example, is a regular cluster (figure 14.4).

In contrast to these, 'irregular' clusters show little or no symmetry. There is no marked concentration towards the centre although subclustering is often present. The membership contains all types of galaxies including perhaps

Fig. 14.4 The centre of the Coma Cluster, where X-ray emission is most intense. The brightest objects are two giant elliptical galaxies and a foreground star (with diffraction spikes). The objects which appear in figure 14.1 as point sources are seen in this picture to be galaxies. The region shown is 13 × 17 arcminutes in extent. (KPNO 4 metre photograph.)

50% spirals. The Virgo Cluster is a familiar example of this type. Although the diameter is about the same as that of a regular cluster, the mass contained in an irregular cluster is only 10% to 30% of that of a regular cluster.

Spectroscopy of the individual galaxies yields both an average redshift (and consequently the cluster distance) and the motion of each galaxy within the cluster. If the cluster is 'virialised' (see box 14.1), the galaxies follow trajectories determined by the gravitational potential of the cluster. The relative velocities within the cluster can be used to calculate the total mass of material present. The quantity measured, however, is a Doppler shift and only the projection of the velocity along the line of sight is determined. After observa-

Table 14.1 Properties of some bright clusters

Cluster	Type	Dominant galaxy	Velocity dispers. (km s^{-1})	Spiral fraction	Richness[a]	Redshift (z)	Distance (Mpc)
A1367	Early	none	830	0.40	117	0.0215	130
Virgo	Early	M87	660	0.55	—	0.0037	20
SC 0627−54	Intermediate	none	—	—	—	0.051	300
A400	Intermediate	cD?	420	0.42	58	0.0232	140
A2199	Intermediate	cD?	810	0.24	88	0.0305	180
Coma	Evolved	none	880	0.13	106	0.0235	140
A2256	Evolved	none	1255	0.24	88	0.0601	360
Perseus	Evolved	NGC 1275	1284	0.10	88	0.018	110
A85	Evolved	cD	1445	—	59	0.0518	310

[a] The number of galaxies not more than 2 magnitudes fainter than the third brightest member of the cluster.

Box 14.1 The virial theorem and the missing mass

There is a theorem, the 'virial theorem', applicable to any physical system of particles bounded in both space and velocity. It can be gas molecules in a box or a system of gravitationally bound galaxies. The theorem states that the time-averaged kinetic energy of the galaxies is equal to 1/2 the average potential energy. Thus average velocity is related to average gravitational potential which is in turn proportional to the total mass.

In order to derive total (or virial) mass of a cluster from observed velocities of galaxies it is necessary to make assumptions about the shape of the potential and about the orbits of the galaxies. If the motion of the galaxies is radial, with each passing through the cluster centre, then measured velocities close to the centre are close to the actual velocities. (Remember only the velocity component in the line-of-sight is measured as a receding or advancing Doppler shift.) On the other hand, if the velocity vectors are isotropic in direction, then the actual velocity dispersion is greater than that measured. Therefore both the general shape of the potential and the trajectories of the galaxies must be assumed before calculating the mass. When this is done a surprising result is obtained which, as a matter of fact, is not very dependent on the assumed orbits of the galaxies.

It is found that the virial mass is always greater than the mass determined from the luminosity of the stars within the galaxies, the 'optical mass'. This mass discrepancy is large. It can be a factor of 100–300 for large clusters! This is the famous 'missing mass'. The conclusion is clear that there is more mass present than is inferred from the optical luminosity alone.

The cluster mass can also be derived from X-ray observations using a model for the distribution and temperature of the hot gas. The X-ray observations, however, have been more valuable in the determination of the shape of cluster potentials, often indicating structure that is not evident in the optical data.

tion of many galaxies, the velocity dispersion (or range of velocities) is calculated and this is used to derive the mass of the cluster. Table 14.1 list properties for some bright clusters.

14.3 X-ray morphology

The first detections of X-rays from clusters were made with rocket-borne instruments launched by the Naval Research Laboratory. In April, 1965, Herbert Friedman *et al.* found X-rays from M87, and in March, 1970, Gilbert Fritz *et al.* discovered X-rays from NGC 1275. These are active galaxies located in the Virgo and Perseus Clusters respectively and the emission was (correctly) associated with the individual galaxies, not with the clusters as a whole. In March, 1969, John Meekins *et al.* identified the Coma Cluster (which contains no active galaxies) as an X-ray source and realised that this emission was much stronger than that expected from the sum of all the individual (normal) galaxies within the cluster.

Data from the Uhuru satellite, launched in December, 1970, rapidly established clusters as an important class of diffuse X-ray sources: the Coma Cluster was discovered to be an extended source by Herbert Gursky *et al.* (AS&E) in 1971. The Perseus Cluster was found to be extended by William Forman *et al.* in 1973. Edwin Kellogg *et al.*, also in 1973, identified 16 Abell clusters as X-ray sources. All-sky surveys by Uhuru, Ariel V, and HEAO-1 soon identified over 150 clusters as X-ray sources.

The first X-ray picture of a cluster was obtained in December, 1975 when Paul Gorenstein *et al.* (AS&E) pointed a small rocket-borne X-ray telescope at the Perseus Cluster for 2 minutes. This showed both NGC 1275 as a point source and diffuse emission from the surrounding cluster.

Table 14.2 The brightest X-ray emitting clusters

Cluster	X-ray source	Uhuru rate (count s⁻¹)	Redshift (z)	Distance (Mpc)[a]	2–10 keV X-ray luminosity
A426 (Perseus)	4U0316+41	47	0.0183	110	14×10^{44}
Ophiuchus Cluster	4U1708−23	30	0.028	168	25×10^{44}
M87 (Virgo)	4U1228+12	22	0.0037	20	2.4×10^{43}
A1656 (Coma)	4U1257+28	15	0.0235	140	9×10^{44}
Centaurus Cluster	4U1246−41	5	0.0107	65	6×10^{43}
A2199	4U1627+39	4	0.0305	180	3×10^{44}
A496	4U0431−12	3	0.0316	190	3×10^{44}
A85	4U0037−10	3	0.0518	310	8×10^{44}

[a] Assuming $H_0 = 50$ km s⁻¹ Mpc⁻¹

During the years 1979 and 1980 the Einstein telescope was pointed at about 400 clusters and almost 300 of these were detected. The spatial extent of emission from the brighter clusters was mapped and the X-ray properties were compared with optical characteristics. Forman and Christine Jones (CfA) have found that clusters can be classified by their X-ray morphology which, since it shows the form of the gravitational potential, is of obvious physical significance.

Spectral measurements, discussed in the following section, show that the diffuse X-ray emission is thermal radiation from hot gas. An X-ray map is a projection of the gas distribution within the cluster. This gas is contained and shaped by the gravitational potential of the cluster. The clusters appear as irregular blobs, sometimes bright at the centre and all fading into the background at the edges. As with galaxy counts, the outer boundaries of clusters are difficult to determine. The central regions, however, are often bright and the density and distribution of gas in these regions can be calculated using models.

As expected, the regular clusters have a relatively strong potential and high concentration of gas at the centre. X-ray luminosity and cluster richness are strongly correlated. The more galaxies within a cluster, the brighter the diffuse X-ray emission. The irregular clusters show weaker central concentration and sometimes emission associated with individual galaxies or subgroups of galaxies. Irregular clusters which appear to be binary, with two concentrations of hot gas, have been discovered.

The clusters with highest central gas concentrations are those with dominant galaxies. These galaxies always appear at the point of maximum X-ray surface brightness. This phenomenon even extends to poor clusters and groups. Those with dominant galaxies are relatively bright X-ray sources. Tables 14.2 and 14.3 list characteristics of a few well-studied clusters of various types.

Let us first consider Abell 1367, shown in figure 14.5. This is a nearby irregular cluster, a loose collection of (mostly spiral) galaxies. It is a relatively bright X-ray source which fills the 1° Einstein IPC field of view, but the X-ray luminosity is low compared with other clusters. Both the distribution of galaxies and hot gas in A1367 indicate a shallow central potential. The cluster is elongated and contains an active galaxy, NGC 3862, in the southeast which shows as a bright unresolved X-ray source. Two other fainter unresolved sources can just be distinguished in the northwest. In addition to these brighter sources there is evidence for clumpiness of X-ray emission within the cluster. A long HRI observation confirmed this and detected 13 sources of X-rays, each approximately one arcminute in extent. Some are associated with individual galaxies with X-ray luminosities of 10^{41} erg s⁻¹. The sources of

Fig. 14.5 An X-ray picture (red) of the irregular cluster Abell 1367 overlaid on the Palomar Sky Survey picture (blue). The region shown is 45 arcminutes in extent. This is a 7 hour exposure with the Einstein IPC. The active galaxy NGC 3862, a member of the cluster, appears as a bright unresolved source at the southeast (lower left) edge of the cluster.

emission are identified as hot gaseous coronae around these galaxies rather than the sum of individual sources within.

Contrast A1367 with the Coma Cluster, shown in figure 14.1, which is rich and often given as an example of a regular cluster. There are 100 bright galaxies in the central cubic megaparsec. If our galaxy were at the centre of the Coma Cluster, we would see 100 large galaxies within a distance of 0.5 Mpc rather than only M31! Our nearest neighbour would probably be 200 kpc from us and we would see, with the naked eye, a fuzzy patch corresponding to the central region of a nearby galaxy every 20°!

The Coma Cluster is probably a rare specimen and more massive than most; less than 5% of the Abell clusters are as rich. The X-ray emission shows a smooth flat central plateau slightly elongated in the east–west direction. The same shape is derived from optical galaxy counts with the major axis approximately marked by two bright central elliptical galaxies, NGC 4874 and NGC 4889. No enhanced X-ray emission from the vicinity of these galaxies was detected. The large number of galaxies, the high velocity dispersion, the bright X-ray centre, and the high X-ray temperature all indicate a strong central potential. As shown in figures 14.6 and 14.7, the X-ray emission can be well fitted by the model described in box 14.2. This assumes a spherical distribution of hot gas with central density 0.0025 ions cm^{-3}, a 'core radius' of 0.3 Mpc (8 arcminutes) and a gas mass of 2×10^{13} $M_{\odot}$, slightly greater than the mass estimated for the luminous stars in the galaxies. The

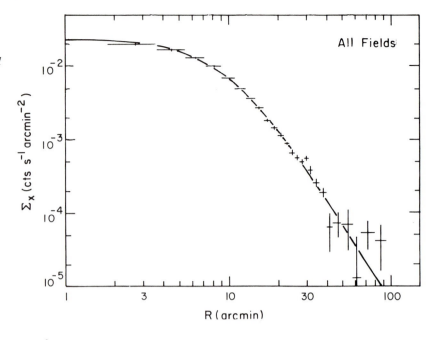

*Fig. 14.6 The observed X-ray
surface brightness of the Coma
Cluster as a function of angular
distance from the centre. The solid
curve shows the brightness
calculated using an isothermal
sphere of hot gas as a model.
Data are from the IPC field
shown in figure 14.1. (Courtesy
of J. Hughes, CfA.)*

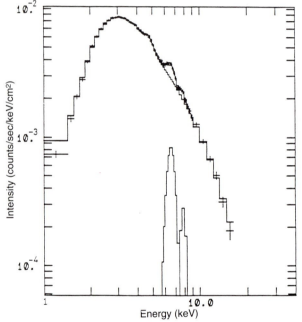

*Fig. 14.7 X-ray spectrum of the
Coma Cluster showing iron line
emission and a continuum, both
indicating a temperature of 10^8 K.
Data are from the Tenma gas-
scintillation proportional counter.
The histogram shows the detector
response to a theoretical spectrum.
The iron Kα and Kβ lines appear
clearly above the continuum and
are also shown separately. The
energy of the iron lines and the
shape of the continuum are both
determined by the temperature.
[Y. Okumura, et al., Pub.
Astronomical Soc. Japan **40**,
639 (1988)]*

value derived for β (explained in box 14.2) is less than 1, implying that the
gas distribution is more extended than that of the galaxies.

Most existing spectral data do not allow us to distinguish between an
isothermal X-ray gas and one in which the temperature varies with radius.
An 'adiabatic' model, in which the temperature decreases with distance from
the centre, is perhaps more realistic, but the isothermal assumption is usually
made for simplicity in calculation. With the exception of 'cooling flows', there
is little observational evidence for a variation of temperature with radius. Jack
Hughes (CfA) has, however, used EXOSAT spectra to show that the temperature of the Coma Cluster gas probably decreases with radius. A model

Box 14.2 The self-gravitating isothermal sphere

This model is commonly used to fit the gravitational potential of clusters. It describes the dark matter which comprises almost all the mass of the system. The nature of this dark matter is unknown. It need not be a gas, but must be distributed throughout the cluster in the manner approximated by the model. The matter is assumed to behave as a gas in hydrostatic equilibrium, with thermal pressure of expansion balanced by the inward directed force of gravity due to the mass of the material.

Two parameters completely characterise the distribution of this material, and consequently the potential: the central density, ρ_0, and the 'core radius', a. The dependence of mass density, ρ, on distance from the centre, r, has been calculated by Ivan King (UCB), and in the central region is

$$\rho = \rho_0 \left[1 + (r/a)^2 \right]^{-3/2},$$

and the two-dimensional projected profile (which we observe) is

$$\sigma = \sigma_0 \left[1 + (r/a)^2 \right]^{-1},$$

where σ_0 is the projected density at the cluster centre.

The X-ray emitting gas floats in, but does not contribute appreciably to, this potential. The X-ray emission is proportional to the square of the gas density, which emphasises the densest part, usually the central region. The observed surface brightness, S, can be described by the form

$$S = S_0 \left[1 + (r/a)^2 \right]^{-3\beta+1/2},$$

where β is the ratio of energy per unit mass in galaxies to energy per unit mass in the gas. For most clusters β is in the range 0.5 to 1.0, indicating that, as radius increases, the density of gas decreases less rapidly than the density of material in galaxies. At the core radius, where $r = a$, the X-ray surface brightness generally drops to 10% – 30% of its value at the centre.

Figures 14.6 and 14.18 illustrate the excellent fits obtained with this model to X-ray data. The core radii derived from such data need not be the same as the core radii of the dark material. The temperature distribution of the gas influences its morphology and determines the connection between the hot gas and the gravitational potential. The model is not expected to fit the outer regions where the gravitational field is weak. The observations are also uncertain at large radii because of difficulty in subtracting the background. The outer boundary of clusters therefore remains an unknown quantity.

Table 14.3 Properties of X-ray-emitting gas in some bright clusters

Cluster	0.5–3 keV luminosity (ergs s^{-1})	Temperature (keV)	Central gas density (cm^{-3})	Core radius (Mpc)	β	Gas mass in cluster ($M_\odot$)	Total mass ($M_\odot$)
A1367	4×10^{43}	3	1×10^{-3}	0.3–0.6	0.53		
Coma	4×10^{44}	7	2×10^{-3}	0.50	0.76	1.6×10^{13a}	2×10^{15}
A2256	5×10^{44}	8	2×10^{-3}	0.45	0.73	4×10^{13a}	6×10^{14a}
Perseus	8×10^{44}	6	17×10^{-3}	0.34	0.57		
A85	4×10^{44}	7	7×10^{-3}	0.26	0.62	3×10^{13a}	2×10^{14a}

[a] Mass within a radius of 0.5 Mpc. The total mass in the cluster is greater.

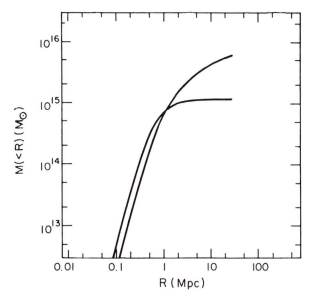

Fig. 14.8 A plot of the radially integrated mass of the Coma Cluster, based on optical and X-ray results. Results from two different models are shown. The scale is 1 Mpc = 24 arcminutes. (Courtesy of J. Hughes, CfA.)

which fits his data uses a temperature of 9 keV in the central region and drops to 3 keV at a radius of 3 Mpc.

Figure 14.8 shows the total mass calculated for the Coma Cluster as a function of radius. This was based on the X-ray profile of figure 14.6 and on the optical measurements. The mass estimates are not very dependent on geometry. All models imply much dark matter at large radii and a total cluster mass in excess of 10^{15} $M_\odot$.

14.4 X-ray spectra

What is the nature of these vast clouds of hot gas which fill the space between galaxies in clusters? What is the composition, density, and temperature of the gas? Are there variations from place to place in the cluster? Why is the gas like it is? The answers are found in spectral observations.

The setting is favourable for spectral measurements. Two common complications which make analysis difficult are absent. First, although the source is physically large (very large indeed!), the gas is thin and is completely transparent to its own radiation. No radiation is self-absorbed and 'hidden' from our view. Second, all elements of the gas are in ionisation equilibrium. The kinetic energy of the positive ions, the kinetic energy of the electrons, and the distribution of ion states are all described by the same temperature in any small element of the gas. The interpretation of spectral data is presently limited only by the instruments.

Continuous variation of temperature throughout most clusters is not excluded by the present data. Most people have found it sufficient, and simpler, to interpret observations with single temperature (isothermal) or two-temperature models. Future observatories will have the capability of measuring spatially resolved spectra accurately for the brighter clusters. This will lead to a refinement of the simple models.

There are two spectral indicators of temperature; the shape of the X-ray continuum, and the energies of characteristic lines. The continuum, at temperatures above a few keV, is almost all from bremsstrahlung. This is generated by the collisions of free electrons with positive ions. The temperature of the gas can be deduced from the shape of the continuum (see box 3.4). At very high temperatures or for pure hydrogen gas there are no X-ray lines and the continuum is the only indication of temperature.

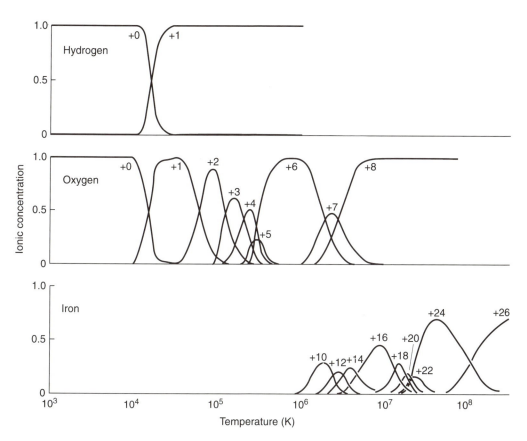

If the gas is of normal composition (similar to that of the Sun) there are characteristic lines in the spectrum. The energy of a line (formed by a transition of a single electron from one energy state to another) identifies the ion from which it is radiated. Since a given ion can only exist in a narrow temperature range, the presence of that particular ion demands a certain temperature if the gas is in equilibrium. As the temperature increases, more and more electrons are removed from the atoms. Ions of low atomic number are the first to be stripped of all their electrons. This is illustrated in figure 14.9 which shows the ion population of hydrogen, oxygen, and iron as a function of temperature.

In September 1975, iron line emission was discovered from the Perseus Cluster by Richard Mitchell *et al.* (MSSL) using proportional counters carried by Ariel V. This established the X-ray emission mechanism as thermal, with a temperature of about 10^8 K. There was also a valuable clue to the origin of the hot gas. If the gas were primordial (created at the time of the 'Big Bang') it would consist of only hydrogen and helium. The detection of iron at roughly one half solar abundance (compared to hydrogen) means that much of the gas has been processed by stars. Thus much of the hot gas has been either stripped from or ejected by the galaxies to form the intracluster medium.

The X-ray spectrum of most clusters is well fitted using a thermal spectrum with a single temperature varying from 2 to 8 keV (20 to 90 million degrees) from cluster to cluster. A good example is that of the Coma Cluster shown in figure 14.7. The data require a continuum with temperature of 8.2 keV (95 million degrees). The energy of the iron line indicates a mixture of FeXXV and FeXXVI. Figure 14.9 shows this is exactly as expected at this temperature.

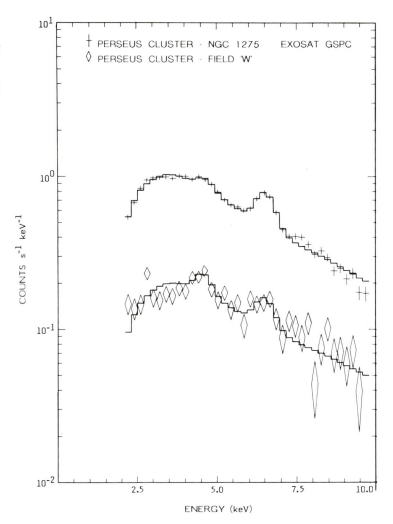

Fig. 14.10 X-ray spectrum of the Perseus Cluster showing line emission from FeXXV and a hard continuum indicating a gas temperature of about 10^8 K. The lower data were taken pointing 30 arcminutes west of the cluster centre. This was done to distinguish cluster emission from that of the central galaxy, NGC 1275. These data are from the EXOSAT gas-scintillation proportional counter. (Courtesy of G. Branduardi-Raymont, MSSL.)

Not all clusters, however, have spectra which can be so fitted. These require an additional low temperature component. Usually there is also strong diffuse emission surrounding a large galaxy at the centre of these clusters. A high density, low temperature gas surrounding a central galaxy is called a 'cooling flow'. Many clusters with bright central galaxies show this phenomenon.

14.5. Radiative cooling in the Perseus Cluster and Abell 85

Let us consider the Perseus Cluster which has an unusually rich (meaning complex and difficult to unravel) X-ray morphology. Figures 14.10 to 14.14 show the observations. The largest structure is due to a mass of hot gas with temperature 7 keV and characterised by a core radius of 8 arcminutes. This is bound by the cluster potential. The parameters are derived by a fit to the X-ray surface brightness between radii of 10 and 17 arcminutes and from the average spectrum of the entire cluster.

At the centre of this, there is a smaller, cooler, cloud of gas with a core radius of 1.2 arcminutes. This surrounds the dominant galaxy NGC 1275. This core radius fits the X-ray surface brightness between 1 and 6 arcminutes. Spectral observations of the centre contain lines which must come from gas appreciably cooler than the average temperature of the entire cluster.

Fig. 14.11 X-ray spectrum of the Perseus Cluster obtained with the Einstein FPCS showing line emission from FeXVII. These results require gas at a temperature of 10^7 K. These data and the results shown in figure 14.10 require gas of at least two temperatures within the cluster. (Courtesy of C. Canizares, MIT.)

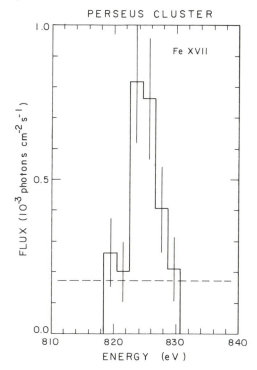

PERSEUS CLUSTER

Fe XVII

FLUX (10^{-3} photons cm^{-2} s^{-1})

ENERGY (eV)

Fig. 14.12 X-ray emission of the Perseus Cluster superposed on the Palomar Sky Survey picture. The red X-ray map is from a 4.5 hour exposure with the Einstein IPC. A bright unresolved X-ray source is centred on the galaxy NGC 1275. Faint emission from the intracluster gas extends at least 10 arcminutes in all directions. This figure is 40 arcminutes on a side.

Lastly, there is an unresolved source of non-thermal radiation coincident with the nucleus of NGC 1275. This is the bright nucleus of an active galaxy. Table 14.4 gives the X-ray characteristics of the Perseus Cluster.

The existence of the inner gas cloud and the spectral evidence for lower temperatures in the Perseus Cluster (and other clusters) form the observational evidence for the existence of cooling flows. When the gas density at the centre of a cluster is high, the radiative energy loss from X-ray emission can have a significant effect on the behaviour of the gas. Energy is radiated, the gas cools, and pressure drops. The gas is further compressed by gravity and somewhat by pressure of the hotter surrounding gas. The gas sinks further into the cluster potential well. The density increases and, since the rate of radiation goes as the square of the density, energy is radiated still more rapidly. This continues until the temperature is low enough for the gas to condense into filaments and presumably eventually stars. This inward motion of the cluster gas is the 'cooling flow'.

Table 14.4 X-ray characteristics of the Perseus Cluster

Component	Size	X-ray luminosity (erg s^{-1})	Temperature (keV)
Hot gas	Core radius 8 arcmin	7×10^{44}	6.4
Cool gas	Core radius 74 arcsec	5×10^{44}	1.0
Nuclear source	Unresolved	1.2×10^{44}	non-thermal

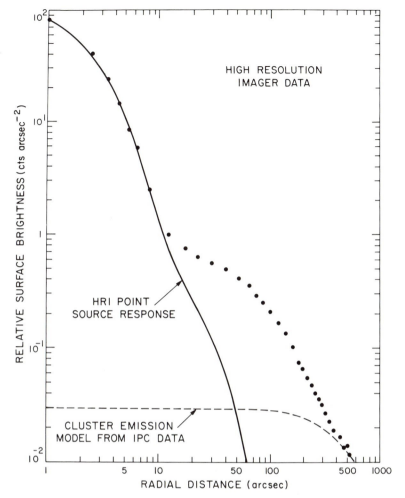

Fig. 14.14 X-ray surface brightness profile of the Perseus Cluster as recorded with the Einstein HRI. The bright nuclear source in NGC 1275 contributes the solid line. The dashed line shows emission from the intracluster gas derived from the IPC observation. The remaining emission, shown by dots, at radii 20–300 arcseconds, is attributed to cooler gas surrounding NGC 1275. It is perhaps associated with the filaments in figure 14.13. [G. Branduardi-Raymont et al., Astrophysical J. 248, 55 (1981).]

At a density of 3×10^{-4} ions cm^{-3} the time required for the hot gas to radiate half its thermal energy is 1.6×10^{10} years. This is comparable to the Hubble time (see box 14.3); and this cooling has no effect on the dynamics of the gas in the cluster. At a density of 3×10^{-3} ions cm^{-3}, the thermal energy is radiated in 1.6×10^{9} years. At a density 10 times higher, the radiation time is 10 times shorter.

The gas density at the centre of the Perseus Cluster is about 3×10^{-2} ions cm^{-3} so cooling occurs on a timescale short compared with the lifetime of the cluster. At radii less than 10 arcminutes the gas density exceeds that required to establish a cooling flow. The inward flow of gas can be calculated but with difficulty. Estimates of this flow for the Perseus Cluster range from 3 to 300 M$_{\odot}$ per year. Consequently there is some disagreement on the importance of the cooling flow. It is, however, clear that there are unusual conditions in the region surrounding NGC 1275.

If the larger estimates are correct, after 10^{9} years this mass inflow will have accumulated to the mass of a good-sized galaxy. There might be an observable accumulation of mass at the cluster centre, perhaps associated with the galaxy located there. This indeed seems to be the case. Figure 14.13 shows optical filaments found in the core of the Perseus Cluster. These filaments are attached to the outer regions of NGC 1275 and are best seen through an Hα filter. They are huge, the size of a galaxy, and are presumably formed from

Box 14.3 The age of the universe and the Hubble time

Looking outward, one looks back into the past, and the further the source, the greater the redshift of the radiation detected. This is due to the expansion of the universe and the rate of this expansion is given by the Hubble constant, H_0. In this book, the value of the Hubble constant has been assumed to be 50 km s^{-1} Mpc^{-1} or, in different units, 15 km s^{-1} per million years of light-travel time.

The farthest one might conceivably see is out to the point where the recession velocity becomes equal to the speed of light, 3×10^5 km s^{-1}. The light-travel time from this point is 2×10^{10} yr. This is the reciprocal of the Hubble constant and is called the Hubble time. Since the speed of light cannot be exceeded, information cannot be received beyond this horizon.

As one looks back in time, before reaching this horizon 20 billion years in the past, one expects to see the moment of creation, the Big Bang. The time elapsed since the Big Bang is the 'age' of the universe, and this age is dependent on the properties of space. If, for example, space is 'flat' the time since the Big Bang is 2/3 of the Hubble time or 15 billion years ago. The age is uncertain because the average density of matter (which determines the 'curvature' of space) is unknown. Present estimates for the age of the universe fall between 10 and 20 billion years.

Fig. 14.15 X-rays from the Virgo Cluster, a mosaic of 40 Einstein IPC fields. Each field is 1 degree square. The brightest galaxy is M87 which is shown in more detail in figure 14.17. (Courtesy of W. Forman, CfA.)

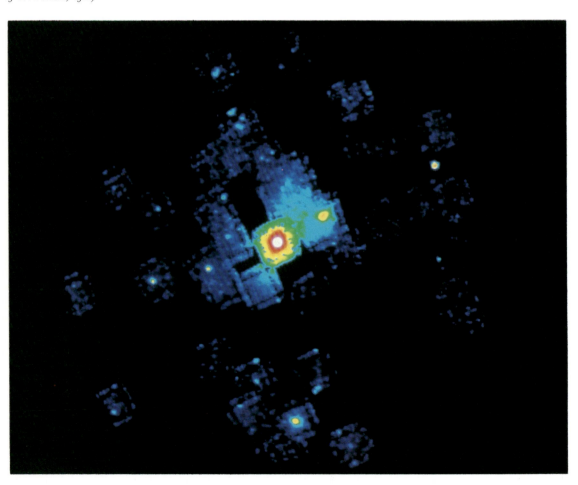

the inflowing gas. Thus NGC 1275, long recognised as a 'peculiar' galaxy, and first described as an 'exploding' galaxy, might actually be an accreting or condensing galaxy.

Abell 85 is another cluster with central radiative cooling. It is similar to the Coma Cluster; rich, relaxed, and with a strong central potential. But, unlike the Coma Cluster, at the centre there is a region of bright diffuse X-ray emission responsible for about 5% of the X-ray luminosity and centred on a very bright galaxy. Here is another possibly-accreting galaxy surrounded by a cooling flow from the hot intracluster gas. Many clusters show X-ray morphology or spectra indicating central cooling flows. In all of these there is a central dominant galaxy. It is not known whether these central galaxies are large because of accretion of intracluster gas or whether they achieved their size through other processes.

14.6 The formation of giant or dominant galaxies within clusters

A 'dominant' galaxy, usually a giant elliptical, is considerably brighter (and more massive) than its neighbours. An excellent and well-studied example is M87 within the Virgo Cluster. Figures 14.15 and 14.16 show a mosaic of Einstein IPC fields centred on M87, which is by far the brightest source in the field. Emission extends 1.5° from the nucleus. Other galaxies within the cluster can be seen as individual X-ray sources but none approaches the strength of M87. There is perhaps faint diffuse emission from the intracluster gas but this is too diffuse to detect reliably in these Einstein data.

Fig. 14.16 Key to figure 14.15 identifying some of the X-ray emitting galaxies in the Virgo Cluster.

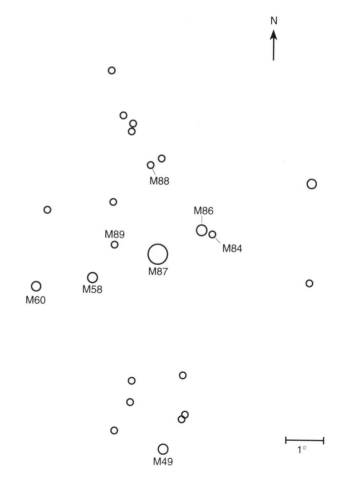

M87 is familiar to us as an active galaxy. It contains a bright radio source with the famous optical jet extending about 30 arcseconds from the nucleus. This nuclear activity has been discussed in chapter 13 and is shown in figure 13.10. The optical galaxy has a measured size of 4 arcminutes on the Palomar Sky Survey and has a faint optical halo extending at least 20 arcminutes from the nucleus. This is surrounded by a vast cloud of X-ray emitting hot gas indicating an even greater extent of gravitationally bound material. The gravitational potential can be derived using the same model used to describe regular clusters. This has been done by Fabricant and collaborators (CfA) with results shown in figures 14.17 to 14.19. In this case the source was large and bright so that X-ray spectra were obtained at different radial distances and the gas temperature was derived as a function of radius. The variation of temperature with position is not large except at the centre where a strong soft component exists. The temperature at the centre is a factor of two below that of the outer regions, an indication of cooler gas.

Fig. 14.17 IPC picture (red) of M87 overlaid on the Palomar Sky Survey (blue). The field is 34 arcminutes square. The central part of M87 is bright at both X-ray and optical wavelengths and is white in this picture. Since both X-ray and optical emission fade gradually into the background, it is difficult to measure the full extent of the galaxy. This figure is meant to give a feeling for the appearance of M87 rather than to accurately depict its size.

Figure 14.19 shows the derived mass distribution. The optical data are from measured velocities of stars in the central region and imply a mass equal to the mass of the stars in the Milky Way within 1 arcminute of the nucleus (table 14.5). The measured X-ray surface brightness, shown in figure 14.18, is very well fitted by the model. The potential derived from the model shows material in the galaxy extending to a radius of 90 arcminutes.

Fig. 14.18 Measured 0.2 to 4.0 keV X-ray surface brightness profile of M87. Data from five Einstein IPC fields were used to extend coverage out to the point where the galaxy could not be distinguished from the background (at a radius of 90 arcminutes). The solid curve is that expected from an isothermal sphere of hot gas with core radius 1.6 arcminutes. [D. Fabricant and P. Gorenstein, Astrophysical J. **267**, 535 (1983).]

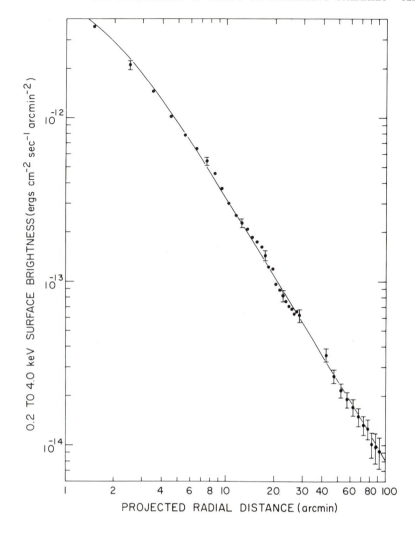

Thus M87 is 10 times as large as our galaxy and 200–500 times as massive. The integral mass to light ratio for M87 (the mass indicated by the potential, divided by the luminous mass seen as stars) increases from about 10 at 1 arcminute to about 200 at 20 arcminutes from the centre. Thus the relative amount of 'dark matter' increases with increasing radial distance. The dark matter forms an appreciable halo surrounding the optical galaxy.

The great mass and the central position of M87 within the Virgo Cluster are not accidental. M87 was probably large when first formed and acquired more mass early in the life of the Virgo Cluster. Even if it had not been formed at the centre of the cluster, collisions between galaxies will have settled this massive galaxy to the centre faster than the lighter ones.

Table 14.5 The mass of M87

Distance from nucleus		Mass
(arcminutes)	(kpc)	$(M_\odot)$
1	4.4	2×10^{11}
90	400	4–10×10^{13}

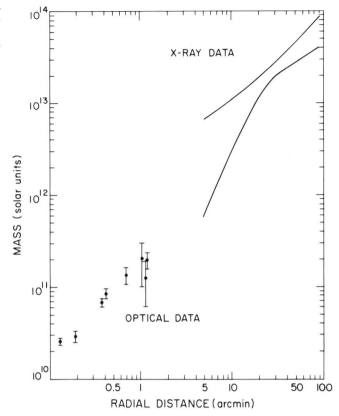

Fig. 14.19 The mass of M87 as a function of radius, incorporating both optical and X-ray results. The allowed mass range indicated by the X-ray data is bounded by the two solid lines. [D. Fabricant and P. Gorenstein, Astrophysical J. **267**, *535 (1983).]*

Additionally, M87 is now accreting cooling intracluster gas, which may have contributed an appreciable portion of its mass.

The brightest galaxies in rich clusters are usually classified 'cD' in a system of nomenclature developed by William Morgan. The 'D' galaxies are 'dustless' systems dominated by amorphous light. The prefix 'c' means supergiant. M87 is a giant elliptical but probably not a cD galaxy.

The cD galaxies are observed to have faint, extensive optical envelopes about a nuclear region which sometimes has multiple components. Some are truly enormous. The cD galaxy in A1413 has low optical surface brightness to a radius of 2 Mpc! Such giant galaxies are only found in cluster centres and, as for M87, cluster and galaxy are intimately connected. The galaxies have grown to their present size by accretion of gas and/or by the merging of smaller galaxies. They can only exist in clusters where conditions are right for such mass accretion. The possibility of the cD galaxies acquiring material from their smaller neighbours has produced the delightful expression 'galactic cannibalism'. The giant galaxies, with insatiable appetites, tear their companions to bits with tidal forces and then feed on the remains, growing fat and prosperous in the process.

Note that the distribution of gas around M87, and other giant galaxies, is determined by the gravitational potential of underlying material. The high surface brightness is caused by the high density of gas trapped by a high mass concentration. The variation of brightness with temperature is a secondary effect.

Other, smaller, galaxies within clusters also show an interaction with the hot intracluster gas. As discussed in chapter 6, many normal galaxies have X-ray halos. This shows the existence of gravitationally bound hot gas and dark matter in outlying regions. Figure 14.20 shows the halo of M86, an elliptical galaxy in the Virgo Cluster. The halo is not symmetrical about the optical galaxy but is stretched out to the north and northwest. This plume is thought to be gas that has been stripped from M86 by ram pressure (or the resistance) of the intracluster gas through which it is moving. The observed velocity of M86, in our direction, relative to the cluster centre is 1500 km s^{-1}. This is high enough to carry the galaxy through the cluster core and into the outer regions in about 10^9 years. While outside the core, interstellar gas will again accumulate in M86, building up until it is removed during the next passage through the core of the cluster.

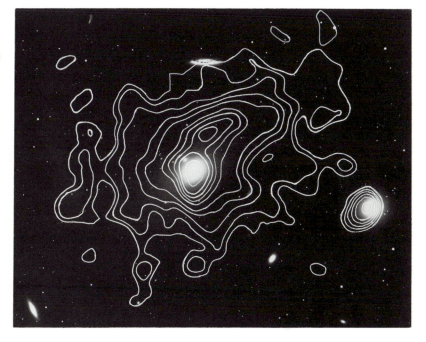

Fig. 14.20 X-ray halos around normal galaxies. Contours of constant X-ray surface brightness from the Einstein IPC overlaid on an optical picture of M86 and M84. Note the large halo of hot gas associated with M86 and the 'plume' extending to the northwest (upper right). These galaxies also appear in figure 14.15. (Courtesy of C. Jones, CfA.)

The galaxy M84, also shown in figure 14.20, has a velocity of less than 100 km s^{-1} within the cluster and is apparently a permanent resident of the core. The ram pressure, although lower than that acting on M86, is continuous and interstellar gas never accumulates. The observed X-ray emission is centred on the galaxy and the extent of the halo is small.

This is another clue to the origin of the hot intracluster gas. Gas stripped from M86 and other high velocity galaxies must diffuse and be incorporated into the intergalactic medium. We know already that this gas has been processed in stars because of heavy elements (e.g. iron) seen in X-ray spectra. The stars evolve in the galaxies and eject heavy elements through winds and supernova explosions. This gas is then removed from the galaxies during their motion within the cluster.

Similar effects of intergalactic gas can be seen in some double radio sources located within clusters. A classical double radio source has two lobes of radio emission located on diametrically opposite sides of an optical galaxy and extending sometimes for hundreds of kiloparsecs. The emission comes from high energy electrons beamed in two jets from the galactic nucleus. Maximum emission usually occurs at the ends of the jets but high resolution observations can often trace the jets back to the nucleus. If the radio galaxy is moving through a cluster containing hot gas, ram pressure of the gas will cause the jets to lag behind the galaxy. There are many examples of trailing radio structure such as that illustrated in figure 13.3.

14.7 Evolution

The different types of clusters can be placed in an evolutionary framework. It is convenient to think of them, and to classify them, in this manner. Bear in mind, however, that data have not yet been gathered at high redshifts. No instrument has yet looked far enough back in time to determine that clusters really do evolve. Present ideas may require modification.

Nevertheless, a young cluster is thought to be irregular. The cluster potential is shallow and intracluster gas is relatively cool and unevenly distributed. Many of the galaxies contain gas and dust. As time passes, the galaxies move in orbits around or through the cluster centre. Some collide with other galaxies and the tidal forces generated in close encounters strip gas from the galaxies. The tidal force of the general potential also helps transfer gas from galaxies to the intracluster medium. This intracluster gas in turn strips gas from rapidly moving galaxies. The galaxies and gas settle toward the centre. The intracluster gas heats up, and the cluster potential deepens. At this point, the cluster has 'relaxed' and the galaxies contain little gas and dust. The cluster properties become those of a regular cluster.

It is important to remember that when the cluster potential changes, this is caused by a change in the distribution of the dark matter. To describe cluster evolution in terms of galaxy motion and interaction implies that the dark matter is associated with the individual galaxies, as indeed at least some of it is.

The speed of evolution depends on the mass. More massive clusters will evolve to a regular state faster than those with lesser mass. The present regular clusters were probably born with a high enough density so that they are now farther along the evolutionary track than the present irregulars. None of the present irregulars for example have the mass of the Coma Cluster. The nearby irregular clusters will not evolve into regular clusters exactly like those already observed. They will evolve to some sort of low-mass regularity.

If a system has reached a stationary state where the distribution of objects does not change appreciably with time, the system is said to be 'relaxed'. If the system starts out-of-balance it will move towards or relax towards an equilibrium state. An example is a system of many interacting particles with different masses. If in equilibrium, energy will be distributed evenly among the different particles. There will be 'equipartition' of energy. If the particles are atoms in a gas, the equilibrium velocity distribution will have a characteristic form called 'Maxwellian'. If there are processes which dissipate energy the system will move from a state of high internal energy to one of lower energy.

A cluster of galaxies is expected to relax in steps. First the general distribution of galaxies and the cluster potential will come to a quasi-final configuration. The process which brings this about is that of 'violent' relaxation (a curious combination of words). This is a collective effect in which galaxies interact with the rapidly changing potential created by their neighbours. This slows rapidly moving galaxies faster than individual collisions and causes the core to collapse on a timescale of about 10^9 years for a rich cluster. The process is called 'violent' because the change of cluster potential is rapid compared to the timescale of other processes. There is not a great deal of banging together of individual galaxies. Although the general potential at this point is now close to its equilibrium configuration, the galaxies are not.

A cluster with many galaxies and in equilibrium should have equipartition of energy among the galaxies. The more massive galaxies should move more slowly than the less massive. Furthermore, the more massive galaxies should have gravitated towards the cluster centre so that the relative number of heavy galaxies is higher at the centre than at large radii. Although attempts have been made to observe these effects, at this time there is no conclusive evidence that any cluster shows equipartition of energy.

The galaxies have not yet come to equilibrium because the timescale for relaxation is large. A galaxy moving at 1000 km s^{-1} will take 10^9 years to pass through the core of a dense cluster and 10^{10} years to travel the entire extent of the cluster. The average time between collisions is about 10^9 years. The 'two-body relaxation time', during which collisions can change the velocity distribution appreciably, varies from about 10^9 years for heavy galaxies in the cores of regular clusters to over 2×10^{10} years for irregular clusters. It is not surprising that most of the galaxies observed in clusters do not appear to be far along the track of dynamic evolution.

In the about 400 images of clusters obtained with the Einstein telescope, there are several which appear to be double. These are not accidental super-positions of clusters at different distances but single clusters with bimodal mass distributions. The total mass calculated is about the same as that in the regular clusters. The double clusters are apparently in an intermediate state of evolution: between unevolved, irregular clusters, and evolved, regular clusters. Numerical modelling studies tend to confirm this. In these studies the motions of hundreds of test-galaxies in a gravitationally bound system are followed with a computer. The test-galaxies, started with random distribution, first form small groups. Then, about 25% of the time, they form a bimodal arrangement before finally relaxing to a cluster with high galactic density at the core.

The X-ray images also suggest that clusters may be divided into two parallel streams of evolution, corresponding to those with and without central dominant galaxies. Clusters with dominant galaxies tend to have smaller core radii and higher central gas densities than those without. Figure 14.21 shows surface brightness contours of clusters belonging to these two families and includes unevolved, intermediate, and evolved clusters.

Perhaps the only way of 'seeing' cluster evolution is to study distant clusters. As redshift (z) increases, the light comes from further in the past. At early times, clusters are expected to be less X-ray luminous since, as a cluster evolves, both the gas content and the central potential should increase. P. Henry (University of Hawaii) and colleagues have compiled a sample of X-ray emitting clusters with $z = 0.2$ to 0.5. They found a luminosity function compatible with that of the nearby clusters and concluded that no evolutionary trend was evident. The sample, however, was not uniform since only the brightest clusters were detected. The test is difficult, even with the best observations, because none of the nearby clusters is fully evolved. All clusters seen are just starting, or at an intermediate stage of, their evolution.

14.8 Present status

With X-ray telescopes, the most massive aggregation of matter that can be seen as a single entity is a cluster of galaxies. The X-ray morphology follows the underlying gravitational potential. The total mass within the cluster can be derived with the help of a model specifying the radial distribution of gas density and temperature. In the central region, generally 10% of the mass is

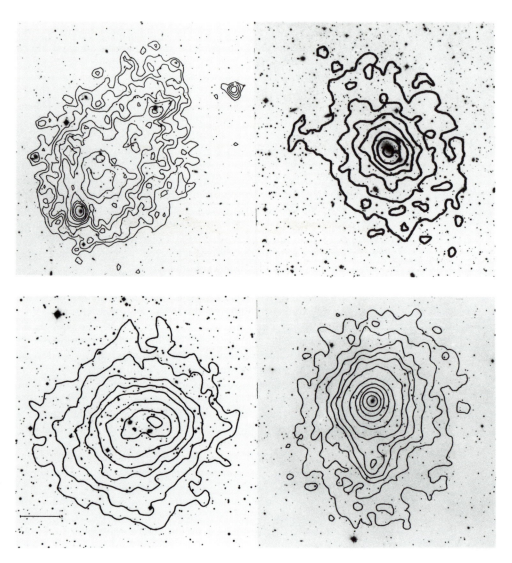

Fig. 14.21 The evolution of clusters. Contours of constant X-ray surface brightness are overlaid on optical pictures from the Palomar Sky Survey. A1367 at top left is unevolved, A2256 at bottom left, and showing greater central concentration of mass, is more evolved. Both clusters at the right have dominant galaxies. A262 at top right is less evolved than A85 at bottom right. (Courtesy of C. Jones, CfA.)

in the form of visible galaxies, 10% is hot intracluster gas, and 80% is invisible. In the outer regions, where measurements are more difficult, the fraction of invisible matter is even higher. The nature of this 'dark matter' is unknown and, needless to say, of the greatest astrophysical interest.

The observations are consistent with an evolutionary scheme in which the density of matter in clusters increases with time. Clusters progress from a loose aggregation of galaxies to a more tightly bound system. The less massive clusters will probably never achieve the density reached by the more massive ones. The galaxies supply the gas which eventually permeates the cluster. About 40% of clusters contain a dominant galaxy at the centre which both affects the evolution of the cluster and might grow appreciably by accreting the cluster gas.

At this point, several models have been developed which agree with the observations. More observations are needed to distinguish between these. Measurements of gas temperature as a function of radius are necessary to better model the gas behaviour and to derive more accurate potentials. Observations of clusters out to distances of at least $z = 1$ are vital to

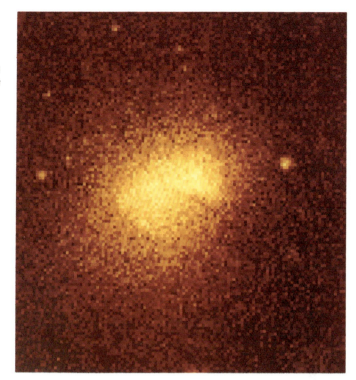

Fig. 14.22 ROSAT PSPC X-ray picture of A2256 which is now seen to consist of two lobes, perhaps in the process of merging. Compare this observation with the contours in figure 14.21. Note the weak serendipitous sources in the field. The picture is 32 arcminutes on a side. (From the ROSAT Archive.)

determine the nature of the evolutionary process. We look forward to future instruments which will produce great advances in both these areas.

15 X-rays from planets

15.1 The production of planetary X-rays

Planets are small and, compared to the 'cosmic' subjects of earlier chapters, extremely weak sources of X-rays. Nevertheless X-rays have been detected from the Earth, the Moon, and Jupiter. These results have been scientifically useful and some even surprising. As in the visible band, all the planets shine with reflected solar energy. Some also have magnetospheres which provide a mechanism for generating a locally powered source of X-rays. Thus planetary X-rays carry information about the reflecting atmosphere or surface, and sometimes about the aurora.

In 1967, solar X-rays scattered from the top of the Earth's atmosphere were detected with rocket-borne proportional counters by J. Harries and R. Francey (Universities of Adelaide and Tasmania) and by Rod Grader and colleagues (LLL). This X-ray glow from the sunlit Earth produced a considerable background in the detectors. At energies below 1 keV, it completely ruined the ability to detect the fainter cosmic sources. After this, rockets carrying soft X-ray detectors were launched at night when the sunlit atmosphere was not visible to the detectors.

Most incident solar X-rays are absorbed through the photoelectric effect by atoms and molecules in the upper atmosphere. Only a small fraction are scattered, either by individual electrons or collectively by all the electrons bound in an atom. These scattered X-rays produce most of the signal in a detector looking down on the top of the sunlit atmosphere. The incident solar spectrum is quite soft so the intensity of scattered X-rays is highest at low energies. There are also X-rays at specific energies from ionised atoms which recombine to emit fluorescent radiation.

The Earth's atmospheric composition up to 100 km is about the same as at sea level: 78% molecular nitrogen, 21% molecular oxygen, and 1% argon. Above 125 km atomic oxygen is the main constituent. Solar X-rays penetrate to an altitude strongly dependent on the X-ray energy. The layer of the atmosphere 'seen' by a detector above it depends on the energy band of the detector. In the 0.2–1 keV band, observable scattering occurs at altitudes of 100–120 km; and in the 1–10 keV band, at altitudes of 70–100 km (table 15.1).

Table 15.1 Solar X-ray penetration of the atmosphere

X-ray energy	(keV)	1	5	20
Minimum height	(km)	100	75	50

Auroral particles are a second source of atmospheric X-rays. Electrons precipitating into the atmosphere generate bremsstrahlung and some fluorescence. This precipitation occurs in the auroral zones. On Earth these form two ovals at 65° to 70° magnetic latitude circling the north and south magnetic poles. Here the magnetic field lines connect with the boundaries of the radiation belts where energetic electrons are stored. The precipitating electrons are not very energetic and optical aurorae occur at altitudes of 100–300 km. None of the electrons penetrate to altitudes less than 80 km.

X-rays generated by these particles, the auroral-electron bremsstrahlung, also have a soft spectrum. These X-rays should be detectable from a point directly above the source. The measurement, however, is unusually difficult with detectors carried by sounding rockets or near-Earth satellites. At the time of the measurement the instrument is embedded in the flux of electrons causing the aurora. These particles generate X-rays within the detector which can overwhelm the atmospheric X-rays of interest.

If viewed from below, most X-rays are absorbed by the atmosphere. There is, however, a hard X-ray component with energy above 25 keV which can penetrate to 40 km. This altitude can be reached by balloons. For several years starting in 1957, balloon-borne detectors were used by Kinsey Anderson (UCB) and others to measure auroral X-rays. The flux, spectrum, and time dependence of precipitating electrons were deduced from these observations.

Fig. 15.1 Solar X-ray flux and X-rays from Earth's atmosphere during the March 28, 1976 solar flare. The right hand scale refers to the solar flux as measured by the satellite GEOS 1 in two energy bands shown by fine solid and dashed curves. The short heavy lines show the counting rate of the Ariel V detector and are labelled by orbit number.

15.2 Earth

The British satellite Ariel V carried a proportional counter with a 3.5° field of view. It was sensitive in the energy range 1.5–10 keV, and was pointed approximately along the satellite spin axis. The spin vector was held constant for the duration of each observation and in this way the detector was pointed at various astronomical targets. Since the spacecraft was in a near-Earth orbit, the

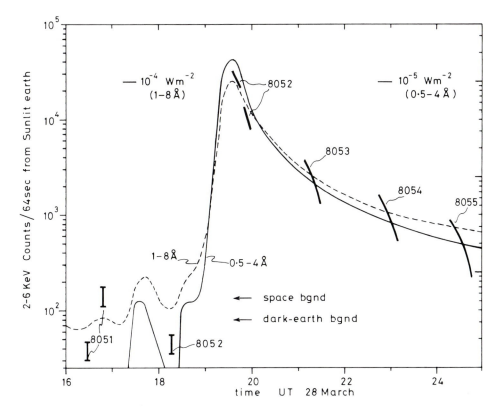

Earth generally passed through the field of view once every orbit. Usually the Earth was dark. No X-rays were detected from the atmosphere day or night.

On March 28 and April 30, 1976, there were two large solar flares. During these flares the X-ray flux from the Sun greatly increased. The Earth, as seen by the Ariel detector, also brightened. On March 28 the Earth was observed (once every 100 minute orbit) for 9 hours (figure 15.1). At the start of the flare the Earth-scattered flux increased by a factor of 1000 and then decayed over the next 6 hours in a way which exactly followed the solar X-ray emission observed by another satellite.

During each orbit the Earth was viewed for 40 minutes. Within this interval the X-ray flux varied much more rapidly than did solar emission. This effect is caused by the varying geometry (see box 15.1). The scattered flux at the satellite depends on the orientation of the Sun, Earth, and detector. Since the satellite spin axis remains fixed in space, the field of view scans over the top of the atmosphere as the satellite circles the Earth. The angle that detected scattered X-rays make with the terrestrial radius vector changes. The total thickness of atmosphere that these X-rays must traverse, travelling from Sun to Earth to detector, also changes. The thicker this layer, the more the X-rays are attenuated. Thus the detected flux depends strongly on the position of the satellite in its orbit (figures 15.2 and 15.3).

Fig. 15.2 Counting rate of the Ariel V detector during orbit 8055 as the sunlit Earth passed through the field of view. The solid line shows the predicted dependence of scattered X-rays on satellite position. The decrease in scattered flux is due to increased absorption in the atmosphere as the geometry changes.

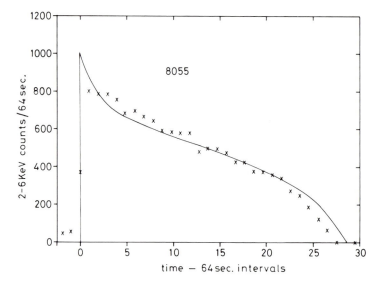

The maximum flux observed occurred when the detector viewed the sunlit limb of the Earth and was looking parallel to the top of the atmosphere. When the scattered flux dropped rapidly to zero the detector was looking at the terminator (the dividing line between day and night). Solar X-rays were absorbed before they reached that part of the atmosphere viewed by the detector.

The Einstein telescope was sensitive to softer X-rays than Ariel V. Figure 15.4 shows scattered X-rays detected by Einstein under these same geometrical conditions. These events are all due to sub-keV photons and the Sun was 'quiet' during this observation. Scattered X-rays from the quiet Sun can be a considerable background problem at sub-keV energies. When the Einstein telescope was pointed close to the sunlit Earth this background would degrade the sensitivity for astronomical targets. Above 2 keV, as for Ariel V, scattered solar X-rays are only a problem during solar flares. Counting rates for these X-rays are discussed in box 15.2.

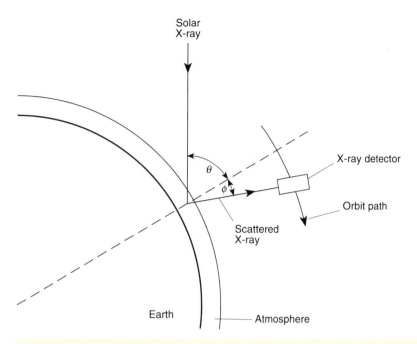

Fig. 15.3 Sun–Earth–satellite configuration during the Ariel V and Einstein observations of scattered solar X-rays.

Box 15.1 The dependence of scattered X-ray flux on geometry

A simple calculation shows the form of the observations described here. Assume that the top of the atmosphere is planar, the X-ray absorption coefficient is the same for both incident and scattered X-rays, there is only a single scattering, and solar flux is constant during the observation. If the angle between the path of the incident X-ray and the terrestrial radius vector is θ and the angle between the path of the scattered X-ray and the radius vector is ϕ, then the scattered flux at the satellite, F_s, is:

$$F_s = \text{constant} \times [1 + (\cos \phi / \cos \theta)]^{-1}.$$

The geometry is shown in figure 15.3, and this function was used to calculate the solid line in figure 15.2. It fits the measurements well.

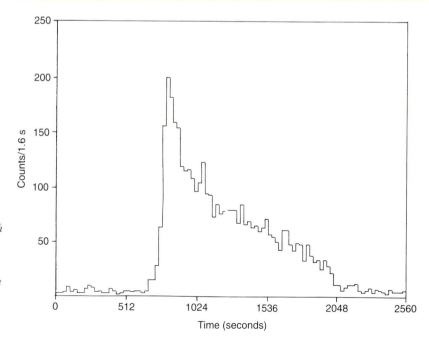

Fig. 15.4 Counting rate of the Einstein HRI December 25, 1979 as the sunlit Earth passed through the field of view. Although the photon energies are considerably lower than those detected by Ariel V, the variation of scattered flux with time is quite similar and depends only on the geometry. (Courtesy of F. R. Harnden Jr., CfA.)

Box 15.2 Counting rates for scattered solar X-rays

Some approximate numbers, useful for calculating the strength of scattered and fluorescent X-rays above 1.5 keV have been derived from the Ariel V observations: during the maximum of the solar flare in figure 15.1, when the incident energy from 1.5 to 12 keV (1–8 Å) was 10^{-1} erg cm^{-2} s^{-1}, the flux from the atmosphere was 550 argon fluorescent photons cm^{-2} s^{-1} ster^{-1}, with an equal number of scattered photons above 3 keV. This is equivalent to 1.5×10^{-3} of the incident solar energy re-emitted into 2π steradians and 11% of this albedo energy is in the argon line.

At lower energies, during quiet Sun conditions, Hugo Rugge (Aerospace Corp.) and colleagues detected atmospheric X-rays with the HEAO-1 detectors. They estimated that half of the photons detected in the 0.2–1 keV band were fluorescent radiation from nitrogen and oxygen.

In the Einstein energy band (0.2–3 keV) the average surface brightness of the sunlit Earth was measured as 0.05 counts s^{-1} arcmin^{-2}. This can be used to estimate the signal expected from other bodies in the solar system (table 15.2).

Table 15.2 Einstein IPC counting rates for planets

Object	Solar flux relative to Earth	Solid angle (arcmin^2)	Predicted rate (counts s^{-1})	Observed rate (counts s^{-1})
Mercury	6.6	0.005	0.0016	—
Venus	1.9	0.03	0.0028	—
Earth	1.0	—	—	0.05 arcmin^{-2}
Moon	1.0	700	35	17 (ROSAT, half Moon)
Mars	0.43	0.07	0.0015	—
Jupiter	0.037	0.49	0.001	0.012
Saturn	0.011	0.08	0.00004	less than 0.003

It has been assumed that the X-ray albedo of all objects is the same as that of the Earth and that Mercury and Venus are at maximum elongation and consequently only half illuminated.

Venus is predicted to give the strongest signal and an observation of duration 10^4 s (3 hours) might have yielded 30 counts. Although detectable, it would have been difficult to observe long enough to collect sufficient counts for spectral studies. Since it was strictly forbidden to point the telescope anywhere near the Sun, Venus was not observed. The only planets observed with Einstein were Jupiter and Saturn. X-rays from Jupiter were detected at a level 10 times greater than expected. It is believed that these were auroral X-rays, not the solar albedo, so there is no reason to doubt the calculated rates listed in table 15.2.

15.3 The Moon

X-rays from the lunar surface have only recently been detected by an instrument in Earth orbit. Einstein and EXOSAT could easily have done so but did not. On June 29, 1990 the observation was accomplished by ROSAT.

This is not a trivial observation. The Moon is a moving target. Either it must be tracked during the observation or the image must be reconstructed to follow the motion after-the-fact. The apparent motion with respect to the stars is both the lunar orbital motion of about 30 arcmin hour^{-1} and, for an instrument in near-Earth orbit, an even larger parallax. An X-ray picture of the full Moon might show bright and dark spots due to differences in surface composition. Such differences in X-ray albedo are expected to be small and long observations in several wavebands or at high spectral resolution will be needed to be useful.

Figure 15.5 shows the ROSAT picture. Jurgen Schmitt and colleagues (MPE) assembled this from a 1500 second interval during which the Moon passed from the edge to the centre of the field of view. The Sun was illuminating the Moon from the right side of the picture and X-rays scattered from the illuminated side were brightest at the limb, as expected. Total luminosity of these X-rays was 7×10^{11} erg s^{-1}, equivalent to only 700 100-watt light bulbs! In another historic first, the diffuse background was observed to be shadowed, showing that it originates (as we all believed) beyond the lunar orbit. However, the count rate observed from the lunar dark side was significantly higher than the detector background. These events are thought to be X-rays generated by bombardment of the lunar surface by solar-wind electrons. This picture of the bright side is a step towards using X-ray imaging for lunar surface diagnostics.

Fig. 15.5 The Moon observed by ROSAT. X-rays scattered from the solar-illuminated half of the Moon are brightest at the limb and fade to nothing at the terminator. The Sun–Moon–detector orientation was as illustrated in figure 15.3. Note that the Moon shadows the diffuse X-ray background so the dark side of the Moon is darker than the surrounding sky. (Courtesy of J. Schmitt, MPE.)

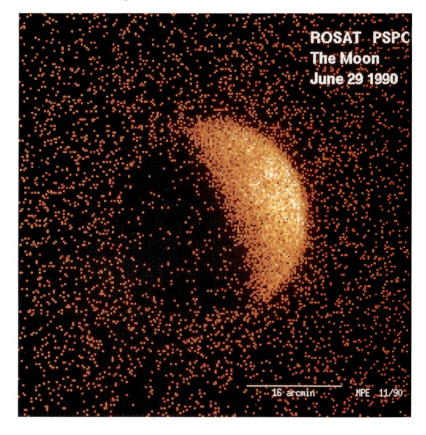

Lunar diagnostics were actually accomplished in 1971 and 1972 by Isidor Adler (GSFC) and associates by taking the detectors to the Moon. They used a special set of X-ray detectors which were carried in lunar orbit by the Apollo 15 and 16 command modules. A map of scattered X-rays from the sunlit lunar surface was assembled from approximately 200 hours of data. These were taken at an altitude (above the Moon's surface) of 100 km. Although restricted to sunlit territory directly under the spacecraft, 20–25% of the lunar surface was examined during the two missions.

The instrumentation consisted of three proportional counters operating in the energy range 0.7– 2.8 keV. Each was pointed at the surface and collimated to a 30° field of view. There was also a solar monitor. Two of the counters were respectively covered with thin foils of magnesium and aluminium. The K absorption edges of these materials formed transmission windows which enhanced the relative response at particular energies. One detector had relatively high efficiency for the fluorescent Kα line from magnesium. The other had a relatively high efficiency for the Kα line from aluminium. The third detector was equally sensitive to fluorescent lines from magnesium, aluminium, and silicon. The relative counting rates of these three detectors were used to determine the relative abundances of the three elements in the lunar material. Since the solar X-rays are easily absorbed, the measurement only applies to a depth of 0.1 mm. This is truly the lunar surface!

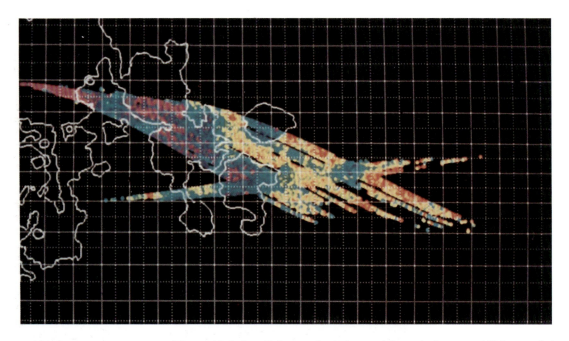

Fig. 15.6 The elemental composition of the lunar surface determined from Apollo 15 and 16. The lunar seas are outlined in white. Mare Crisium is just above the centre of the map and only the lower half has been surveyed. The blue colour within the sea indicates a low aluminium concentration and yellow and red in the surrounding highlands shows a high aluminium concentration. (Courtesy of J. Trombka, GSFC.)

Figure 15.6 shows the results. The spatial resolution was 100 km and the following properties of the lunar surface were inferred: there exists a global lunar crust characterised by a high aluminium to silicon ratio. This occurs in the highland areas both on near and far sides. The highland areas show high aluminium and low magnesium concentrations (yellow and red). The reverse is true in mare areas (blue). The irregular maria show a higher aluminium to silicon ratio than the circular maria. Comparison with optical pictures showed that areas of high optical albedo correspond to areas of high aluminium to silicon composition.

Since two of the Apollo landing sites were surveyed, lunar samples returned to Earth were used to calibrate the system. The X-ray data were thus tied to actual samples of material, making it possible to comment on the global distribution of material over the lunar surface.

15.4 Jupiter

In 1955, Jupiter was identified by Bernard Burke and Kenneth Franklin (Carnegie Institution) as a strong source of nonthermal radio emission. This was soon recognised to be synchrotron radiation, thus establishing the presence of a magnetosphere with a large population of energetic electrons. By analogy with Earth, these electrons were expected to precipitate along field lines and to generate X-rays when they collided with atoms in the Jovian atmosphere. There was also speculation that bombardment of the inner Galilean satellites might produce characteristic X-rays from their surfaces.

In 1973–74 the spacecraft Pioneer 10 and 11 measured the extent of the radiation belt surrounding Jupiter. The zone of stable trapping extends out to 25 times the radius of the planet. The zone encompasses the orbits of Almathea (a small moon), Io, Europa, and Ganymede. The trapped particle flux was found to decrease at the orbital distance of Almathea and of Io. Io, in fact, almost completely removes low energy protons from the radiation belts.

Radiation from particle bombardment of Jupiter's moons has not yet been detected. Auroral radiation from Jupiter itself, however, was detected in the ultraviolet by the Voyager spacecraft. Subsequent measurements with the IUE spectrometer have shown the radiation to consist primarily of hydrogen Lyman alpha at 1216 Å and the 1175–1650 Å Lyman and Werner bands of molecular hydrogen. Hydrogen is, of course, the major constituent of the upper atmosphere of Jupiter. X-rays from Jupiter were first detected on April 13, 1979 by Albert Metzger (JPL) and colleagues using the Einstein IPC. A subsequent observation with the HRI showed that the emission was from the polar regions and consequently auroral in nature. This satisfying result ended a 15 year search which started with the first observations of extrasolar X-rays. No X-rays were observed from the Galilean moons which, at several arcminutes separation, would have been easily resolved from Jupiter.

Jupiter was observed with the Einstein detectors four times. Each time was chosen to be when retrograde motion was starting or ending. In this way the apparent motion of the planet during the observation (typically of 3 hours duration) was kept to a minimum. The flux was established to be relatively constant with time, to have a very soft spectrum, and to originate approximately equally from both north and south polar regions. Because no emission was seen from equatorial regions, the emission cannot be solar-scattered X-rays.

The luminosity of the planet in the Einstein band was 4×10^{16} erg s^{-1}. (This is the next-to-least powerful extrasolar X-ray source detected to date. Earth's moon is ten thousand times less luminous.) The emission from each polar region appears diffuse with possible contributions from high above the planet's atmosphere. Since the planet was a moving, spinning target, and since only 35 photons were detected from each pole, the spatial extent of the source cannot be well determined from these data (figure 15.7).

The energy carried by electrons in the magnetosphere is known from the Pioneer observations and from the radio observations. There is not enough power in these electrons to produce the auroral X-rays. This leads to the hypothesis that these X-rays are not generated by electrons but by positive ions precipitating from the magnetosphere. Characteristic X-rays from oxygen and sulphur can produce the observed soft spectrum.

The origin of these elements is thought to be the volcanic plumes of Io, the innermost moon! The atoms are first spewed into space by the volcanos. They are then ionised and trapped by Jupiter's magnetic field. Next there must be acceleration within the magnetosphere, and finally scattering into trajectories which carry them into Jupiter's atmosphere. The volcanic plumes were discovered by the Voyager spacecraft and are probably familiar to the reader. Also a plasma torus in the vicinity of Io's orbit has been mapped with Earth-based telescopes. Although the electromagnetic forces which accelerate particles in the magnetosphere are not well understood, the basic source of power for the high energy ions and subsequent Jovian X-rays is probably the rotation of Jupiter.

An Einstein observation of Saturn yielded only an upper limit for X-ray

Fig. 15.7 Einstein observation of X-rays from the polar regions of Jupiter. The circle shows the position of Jupiter's 40 arcsecond diameter disc at the midpoint of the observation. During this 6 hour observation Jupiter travelled about one diameter along the path of its orbit which is indicated by solid lines. (Courtesy of A. Metzger, JPL.)

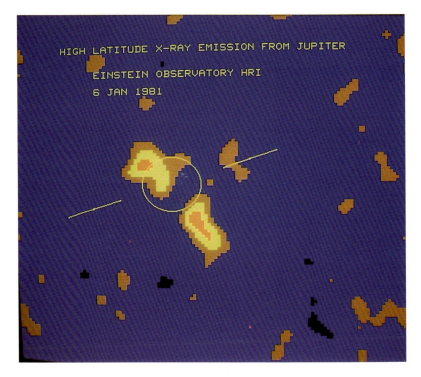

emission corresponding to an X-ray luminosity 0.8 that of Jupiter. This was not a surprising result. Einstein did not observe Mercury, Venus, or Mars.

Future X-ray observatories will carry instruments 10 times more sensitive than the Einstein detectors. These new missions hold great promise for planetary X-ray astronomy. Instruments will be able to map the spatial emission within the polar regions of Jupiter and to detect characteristic lines in the spectrum. These data together with advanced ultraviolet maps and spectra will characterise the nature of Jupiter's auroral emission.

16 The diffuse X-ray background

16.1 Early observations and general characteristics

It is difficult, at first, to be excited about the X-ray background. At all except the lowest energies, it is featureless. X-rays come uniformly from all directions to form a high energy glow between the stars. There is neither spatial structure nor sharp spectral feature to indicate production mechanism, distance, or origin. For early observers, the background was often just one more obstacle to be overcome in the search for faint sources.

Nevertheless, in spite of appearances, the X-ray background is a strong signal, and must be understood. Among other things it probably contains information about X-ray emission from the universe at early times. Because of this, many consider the background to be the most important topic in X-ray astronomy. It has been the subject of many scientific papers. Its exact nature, however, remains elusive. All the evidence does not point in the same direction. In this chapter we will review a little history, show the most extensive observations, and summarise current knowledge.

The X-ray background was not anticipated before discovery. It was initially recognised in data from the 1962 rocket flight which first detected Sco X-1. The observers, Riccardo Giacconi and colleagues (AS&E), concluded that the background was of 'diffuse character' and due to X-rays of about the same energy as those from Sco X-1.

This rocket flight was the first successful attempt to detect X-rays from sources other than the Sun or Earth. It used an uncollimated detector viewing about 10 000 square degrees of the sky. With such a detector, the observed 'diffuse' signal could have been generated by a few moderately strong 'point' sources spread over the sky. There was, in fact, an enhancement observed from the bright galactic sources in the Cygnus region. The next observations, however, used detectors collimated to observe only 100 square degrees of sky. These showed the background to be uniform to about 10% and, within the accuracies of these detectors, diffuse.

Signals observed during these early observations were clear and unmistakable indicators of the presence of the background. Rocket payloads used a nose cone or door to protect the X-ray detector. Without this protection the heat and pressure generated during the rapid ascent through the atmosphere would easily destroy the fragile detector windows. When above the atmosphere, this shield was ejected. At this time the counting rate of the detector always increased. When the payload fell back into the atmosphere, if the detector survived, the count rate would decrease. If, during the flight, the payload motion caused the detector to scan over the dark Earth, the rate always decreased.

All observations, without exception, showed a few bright sources embedded in a uniform X-ray glow. The night sky at X-ray wavelengths was uniformly bright! Sources appeared superposed on this background rather like stars viewed with the naked eye on a night when the Moon is full. At this time the faint stars disappear into the background of moonlight scattered from our atmosphere.

Because the observed emission was uniform, the possibility of a cosmic origin was, and still is, seriously considered. Right from the beginning, this phenomenon has been called the 'diffuse X-ray background'. It is now known that a large fraction of this background is due to faint unresolved sources. Perhaps all the background is due to discrete sources, but this has not yet been conclusively demonstrated. Some truly diffuse emission of cosmic origin is still a possibility, implying a vast amount of previously unseen matter in the universe. It might even require a new energetic phenomenon, pervasive at some past epoch. The study of the background, therefore, has been a vigorous field of research.

Attempts to understand this background have centred on three observational efforts: (1) the generation of precise maps of the background with detectors having 10–100 square degree fields of view, (2) measurements of the background spectrum over a large range of frequency, and, (3) the use of high spatial resolution detectors capable of resolving the background into a finite number of faint sources – or capable of at least determining the contribution of faint sources to the background.

Thoughts about cosmic background radiation are of course strongly influenced by observations at other wavelengths. Figure 16.1 shows the measured isotropic sky flux as a function of frequency from the radio band to gamma-ray energies. The dominant feature, containing 95% of the energy, is in the microwave–far-infrared band and has a spectrum closely resembling that of a blackbody at a temperature of 2.7 K (box 16.1).

The optical and UV points shown in figure 16.1 are theoretical upper limits. Our galaxy is opaque to UV radiation and local conditions make it difficult to detect faint diffuse optical radiation. The galaxy is, however, transparent to

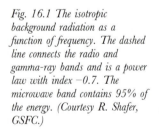

Fig. 16.1 The isotropic background radiation as a function of frequency. The dashed line connects the radio and gamma-ray bands and is a power law with index −0.7. The microwave band contains 95% of the energy. (Courtesy R. Shafer, GSFC.)

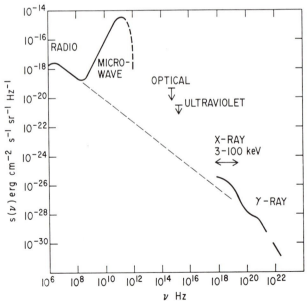

Box 16.1 The microwave background

The microwave background was discovered by Arno Penzias and Robert Wilson in 1965, and explained in an accompanying paper by Robert Dicke, James Peebles, Peter Roll and David Wilkinson. It is universally accepted as 'relic radiation from the Big Bang'. It is the most convincing piece of evidence we have that, in the past, matter in the universe was much denser and hotter than it is now − a gross understatement. The 'Big Bang' theory postulates that our entire observable universe originated in a configuration of unimaginably high density and temperature. All matter and energy was in the form of a 'soup' of elementary particles. Since then the density and temperature have been steadily decreasing with time as the universe expands.

At an age of about one million years, particles that make matter as we know it had been formed. The universe at this time consisted of a sea of ionised material, mostly electrons, protons, and helium nuclei. This material, a plasma, was opaque to electromagnetic radiation. Photons could only travel short distances before colliding with free electrons. As the temperature decreased, a point was reached where the photons were not energetic enough to remove electrons from bound states with the protons which attracted them. At this time hydrogen was formed. All the electrons were captured and locked into hydrogen atoms. Since hydrogen is transparent, the photons were suddenly able to stream through space for great distances.

We are now able to look through space back to the time when the universe was opaque. The photons we see are the UV–optical radiation that was freed by recombination, but redshifted by a factor of 1000 into the microwave range.

X-rays of energy greater than 1 keV. (This is true except for observations within a few degrees of the galactic plane where photon energies must sometimes be greater than 10 keV to penetrate the gas and dust in the disc.)

Since above 1–2 keV, the universe (except for clumps of cold material in galaxies) is essentially transparent, the discovery of an isotropic X-ray flux was exciting. Perhaps here was a high energy phenomenon somehow analogous to the microwave background. The high degree of isotropy implies that the source of radiation is at a great distance. The brightness implies that this was not a rare phenomenon. The fact that the photons are X-rays implies a high energy process, certainly much higher than the energy of hydrogen recombination.

Two possibilities for a cosmic component are: a tenuous hot gas filling the space between clusters of galaxies, and inverse Compton photons from cosmic rays interacting with the microwave background. (In this process a very high energy electron collides with a low energy photon and transfers a small fraction of its energy to the photon. The photon gains enough energy to make it an X-ray.) It would be truly exciting to find that part of the background might come from the distant past, before galaxies were formed, but observational proof of this is tricky.

16.2 The HEAO-1 background map, 3–50 keV

The satellite HEAO-1 carried a set of detectors cleverly designed to determine the characteristics of the X-ray background. Six detectors with 400 cm^2 area windows provided coverage in three energy bands from 0.1 to 60 keV. Special care was taken to reject internal and charged-particle induced background with internal anticoincidence requirements (as discussed more fully in chapter 2). Each detector had a collimator which defined two fields of view, one having twice the solid angle of the other. The diffuse signal was twice as great in the detector chambers viewing through the large aperture part of the collimator. The internal background, however, was the same as in chambers viewing through the narrow aperture. Thus Elihu Boldt and

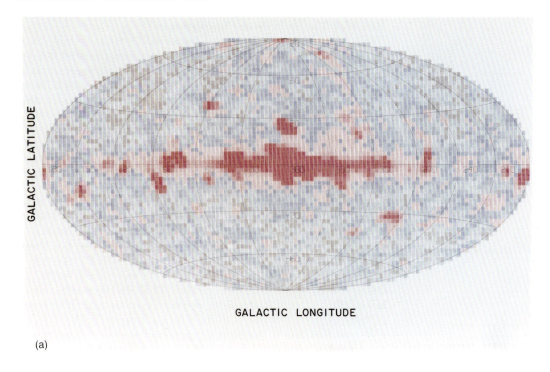

(a)

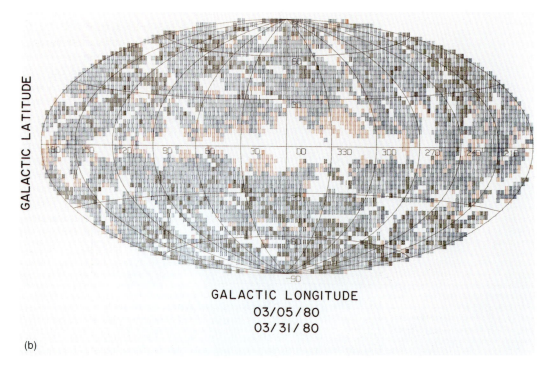

(b)

Fig. 16.2 (a) The surface brightness of the X-ray sky mapped by HEAO-1. Pixels are 3° by 3°, energy range is 2 to 60 keV, and the map is in galactic coordinates. The sky brightness is colour coded black (lowest), blue, pink, red (highest). The brightest emission is from discrete sources lying in the galactic plane and concentrated toward the galactic centre. (b) Resolved sources have been removed. Each pixel is now coded with lines representing a small fraction of the mean intensity. Pixels with few or many lines show fluctuations below and above the mean. Some dark pixels contain unresolved point sources and, although it is hard to see, the emission is brightest close to the galactic plane. (Courtesy F. Marshall, GSFC.)

his colleagues (GSFC) attacked the instrumental problems which usually limit the accuracy with which diffuse signals can be measured. The background map produced from observations with this instrument is the best currently available and is likely to remain so for some time.

In three years of operation, HEAO-1 performed 2.7 complete scans of the sky. Although coverage was complete, more time was spent looking at the ecliptic poles than at the ecliptic plane. Rotation of the spacecraft fixed the scan path to be a great circle perpendicular to the Earth–Sun line. This path crossed the poles once every rotation but precessed over the rest of the sky once every 6 months.

The HEAO-1 GSFC detectors made an all-sky X-ray map (figure 16.2(a)) with 3°×3° and 3°×6° collimators. The map is dominated by bright sources clustered in the central region of the galactic plane. The first step in the study of the background was to remove the effect of these sources. This was not difficult since the response of the detectors to a point-like source passing through the field of view was well calibrated. Source strengths and locations could also be accurately determined. After making a catalogue of detected sources, the signal due to these was subtracted from the data. The 'source-free' map shown in figure 16.2(b) was the result. It is coded to illustrate small fluctuations so these show more clearly than in figure 16.2(a). Coverage of the sky is good but spotty because of regions hidden by the bright sources. Almost all of the events comprising this map are due to the X-ray background and to a few large-scale diffuse features.

16.3 HEAO-1 diffuse galactic emission, 2–60 keV

Except for a few obvious fluctuations, these data have been resolved into two components: the isotropic background, and emission associated with our galaxy. Figure 16.3 shows the X-ray surface brightness as a function of galactic latitude over different ranges of galactic longitude. Note that the diffuse

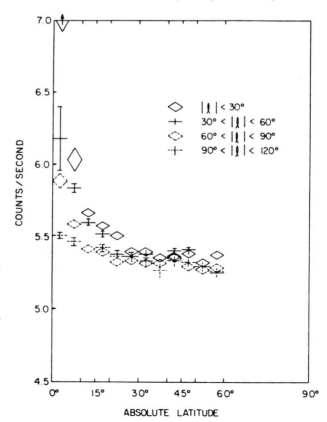

Fig. 16.3 The galactic component of the HEAO-1 background. The background X-ray intensity is plotted as a function of galactic latitude for four different bands of longitude. Emission is highest close to the plane and strongest in the direction of the galactic centre. [From D. Iwan et al, Astrophysical J. 260, 111 (1982).]

emission (strong sources have been subtracted) is brightest in the galactic plane, and that the anisotropy is strongest at low latitudes. The relative contribution of the galactic component at the poles is only about 2%.

This galactic emission has been fitted by De Ann Iwan and colleagues (GSFC) with a model which assumes that the radiation is from a disc-shaped region aligned with the galactic centre and plane. This disc contains a uniform density of sources or emitting material. Dimensions of the disc can be derived from the data. Using 10 kpc as the distance to the galactic centre, the thickness of the disc is 7 kpc and its radius is 28 kpc. Thus our galaxy, viewed from afar with an X-ray telescope, would be an array of about 20 bright sources (associated with the inner spiral arms) contained in a rather squashed, apparently diffuse, halo with radial extent almost twice that of the region containing the bright sources.

The X-ray luminosity of this halo is about 8×10^{38} erg s^{-1}, 30% of that of the array of bright sources. This diffuse emission is therefore important. The spectrum is softer than that of the isotropic cosmic background. A temperature of 1×10^8 K (9 keV) produces a reasonable fit to the galactic-halo data.

Three possible sources of this halo emission have been investigated. (1) A halo of subdwarf M-stars is promising and the stellar contribution at low energies was calculated from Einstein survey data. The Einstein telescope, however, obtained no data above 4 keV so the overlap with HEAO-1 results is small. (2) Inverse Compton emission from galactic cosmic ray interactions with the microwave background should provide a diffuse source with about the right shape. A 1 GeV electron colliding with a microwave photon will raise the photon energy to the X-ray range. However, if the electron flux and spectrum in the halo is the same as that observed in the vicinity of the Earth, this mechanism fails by an order of magnitude to account for the observed halo X-rays. (3) A very hot gas (temperature about 10^8 K and density about 2×10^{-4} particles cm^{-3}) could easily produce the X-rays. Energy could be supplied to such a halo by supernova explosions but the temperature is so high that the gas would not be confined by gravity and would soon escape. A magnetic field could contain the gas but the field required is stronger than currently believed to exist in the galactic halo. Although there are severe problems explaining the X-ray emission with any one of these models, they all must contribute to the halo at some level.

After subtraction of this galactic component, an isotropic X-ray background remains. This result, if displayed on a sky map, has a singularly unspectacular appearance . The HEAO-1 data, binned in $9° \times 9°$ pixels, after subtraction of the galactic component has a median standard deviation of 1.5% from the mean intensity.

Further analysis of these data has centred on the spectra and on the observed fluctuations.

16.4 HEAO-1 fluctuations

If the background emission is from a universal process which occurred in the distant past, the radiation is expected to be isotropic. Isotropy alone, however, is not sufficient to prove that such is the origin of the radiation. There is, in fact, an important clue in the data which shows that this is not strictly the case. There is a granularity in the background, above the expected statistical fluctuations, implying that faint unresolved sources contribute part, perhaps all, of the isotropic background.

Analysis of these data is difficult. First the obvious sources must be subtracted, then the galactic component. After excluding regions dominated by these effects, the expected fluctuations in the remaining data must be precisely evaluated. The HEAO-1 data were divided into $9° \times 9°$ cells, each containing a finite number of counts. If the source of the radiation producing these counts is truly uniform, the average deviation from the number of counts in the individual cells is exactly specified by the laws of chance. The expected fluctuations must first be precisely calculated. These are then compared with the data to identify deviations from random behaviour. The measured excess can be used to derive some characteristics of the unseen sources.

Such analyses have been done using survey data from Uhuru, Ariel V, and HEAO-1. The scientists involved: Andrew Fabian (IOA), Daniel Schwartz (CfA), John Pye and Robert Warwick (Leicester), Elihu Boldt and Richard Shafer (GSFC) and colleagues, have, after diligent labour, reached similar conclusions. These are illustrated here by the HEAO-1 result. Figure 16.4 shows that the 'diffuse' component really has a larger spread in count rate than expected from a uniformly bright sky. This level of granularity can be used to set limits on the brightness of sources too faint to be seen as individuals. Furthermore, if the sources are assumed to be distributed in intensity in a particular way, limits can be placed on the source population at a level two orders of magnitude below the faintest individual source seen.

Fig. 16.4 Fluctuations in the HEAO-1 diffuse background intensity. The histogram shows the distribution of sky brightness in 4° pixels centred on the average counting rate. The long-dashed curve shows the statistically expected distribution. Fluctuations in the data are greater than expected and have the width of the short-dashed Gaussian curve. (Courtesy R. Shafer, GSFC.)

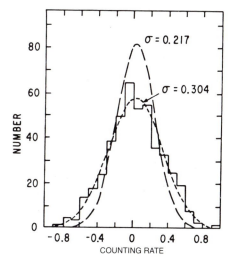

The population of sources needed to make up the diffuse background can be drawn on a log$N$–log$S$ plot (see box 16.2) and compared with observations, which so far have been of sources very much brighter than those expected to make up the background.

Box 16.2 The log$N$–log$S$ relation

In most astronomical surveys the number of sources increases rapidly as the flux decreases. The majority of sources known are faint. This is certainly true for the X-ray sources. Because these faint sources are not far above the threshold of detection, not enough photons have been collected to measure spectra or time variability. Only the location, the approximate strength, and sometimes identification with an optical counterpart are known. The study of faint sources, therefore, proceeds using the number density and identifications. The number density is commonly shown on a plot of the total number, N, of sources as a function of threshold flux or source strength, S. Logarithms are used to give a large range on one graph.

If the universe were Euclidian and if all sources were distributed uniformly through space, the plot of log$N$ vs log$S$ would be a straight line with slope of $-3/2$. Again N is the total number of sources in a given angular range having flux at the detector equal to or above *S*. The slope is −3/2 because the flux of radiation from any source falls as the inverse square of the distance and the number of sources increases as the cube of the distance.

If all the sources were contained in a disc, as expected for those associated with the spiral arms of our galaxy, the number of sources within range of the detector goes as the square of the distance. In this case the expected slope of the log*N*–log*S* plot is −1, rather than −3/2.

This relation has been used to determine the relative number of galactic and extragalactic sources in early X-ray surveys. X-ray sources at high galactic latitudes (looking out of the galactic plane) were found to follow the −3/2 slope and were thus predicted to be mostly extragalactic.

Figure 16.B1 shows a log*N*–log*S* plot for extragalactic sources incorporating all the HEAO data. Open circles are the result of the HEAO-1 survey illustrated in figure 16.2(a). The Einstein deep survey result is a solitary filled circle, and the MSS a filled circle anchoring a dark solid line. The slope of this line is well determined and is −3/2, exactly that expected for a uniform population of extragalactic sources in Euclidian space.

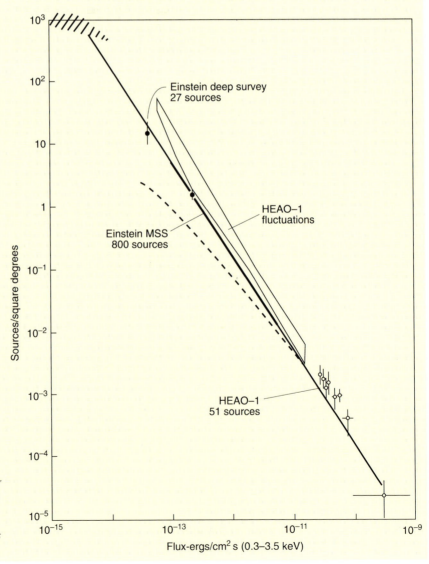

*Fig 16.B1 A log*N*–log*S* plot for extragalactic sources. This shows the total number, N, of observed high-latitude sources having flux equal to or greater than flux S as a function of S.*

The long box extending to low flux levels shows the result of an analysis of fluctuations in the HEAO-1 background data. The analysis assumes that these faint sources are distributed on this plot as a power law, and the box shows the allowed limits in which the line can be drawn.

The vertical scale is in sources degree^{-2}. The level of the lowest point corresponds to only one source in the entire sky. The horizontal scale units are those of the Einstein detectors. The HEAO-1 points were obtained in a higher energy range and have been transferred to this plot by assuming that all sources have the same power-law spectrum. A different assumption will lead to a different normalisation and the HEAO-1 and HEAO-2 data sets can be made to overlap exactly if the spectrum is chosen to do this.

The dashed curve shows the expected contribution of clusters, normalised to a 50% contribution in the HEAO-1 sample. The hatched area in the upper left corner shows an area in which the total contribution of sources might account for 100% of the X-ray background.

It is the sample of extragalactic sources which might reveal the composition of the diffuse background. However, if the faint sources are far away and the redshift is large, the distribution will no longer follow the simple $-3/2$ power law expected from classical considerations. Space is not Euclidian. Because of the 'curvature' of space, our intuitive concept of distance and volume is misleading when we consider very distant objects.

The universe is also expanding, and at large distances galaxies are moving away from us with velocities approaching the speed of light. At redshift, $z = 2$, for example, the recession velocity is $0.8c$. A clock in a system moving with respect to us at relativistic velocity appears to us to be running slow. This relativistic 'time dilation' causes the energy radiated by a source during one of our seconds to be less than the energy radiated during a second measured by a clock moving with the source. This is one effect which makes distant rapidly-moving sources appear fainter than expected.

Also, at high redshifts, the energy of the photons decreases considerably travelling from source to detector. At a redshift of $z = 2$, for example, a 3 keV detected photon had an energy of 9 keV when it left the source. Distant sources are thus sampled at a higher energy part of their spectrum. If a source has a high energy cut-off, it will be undetectable at large distances because the radiation will be redshifted below the threshold of the detector.

If faint X-ray sources have high redshifts, these effects will cause the log$N$–log$S$ plot to fall below a plot with $-3/2$ slope. This is shown by the dashed curve in Fig. 16.B1.

Evolution is also important. The sources themselves are expected to evolve with time. The quasars seen nearby may have been much more luminous in the distant past, or may have turned on in a particular epoch. If observations can be extended to faint enough sources, source evolution should produce a departure from the expected form of the log$N$–log$S$ relation.

16.5 Imaging the background, 0.3–3.5 keV; The Einstein deep and medium-sensitivity surveys

The deep surveys were considered by some to be the most important observations made with the Einstein telescope. The fields were certainly among the most curious. The telescope was deliberately pointed at places known to be devoid of interesting targets. A great deal of observing time was used looking, to the limit of the telescope's ability, at a hitherto blank field!

The purpose of these long observations was to resolve the diffuse background into individual sources. It was rather like looking at a cloud and trying to see the individual water droplets. The regions selected for study were all at high galactic latitude with low hydrogen column density and with no known bright X-ray, optical, or radio sources. These fields were imaged with both the IPC and HRI. Exposure times varied and ranged up to 21 hours with the IPC and up to 24 hours with the HRI. A typical result is shown in figure 16.5.

The IPC was the more sensitive instrument but suffered from source confusion, a common difficulty with deep observations. If the sensitivity is high enough to detect typical sources contributing to the background, they will appear packed close together. Individual sources will not be well separated from one another. Instrumental artefacts within the IPC also made the analysis difficult. Small variations in detector gain put bright stripes and patches into a uniformly illuminated field. Although slight and usually negligible, these nonuniformities appeared clearly in the deep-survey exposures. Thus, although the IPC could in theory detect faint sources, the superior spatial resolution of the HRI was necessary to confirm, separate, and locate the individual sources. Using both detectors almost 100 sources were detected in eight deep fields. Corresponding optical objects are faint, however, and identification of the sources is a slow process.

Fig. 16.5 The Einstein IPC deep survey field in Pavo after background subtraction and correction for vignetting in the telescope. There are 15 sources significantly above background in this field. The HRI was used to resolve source confusion. (Courtesy F. Primini, CfA.)

As a start, a set of 25 IPC sources were selected for detailed analysis. The density of these is indicated on the $\log N$–$\log S$ plot in Fig. 16.B1. At the lowest flux level, 4×10^{-14} erg cm^{-2} s^{-1}, the density is 15 sources degree^{-2}. Eighteen of the sources have been identified. Seven are nearby stars, nine are quasars,

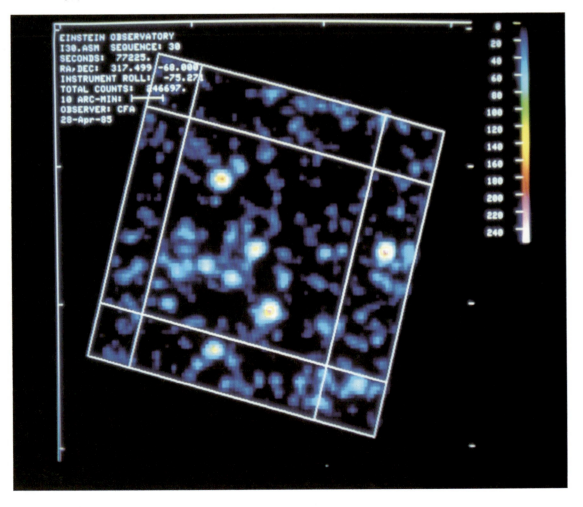

one is a galaxy, and one is probably a distant cluster. The remaining seven are known to be extragalactic because of the high ratio of X-ray to optical flux, but are so far unidentified. Sources such as these are expected to contribute at least 25% of the diffuse background in the range 1–3 keV.

The relative contribution of different types of sources agrees with that found by another Einstein survey, the medium-sensitivity survey (MSS); and, since the statistics are better, this latter survey gives a better indication of the kinds of sources likely to contribute to the background.

In contrast with the deep surveys in which the Einstein telescope was pointed at an empty field, most other observations were aimed at particular sources of interest – stars, radio sources, known quasars, etc. If the target was an X-ray source, it was positioned to appear at the centre of the field. In about half these observations other sources also appeared within the field of view. These 'serendipitous' sources were completely accidental and of unknown origin. For several years, Isabella Gioia and Tommaso Maccacaro (CfA) systematically searched the IPC data and compiled a list of serendipitous sources found at high galactic latitude. They found 835 previously unknown sources.

Working with John Stocke (University of Arizona) and others, they have identified over 800 of these objects. Of these, 53% are quasars and AGN, 13% are clusters, 5% are BL Lac objects, 27% are foreground stars, and 2% are normal galaxies. This is the same mixture as the more limited results from the deep surveys. As a matter of interest, fluxes from the extragalactic objects identified in this MSS are 2–50 times greater than from the deep-survey sources and ~100 times less than those found in the HEAO-1 survey of figure 16.2(a).

These data on source populations have interesting implications for the X-ray background. Of the 51 high latitude sources in the HEAO-1 survey, 51% were clusters, whereas only 13% of those in the MSS are clusters. This trend

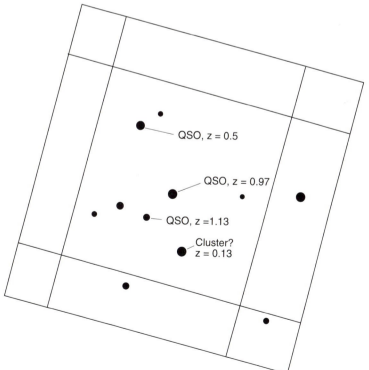

Fig. 16.6 A chart outlining the identification of the brightest sources in the Pavo deep survey field shown in figure 16.5.

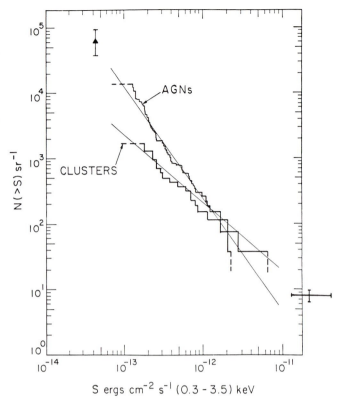

Fig. 16.7 A logN–logS *plot showing results for identified Einstein MSS sources. For this plot, 60 clusters and 200 quasars and Seyfert galaxies have been used. Although half of the bright high-latitude sources are clusters, their contribution to the deep survey (data point) faint sources is small. (Courtesy I. Gioia, CfA.)*

is as expected for sources which do not evolve with time. Both the redshift and the curvature of space cause the logN–logS curve for clusters to flatten as shown in figure 16.B1. As more MSS sources are identified, this expectation is being confirmed, as shown in figure 16.7. Thus X-rays from clusters of galaxies will not contribute to the background unless there is a dramatic reversal of the observed trend.

The deep-survey fields, after extraction of the obvious sources, still contain information. Fluctuations in the deep survey background can be analysed (in the same way as the HEAO-1 background fluctuations) to give information about sources below the detection threshold. Thomas Hamilton and David Helfand (Columbia University) have done this by measuring the granularity in the deep-survey background. They found that unresolved sources must follow a logN–logS curve of lesser slope down to at least 8×10^{-15} (a factor of five below the flux cut off of the deep survey). They also found that unresolved sources must contribute 35–55% of the diffuse background.

The first ROSAT observations confirmed this. Guenther Hasinger (MPE) and colleagues have detected sources down to a flux of 1×10^{-14}. They found ~70 sources degree^{-2}, implying that the slope of the logN–logS curve starts to flatten at a flux below 2×10^{-14}. ROSAT has brought us closer to the goal of 'imaging the background'.

16.6 The contribution of quasars and active galactic nuclei to the background

In order to determine their X-ray properties, the Einstein telescope was pointed at more than 350 quasars which had already been discovered through their radio and optical emission. About 250 of these were detected and X-ray luminosities have been calculated. This sample of objects, by the very nature of the discovery technique, must have abnormally high radio and/or optical emission. The other data set available consists of quasars discovered in the X-ray band which, by similar reasoning, must be X-ray bright. In spite of these biases, a careful analysis shows that most quasars that are bright in the optical or radio bands are also among the more luminous X-ray sources. X-ray emissivity seems to be a universal property of quasars and the X-ray emission is roughly correlated with emission at other wavelengths. Since the spectral lines which signal that an object is a quasar also give the redshift, the distances (assuming $H_0 = 50$ km s^{-1} Mpc^{-1}) and absolute luminosities of quasars are known.

Harvey Tananbaum, Belinda Wilkes, and Yoram Avni (CfA), Bruce Margon (University of Washington), and others analysed the Einstein data on quasars and concluded that it is possible that almost all of the diffuse X-ray background is due to unresolved distant quasars. This is consistent with the MSS result. Keeping in mind that the data do not demand this conclusion, let us assume for the moment that this hypothesis is correct and ask which quasars contribute most to the background.

Even though there is a great deal of individual variation, the X-ray and optical data lead to the conclusion that the most luminous quasars do not contribute appreciably to the background, nor do the distant quasars at $z > 2$. Most of the quasar contribution must come from quasars at $z = 1$–1.5 and with optical magnitudes, V, about 20. (As a matter of interest, the optically-brightest quasar, 3C273, has $V = 12.8$ and $L_x = 2 \times 10^{46}$ erg s^{-1}).

The typical background quasar is not bright enough in X-rays to be observed as an individual. The sample of known X-ray emitting quasars, therefore, does not contain many representative quasars from the hypothetical background-producing shell at $z = 1$–1.5.

The above remarks concern quasars and active galactic nuclei having X-ray luminosities between 10^{42} and 10^{48} erg s^{-1}. Martin Elvis (CfA) pointed out that the nuclei of many 'normal' spiral galaxies show the same evidence of activity (rapid variability, X-ray emission, compact radio source) as AGN but at a lower level, with luminosities 10^{40} to 10^{42} erg s^{-1}. Some specific examples were discussed in chapter 6. A calculation based on X-ray data and observed Hα luminosities of galaxies shows that these low level 'active' nuclei (or 'microquasars') could account for 15% of the X-ray background.

16.7 HEAO-1 spectrum of the background, 3–50 keV

A completely different point of view is that a large part of the X-ray background is truly diffuse. The present data allow this and many astrophysicists find it an attractive hypothesis. The principal evidence in favour is the spectrum of the background above 3 keV, one of the best measured in X-ray astronomy. These excellent data are completely contrary to expectations if the background consists only of radiation from distant quasars.

The large area HEAO-1 detectors, in 6 months observing, collected over 10 million counts which were useful for broad-band spectral analysis. Figure 16.8 shows this spectrum. Between 3 and 50 keV it has exactly the shape expected of thermal bremsstrahlung, radiation from hot gas at a temperature of 5×10^8 K (40 keV). The agreement with the theoretical spectral form is excellent. There is no way a single power law, the type of spectrum emitted by nearby active galactic nuclei and quasars, will fit the data.

Early observations in the 2–10 keV range indicated a power law with index 0.6, but there is a 'break' at about 40 keV after which the slope of the spectrum steepens considerably. The spectral shape leads immediately to the hypothesis that hot gas accounts for the background emission in this energy range. No line emission has been seen but little is expected from material at so high a temperature. Also, if primordial, the gas might consist of only

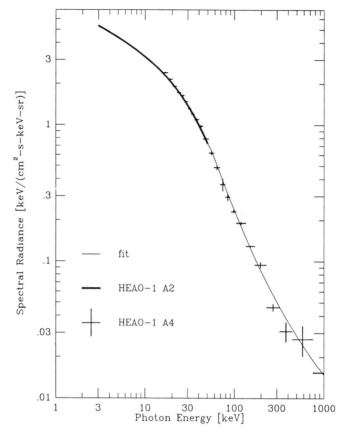

Fig. 16.8 The spectrum of the X-ray background. The solid line from 3 to 50 keV is from HEAO-1 proportional counters. The accuracy of this spectrum is remarkable. On this plot, the uncertainties due to counting statistics of this detector are essentially zero. Crosses are data from the HEAO-1 scintillation counters and the fainter line is an empirical fit. (Courtesy D. Gruber, UCSD.)

hydrogen and helium and there would be no lines in this energy range at any temperature.

We know the physical characteristics of gas within galaxies and within clusters and it is not this hot. The only place left where this very hot gas might be found is the space between clusters of galaxies. A uniform space density of about 10^{-6} ions cm^{-3} would produce the observed signal. It would fill most of the volume available in the universe!

The mass of this gas is 0.3 of that required to 'close' the universe. Here perhaps is much of the material required to stop the present expansion of the universe and start a contraction back to a high density state, appealing to those who like symmetry and who are willing to wait for results!

The energy content of the gas, however, poses theoretical problems. The energy required to heat the gas to a temperature of 5×10^8 K or higher is too large to be supplied by known sources. The pressure of the hot gas is also expected to produce effects on the gas within clusters (the subject of chapter 14) which are not seen. The gas within clusters would in fact soon evaporate if the cluster were surrounded by such a very hot gas. It is not clear that these problems are insurmountable. Clumping of the very hot gas, for example, would make these difficulties less severe.

Theoreticians have not yet ruled out the existence of at least a contribution to the X-ray background from very hot gas. The mass involved could be enormous, more than all other known or implied mass in the universe, and this is exciting.

Further support for the hot-gas hypothesis comes from spectra of nearby extragalactic sources, which have been well measured. Can the background spectrum be reproduced by a collection of these sources at various redshifts?

The answer is no. Clusters of galaxies are the only known thermal sources and these have temperatures of from 2 to 8 keV – much too low to reproduce the X-ray background spectrum. Active galaxies and quasars form the majority of extragalactic sources and these have power-law spectra with average energy index 0.7 (chapter 13). These cannot make up the background spectrum since, even when redshifted, power-law spectra are straight lines on the plot in figure 16.8.

If the X-ray background is due solely to discrete extragalactic sources, the spectra of these sources at an early epoch must have been very different from the spectra observed now. If the sources are all active nuclei and quasars located at z approximately 2, the strength of the background requires 30% of all galaxies to have active nuclei. The spectra must be power-law with breaks (sudden drops in intensity) at about 100 keV. The redshift will then produce the observed spectrum of figure 16.8. Since quasars are strongly evolving sources, they are good candidates for these proposed contributors to the background. A new class of sources, not yet discovered, is also a possibility.

16.8 The galactic soft X-ray background, 0.2–2 keV, the big picture

The background observed at low X-ray energies is quite different. There is both spatial and spectral structure. The bulk of this radiation comes from sources within our own galaxy. It is not necessary to consider strong sources at high redshifts to explain it.

Diffuse background observations below 1 keV are difficult. The Einstein telescope and IPC, a 'sensitive' instrument for sources, was not at all useful for this. The signal-to-noise ratio for a telescope is excellent for point-like sources because X-rays from the source are concentrated into a small part of the detector with consequent small internal background. There is no such concentration of the diffuse background. The ROSAT system, however, with its low-internal-background PSPC (position-sensitive proportional counter) detector, was well suited for background observations.

Data showing the global background have been gathered by rocket-borne detectors. The observations were painstakingly conducted over a period of 8 years by William Kraushaar, Dan McCammon, Wilton Sanders, Alan Bunner, and others at the University of Wisconsin. Data from 10 rocket flights were used to make the maps shown in figure 16.9. These were launched from White Sands, New Mexico and from Woomera, Australia. Each flight scanned 1/8 of the sky with thin-window detectors collimated to a 6.5° FWHM (full width at half maximum) field of view.

The detectors were large-area proportional counters with thin plastic windows. They are rather 'old fashioned' now but were excellent for this particular purpose. The principal difficulties were solar and Earth-associated backgrounds. Scattering of solar X-rays from the sunlit atmosphere can easily overwhelm X-rays of astrophysical interest. Measurements were done at night with the Sun well below the horizon. Windows, if thin enough to transmit soft X-rays, will also transmit low energy electrons and UV. Electrons are sometimes present at the top of the Earth's atmosphere (and are overwhelming over large portions of a satellite orbit). Signals from UV interactions in the detector must be distinguished from those of X-rays through pulse height analysis (explained in chapter 2).

Although the 5 minute observing time of a rocket flight is pitifully short compared to any satellite detector, the rocket has one big advantage. This is the ability to perform observations at low altitude at specific locations where the Earth's magnetic field shields the equipment from energetic charged particles. The University of Wisconsin detectors never exceeded 200 km in altitude so were always too low to encounter trapped particles. Sporadic fluxes of precipitating electrons, however, were still a worry. As a precaution, ceramic magnets were built into the collimators so that low energy electrons (< 80 keV), if present, would be deflected into the walls of the collimator instead of passing through the window into the detector.

The X-ray events were sorted into several energy bands by means of boron and carbon filters and electronic pulse height analysis. Unlike the HEAO-1 results in figure 16.2, the appearance of the soft X-ray sky varies greatly as the energy changes. The maps shown here illustrate the soft X-ray background in three energy bands. Point-like sources have been removed.

Three distinct phenomena appear. The first is an absence of features and has already been illustrated in figure 16.2. The background above 2 keV is isotropic and featureless except for well-known sources. The rocket results in this band are not shown. They are the same as the HEAO-1 results but much less precise.

Fig. 16.9 The soft X-ray sky presented in galactic coordinates as mapped with sounding rockets. Bright point-like sources have been removed. (a) The 'M-band' background. The large diffuse feature is probably a nearby superbubble, a region full of hot gas. (b) The 'C-band' X-ray background. (c) The 'B-band' X-ray background. (d) The 'C-band' background using data taken by the SAS-3 satellite. Note the agreement with (b). (Courtesy W. Sanders, University of Wisconsin.)

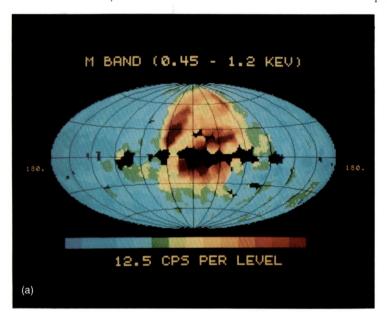

(a)

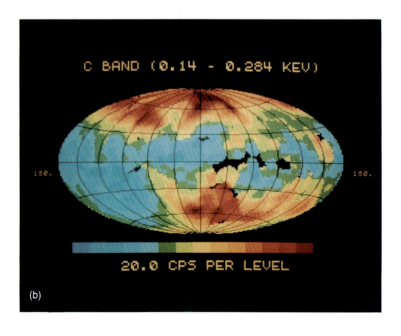

(b)

In the 'M' (middle) band, 0.45 to 1.2 keV, the X-ray sky is still approximately isotropic but three large-scale features appear. The map is dominated by an irregular feature 110° in diameter located roughly in the direction of the galactic centre. It coincides with a radio feature called 'Loop I'. The brightest part of the rim of this loop, in the radio band, is called the 'North Polar Spur'. The two smaller features are the Eridanus–Orion enhancement, very faint at galactic coordinates 200°, −35°; and the Cygnus superbubble at 80°, +5°. These patchy features are the result of hot gas filling cavities in the interstellar medium. The energy to heat the gas comes from supernovae and winds from early-type stars. Chapter 5 describes the Cygnus superbubble.

The 2–10 keV spectrum can be extrapolated to predict the extragalactic diffuse contribution in the M band. This contribution should be strong. At the poles, 50% of the flux is predicted to be of extragalactic origin. This extragalactic component should be completely absorbed by interstellar gas in the direction of the galactic plane. Since figure 16.9(a) shows no minimum at the galactic equator, local sources must contribute almost all of the equatorial flux.

The Einstein observations of stars, which are certainly local, were used by Robert Rosner and colleagues (CfA) to calculate the stellar contribution to

Fig. 16.9 (c, d). For legend see opposite.

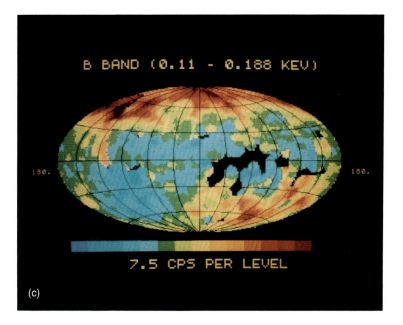

(c)

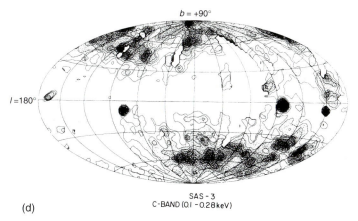

(d)

the M-band X-ray background. The result was that 20% of the flux probably comes from dwarf G, F, K, and M stars, with dM stars producing 80% of this stellar contribution. A second group of local contributors in the plane could be distant, unresolved galactic superbubbles. Thus the M-band background is explained although the exact nature of all the contributing sources is not well understood.

At the lowest energies, the diffuse sky in the 'B' and 'C' bands is distinctly different to that in the 'M' band. (B and C stand for boron and carbon filters used with absorption edges limiting energies to less than 0.18 and 0.28 keV respectively.) The brightest regions are now associated roughly with the galactic poles. This is most easily interpreted as absorption of an extragalactic flux by the neutral hydrogen of our galaxy. The column density of neutral hydrogen is highest in the galactic plane. Although there is a pronounced anticorrelation between X-ray and 21 cm (radio emission from hydrogen atoms) sky brightness, the picture is not this simple. The hydrogen column density in the plane is large enough to absorb all the X-rays in the plane. Since there is appreciable emission coming from that direction, there must also be local sources of these very soft X-rays. The flux of soft X-rays, in all directions, is also considerably higher than that calculated by an extrapolation of the smooth high-energy background spectrum.

As a matter of interest the C band soft X-ray background was also mapped by a small detector on the satellite SAS-3. This carried a small X-ray mirror and detector and much of the time was below the radiation belts in a near-Earth orbit. The field of view was 3° and maximum effective area was only 4 cm². Frederick Marshall and George Clark (MIT) assembled the background information by carefully excluding data taken in regions subject to solar X-ray or charged particle contamination from May 1975 to November 1976. This map (figure 16.9(d)) was essentially the same as that produced by the rocket-borne detectors, 100 times larger and observing only a few minutes. This gives confidence in the result.

The most likely source of the soft X-rays is thermal emission from diffuse hot gas at a temperature of about 1×10^6 K. X-rays from such a gas will be almost completely in the form of spectral lines, with energies characteristic of the ions which radiate. Spectral observations of the diffuse background can therefore both confirm the thermal origin and measure the composition and temperature of the gas. An instrument with higher resolution than the usual proportional counter, however, is needed to distinguish lines at these low energies.

Such an observation was done in March 1980 by Herbert Schnopper (CfA), R. Rocchia (Saclay), and colleagues. A small rocket carried a silicon (lithium) spectrometer above the atmosphere for 4 minutes and soft X-rays were collected from a region 60° in diameter containing the North Ecliptic Pole and the North Polar Spur. Figure 16.10 shows the spectrum. As expected, the observed structure demands the presence of emission lines and a thermal source. The best-fit spectrum, however, requires two temperatures. The radiation is probably from 1×10^6 K gas in the interstellar medium and 4×10^6 K gas in the North Polar Spur. This is in excellent agreement with an earlier observation of the North Polar Spur obtained in September 1977 with a gas scintillation counter by H. Inoue and colleagues (University of Tokyo).

Where exactly is this 10^6 K gas which produces the soft X-ray background in 'interstellar space'? Radiation from a hot galactic corona is one possibility. As discussed in chapter 6, some other galaxies have X-ray emitting halos which extend beyond the optical boundaries. This could well be true for our galaxy but, even if so, the lack of complete absorption in the plane rules out this as the only source of soft X-rays.

The favored explanation is a 'local' source, with emission coming from within 100 pc of the Sun. The Sun is known to be located in a region where

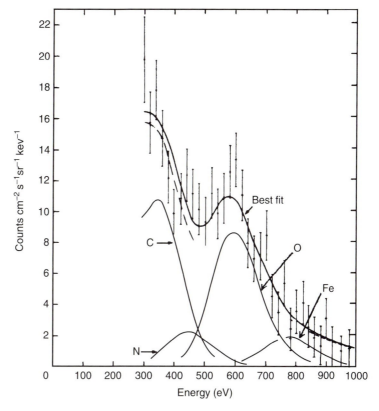

Fig. 16.10 Spectrum of the soft X-ray background compared with predicted contribution of line radiation from the hot gas. The principal contribution is from carbon and oxygen ions. The width of the lines is due to the detector response. [H. Schnopper, et al., Astrophysical J. **253**, 131 (1982).]

the surrounding density of neutral hydrogen is lower than usual. Apparently this is so because this volume is full of hot ionised gas. The Sun is within a bubble of hot gas in the interstellar medium. A diameter of about 100 parsecs and a density of 10^{-2} electrons cm^{-3} would produce the necessary emission. The observed anticorrelation with neutral hydrogen is explained if the bubble is elongated perpendicular to the galactic plane. As illustrated in figure 16.11, in a direction where the path length in hot gas is long, the path length in surrounding cool gas is small. There is other evidence for a very hot com-

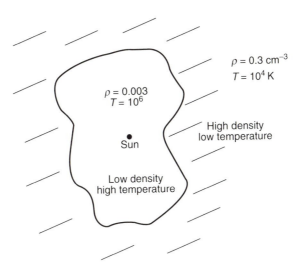

Fig. 16.11 A possible explanation for the C-band background. A bubble of hot gas surrounds the Sun. The extent of the bubble is about 100 parsecs and is greatest in a direction perpendicular to the galactic plane.

ponent of the ISM so this cloud of 10^6 K gas is not unexpected. Perhaps we live within an old supernova remnant! The nearby stellar explosion would have been quite a spectacle to creatures living on Earth ten million years ago.

16.9 The galactic soft X-ray background, detailed structure

The ROSAT PSPC, with low internal background, has produced detailed maps of diffuse soft X-rays. Figure 16.12 shows a portion of the all-sky survey. Diffuse structures in both emission and absorption are seen on scales ranging from a few degrees down to a few arcminutes. The ROSAT survey is expected to reproduce the gross structure of figure 16.9 and individual fields have the potential to reveal arcminute structure.

Fig. 16.12(a) A 15° square portion of the ROSAT all-sky survey containing the galactic centre. (Courtesy J. Trümper and the ROSAT team, MPE.)

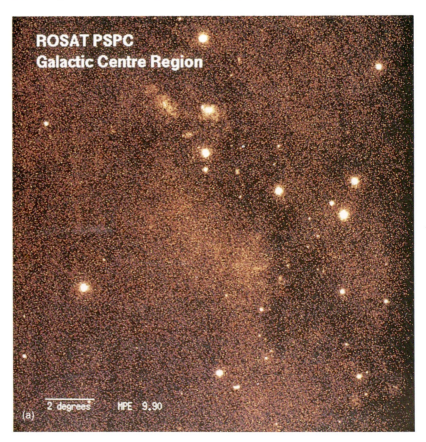

One noteworthy goal was accomplished quickly by ROSAT. Shadowing of the diffuse background by two HI clouds was observed by Steve Snowden and colleagues (MPE) and by David Burrows and Jeff Mendenhall (Pennsylvania State University). Both clouds are in Draco at high galactic latitude and are about $\frac{1}{2}°$ in extent. The C-band X-rays (1/4 keV) were attenuated a factor of 2 by absorption in the clouds. Since the distance to the clouds is ~300 pc, this implies that, although much of the background must come from the Local Bubble, there is an appreciable contribution from 10^6 K gas farther away, in the galactic halo.

16.10 Conclusions

The background spans a large energy range, from 0.1 keV (figure 16.9) to above 1 MeV (figure 16.8). It was discovered in the 2–10 keV energy range and here hopes of finding a new cosmological phenomenon have been strongest. The uniformity of the HEAO-1 background map and the apparently thermal spectrum keep these hopes alive. Something new is needed to explain this spectrum – a new class of object, a strong evolutionary change in a known class of sources, or a diffuse state of matter at an earlier age of the universe.

Much information about faint sources in the energy range 0.3–3 keV has been gathered. Considering just source counts and fluxes, it has been concluded that at least 50% of the background in this range is due to faint sources and that almost all of these are active galactic nuclei, a broad classification which includes quasars. Clusters of galaxies, as we know them, are not a major contributor. Observations 10 times as sensitive as the Einstein deep

Fig. 16.12(b) Key to Fig. 16.12(a) identifying bright sources and other features.

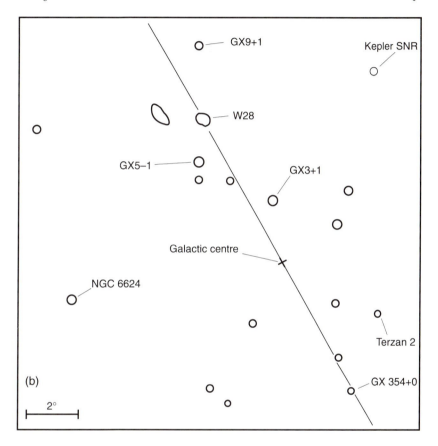

survey are needed before all the sources comprising the background in this energy range can be observed as individuals. Then the hypothesis that 100% of the background is from discrete sources can be fully tested.

The background below 1 keV is largely galactic in origin. There is broad spatial structure, changing rapidly with energy. There is also clearly a galactic component at higher energies, extending to at least 10 keV. The galactic component comes in part from stars and from diffuse clouds of hot gas energised by supernovae and by strong stellar winds. At the lowest energy surveyed, some of the background comes from a local bubble of hot gas.

To understand the background, therefore, requires a knowledge of sources and emission mechanisms located over a remarkable span of distance; from the solar neighbourhood to a distance of at least 3000 Mpc where the redshift of emitted radiation is a factor of 2.

17 Some major observatories of the 1990s

This chapter makes an attempt to cover both existing spacecraft and those planned for the future. There is obvious uncertainty predicting future observational facilities. More X-ray observatories have been proposed than there is money to support. Furthermore, as we know only too well, the business of transporting and establishing an observatory in space is risky. This chapter includes the missions listed in table 17.1, for which building of hardware has started, or for which planning has progressed far enough to define most aspects of the observatory. As in optical astronomy, there is need for both large and small telescopes. Although, in general, bigger is better, there are many important observations for which large telescopes will never be used and for which simpler less expensive instruments are ideal.

Imaging spectroscopy is the goal of many new instruments. This is not surprising. Imaging is necessary for low-background observation of weak sources and spectroscopy is the key to the nature of the sources. The reader will note that the description of many instruments and missions sounds the same. Much of this chapter is best treated as reference material, particularly the tables which are meant for specialists.

It is not a bad thing to have many telescopes with similar capabilities. One instrument cannot look at every target of interest. Optical astronomy has many telescopes, small and large, and they are all put to good use.

17.1 Ginga and Mir-Kvant

Before considering future missions, let us salute two spacecraft, both actively observing in the late 1980s, which contributed some of the first X-ray observations of the 1990s.

The Japanese spacecraft Ginga (figure 17.1) was launched in February 1987 and was under the direction of Yasuo Tanaka and colleagues (ISAS). Its primary instrument was a conventional, but very large area, proportional counter array, the LAC. This was a collaboration between ISAS and the University of Leicester and was used to monitor X-rays from moderately strong sources. Also on board were an all-sky monitor and a gamma-ray burst detector (table 17.2).

Mir-Kvant is a Soviet space station with an attached astrophysics module (figure 17.2). This carries four X-ray detectors: a wide-angle camera using the coded-aperture mask technique to achieve images (Space Research Laboratory, Utrecht and University of Birmingham); a gas scintillation proportional counter for spectroscopy, (ESA/ESTEC); an array of scintillators, for detecting X-rays up to 200 keV (MPE, Garching and University of

Table 17.1 *Recent and future X-ray missions*

Mission	Wavelength	Country	Launch date
Granat	X-ray, gamma	USSR, France, Denmark	Dec 1989
ROSAT	X-ray, XUV	Germany, USA, UK	June 1990
ALEXIS	X-ray, EUV	USA	Apr 1993
DXS	X-ray	USA	Jan 1993
ASCA	X-ray	Japan, USA	Feb 1993
Spectrum-X-Gamma	X-ray, EUV, gamma	Canada, Denmark, Finland, France, Germany, Hungary, Italy, Poland, Russia, Spain, Switzerland, Ukraine, UK, USA	1995
SAX	X-ray	Italy, Netherlands	1994
XTE	X-ray	USA	1996
AXAF	X-ray	USA	1998
XMM	X-ray	ESA	1998

Tubingen); and a high energy phoswich array sensitive to 800 keV (Space Research Institute, Moscow). Table 17.2 lists the characteristics of these instruments.

Both observatories detected hard X-rays from SN 1987A [figure 3.4 (a) and (b)] a once-in-a-lifetime event. The observation was difficult because of the proximity of the supernova to several other sources, particularly the bright source, LMC X-1. Nevertheless, a light curve and X-ray spectra were extracted. This supernova nicely illustrates the obvious fact that astronomical events follow their own timetable, not human schedules. It was fortunate that these two observatories were operating at this time and that the onset of X-ray emission from SN 1987A was recorded. We hope that future observatories will be long-lived and available when needed.

Fig. 17.1 The Japanese satellite Ginga (galaxy). The principal instrument, a group of eight large area proportional counters, fills the left side of the boxlike spacecraft. Star trackers are behind the dark circular aperture and four solar panels are deployed to face the Sun. (Courtesy of ISAS.)

Table 17.2 Instruments on Ginga and Mir-Kvant

Satellite	Detector	Energy band (keV)	Field of view	Purpose	Country or organisation
Ginga	LAC	1.5–30	0.8°×1.7°	timing, spectra	ISAS, Univ. of Leicester, Rutherford Appleton Lab
	ASM	1.5–30	1°×180°	survey	Osaka Univ.
	gamma burst	2–400	4π	survey	ISAS, LANL
Mir-Kvant	TTM	2–30	8°	imaging	SRL Utrecht, Univ. Birmingham
	GSPC	2–50	3°	spectra	ESA/ESTEC
	HEXE	15–200	1.6°	spectra	MPE Garching, Univ. Tubingen
	Pulsar X-1	20–800	3° (2π)	spectra (gamma burst)	IKI Moscow

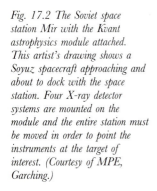

Fig. 17.2 The Soviet space station Mir with the Kvant astrophysics module attached. This artist's drawing shows a Soyuz spacecraft approaching and about to dock with the space station. Four X-ray detector systems are mounted on the module and the entire station must be moved in order to point the instruments at the target of interest. (Courtesy of MPE, Garching.)

17.2 Granat – high energy images

The first major X-ray observatory of the 1990s was launched by the Soviet Union. Granat (meaning pomegranate – a fruit with several chambers and many seeds) was placed in orbit in December 1989. Granat is searching for hard X-ray sources and gamma-ray bursts. It carries 2.3 tonnes of scientific instruments (listed in table 17.3) and is in a highly elliptical orbit (4 day period) allowing long, uninterrupted, viewing times.

Granat (figure 17.3) carries two large X-ray telescopes, but the imaging technique is a radical departure from those of ground-based astronomy. The two principal instruments consist of high energy detectors viewing a small region of the sky through coded masks. Sources cast shadows of the coded pattern on the detectors and a computer later decodes the detector output to derive position and strength of the sources.

The first telescope, 'Sigma', is a pioneer. It is mapping selected regions in the 35–1300 keV range, a region so far only observed with scanning-collimator detectors. The second telescope, 'ART-P', is a four module system sensitive from 4–60 keV and co-aligned with Sigma. The ART-P energy range extends from that of conventional X-ray astronomy to the unknown territory of higher energies. The two telescopes together make maps of the high energy sky. The lowest energies will overlap with images to be obtained with the reflecting telescopes of ROSAT, ASCA, AXAF, XMM, and Spectrum-X-Gamma.

Fig. 17.3 The Soviet Granat spacecraft. Payload weight is 2.3 tonnes. This mission is devoted to mapping high energy X-ray emission from selected regions with two coded-mask telescopes and to the study of gamma-ray bursts. (Courtesy of R. Sunyaev, SRI.)

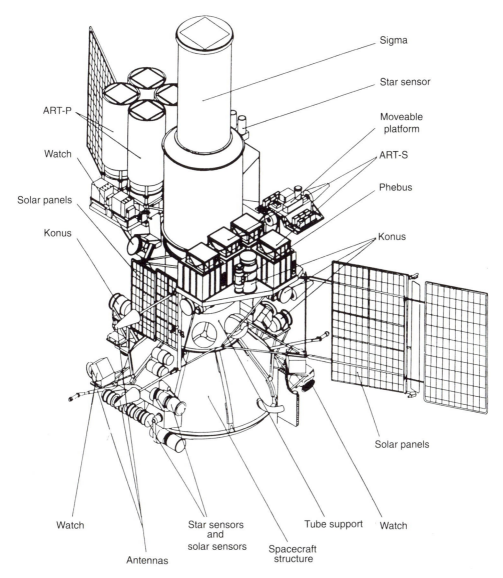

The higher energies are important. The higher the X-ray energy, the more extreme the observed characteristics of the source. The highest energies probably are generated the closest to the central engine of active (and perhaps normal) galactic nuclei and may carry the clearest signature of their characteristics. This is also true for accretion-powered compact sources which contain neutron stars and stellar-sized black holes.

Table 17.3 X-ray instruments on Granat

Detector	Energy band (keV)	Field of view	Purpose	Effective area (cm²)	Spatial resolution	Spectral resolving power, $E/\Delta E$
Sigma	35–1300	4° full-coded 15° partial	coded mask imaging	800	13′	7 @ 60 keV
ART-P	4–60	1.8°	coded mask imaging	1250	5′	5 @ 6 keV
Watch	6–120	4π	all sky monitor	45		
Konus	20–800	4π	gamma burst 7 detectors	315 each	1°	7 @ 600 keV
Phoebus	10^2–10^5	4π	gamma burst 6 detectors	100 each		10 @ 600 keV

Above 5 keV, absorption in the galactic plane should be minimal. A high energy survey of our galaxy will reveal sources in spiral arms on the far side of the galactic nucleus. The X-ray sky above 10 keV is largely unknown having only been explored by scanning instruments on HEAO-2 and 3 and on balloons. A few bright sources have been identified but that is all. There will be surprises when these new instruments reveal fainter high energy objects.

The sky is also being monitored for gamma-ray bursts. A novel feature of Granat is a mechanical device to move directional instruments rapidly to view gamma-ray burst sources within 1 s of their detection. This arrangement should obtain the first sensitive X-ray spectra and time histories of these still completely unexplained phenomena.

17.3 ROSAT – a sensitive survey

The ROentgen-SATellit, built by Germany under the direction of Joachim Trümper and associates (MPE, Garching), was launched in June 1990 on an American Delta rocket. It is appropriately named after Wilhelm Conrad Roentgen, the discoverer of X-rays, and it is much like Einstein. The principal instrument, an X-ray telescope, is built for high sensitivity below 2 keV photon energy and the effective area for X-ray collection is twice that of Einstein. ROSAT, however, has no sensitivity above 2 keV. With larger angles of incidence, it is easier to achieve large collecting areas, but the high energy response is sacrificed. The nested reflectors focus X-rays on either of two imaging detectors.

The primary detector is a gas-filled position-sensitive proportional counter, the PSPC, which, when at the focus of the telescope, has a field of view approximately 2 degrees in diameter. The first six months of the mission were used to conduct an all-sky survey with the PSPC at the focus. If one-half of the time produces useful data (all that can be expected with this orbit), and if the telescope scans the sky uniformly, every celestial object is viewed for about 1000 seconds. It is expected that 5×10^4 X-ray sources will be detected and located by this survey. This will increase the number of known X-ray sources tenfold and will locate 100 times as many sources as were found in the HEAO-1 survey, the most sensitive previous all-sky scan. This is mind-boggling to those of us accustomed to spending weeks, even months, identifying and determining the characteristics of just one cosmic X-ray source. We hope to find new, unexpected, classes of moderately bright sources.

The second focal plane detector is a high resolution imager, the HRI, similar to that carried by Einstein, and supplied by Stephen Murray and associates (CfA). The photosensitive surfaces are coated with caesium iodide to improve the efficiency of converting soft X-rays to photoelectrons. At 1 keV the effective area of this detector is calculated to be approximately four times more sensitive than the Einstein HRI.

A 580 kilometre circular orbit at 56° inclination allows the satellite to pass over a ground station at Weilheim, Germany (latitude 48°) where all data are received. The maximum time interval between readouts is 18 hours.

After the initial survey, operation was changed to a pointed mode with time shared between Germany, the United States, and the United Kingdom. Those countries are putting a major effort into operating ROSAT as a guest observer facility. Data are being disseminated in a form useful to general observers and software for data analysis is compatible with most computation facilities, a praiseworthy departure from past practices. Because of the high sensitivity at low energies, ROSAT observations are a substantial improvement over those of Einstein and EXOSAT for many astronomical sources. More photons can be collected from extragalactic sources and from nearby sources in the Milky Way. However, objects that require arcsecond resolution or better or which are strongly absorbed by intervening material are not well-suited for study by ROSAT.

ROSAT also carries a wide field camera supplied by a UK consortium led by Kenneth Pounds. This instrument was designed to point parallel to the X-ray telescope and map the sky in the extreme ultraviolet at the same time that it is mapped in X-rays by the German telescope. It incorporates a mirror with mean grazing angle 7.5°, geometrical area of 475 cm², and an HRI-like detector. The sky is largely unexplored in this range. Only a few strong sources are known, mostly nearby white dwarfs. The excitement in this detector is the potential for surprises, always forthcoming when a new part of the spectrum is first observed. Limited spectroscopy is also possible with a wheel carrying thin filters of beryllium, aluminium, tin, and plastic.

Note the advantage to the instrument builder of working with large wavelengths. The larger the wavelength, the larger the maximum possible reflection angle, and the shorter the focal length of the telescope. Figure 17.4 shows the small size of the wide field camera telescope compared to the X-ray telescope.

ROSAT's launch coincided with the time of the solar maximum. Because the upper layers of the atmosphere are heated more during solar maximum,

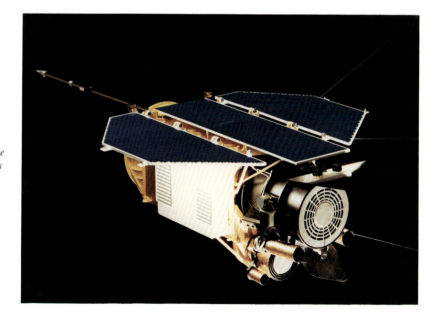

Fig. 17.4 ROSAT. The telescope is pointed to the right and X-rays enter through the concentric-ring aperture. Focal plane instruments are at the other end of the spacecraft. The wide field camera aperture is directly below that of the telescope. The first 6 months were devoted to an all-sky survey which detected and located 50 000 sources. (Courtesy of J. Trumper, MPE.)

Table 17.4 Instruments on ROSAT

Detector	Energy band (keV)	Field of view	Purpose	Effective area (cm²)	Spatial resolution	Spectral resolving power, $E/\Delta E$
PSPC	0.2–2	2°	survey	250 @1 keV	30″	2.3 @1 keV
HRI	0.2–2	0.6°	high res. imaging	160 @1 keV	4″	none
WFC	0.04–0.21	5°	survey	15 @100 Å	2′	filters

the atmospheric density at satellite altitude is high. Since drag on the spacecraft shortens the lifetime in orbit, this was the subject of intense calculation. The prediction was a lifetime of 10 years, longer than the estimated lifetime of the instruments, so this is not a distressing situation.

The first data from ROSAT have been outstanding. The increased sensitivity and the low background of the PSPC have made observation of very faint diffuse emission possible. Some highlights are observation of: structure in the soft background with complexity equal to that of diffuse optical emission from the Milky Way, shadows of interstellar HI clouds, structure in emission from clusters of galaxies, and new supernova remnants. PSPC spectra are good, and the astronomical community is starting to digest 50 000 new X-ray sources from the survey.

17.4 BBXRT – high throughput – the Astro shuttle payload

Work is progressing on several 'high throughput' concepts in which large arrays of reflectors are used to gather X-rays from faint sources. Use of this term usually means spatial resolution has been sacrificed in order to achieve large collecting area. Since a spatial resolution of approximately 1 arcminute is sufficient to isolate many faint sources, the requirement for arcsecond spatial resolution can be relaxed, allowing the mirrors to be fabricated easily and cheaply. With 1 m² (or half the area of a door) of effective area, spectroscopy and timing studies of faint sources are possible in reasonable observing times. The accretion-powered binaries in the spiral arms of M31 are examples of possible targets for a high throughput mission (see figures 6.6 and 6.7). To observe the sources closest to the M31 nucleus, however, a high-resolution mirror, such as that of AXAF, is needed. Table 17.5 compares the characteristics of several X-ray mirrors.

Table 17.5 X-ray telescope mirrors

Mission	Einstein	EXOSAT	ROSAT	BBXRT	AXAF-I	XMM
aperture diameter (cm)	58	28	83	40 one module	120	70 one module
mirrors	4 nested	2 nested	4 nested	118 nested one module	4 nested	58 nested one module
geometric area (cm²)	350	80	1140	1400 2 modules	1100	6000 3 modules
grazing angles (arcminutes)	40–70	90–110	83–135	21–45	27–51	18–40
focal length (m)	3.45	1.09	2.4	3.8	10	7.5
mirror coating	Ni	Au	Au	Au	Ir	Ir and Au
highest energy focused (keV)	4.5	2	2.4	12	10	10
on-axis resolution (arcseconds)	4	18	4	75	0.5	20

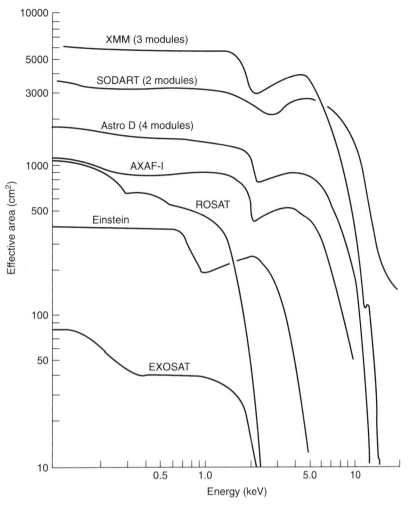

Fig. 17.5 Mirror effective area of past and future X-ray telescopes. The shape of these curves is determined by the reflection angle and the mirror coating. The efficiency of the focal plane detector is not included so these curves are upper limits to telescope efficiencies.

The BBXRT (broad band X-ray telescope) was developed by Peter Serlemitsos and colleagues (GSFC) for flight on the Space Shuttle. Mass-produced thin foil conical mirrors were used as reflectors. Because the mirrors were thin, the shadow cast by each mirror element was minimal. Many elements were arranged concentrically to maximise the reflecting area. The BBXRT mirror assembly consisted of 118 individual mirrors!

BBXRT was flown in December 1990 as part of the Astro mission. Nine days were devoted to the operation of four astronomy instruments carried in the bay of the shuttle. In spite of profound difficulties with the pointing sys-

Table 17.6 BBXRT and ASCA

Satellite	Detector	Energy band (keV)	Field of view	Purpose	Effective area (cm²)	Spatial resolution	Spectral resolving power, $E/\Delta E$
BBXRT	SSS	0.3–12	0.3°	spectra	760 @1.5 keV	5′	11 @1 keV 60 @6 keV
ASCA	CCD	0.4–10	0.4°	spectral imaging	400 @1.5 keV	3′	50 @6 keV
	IGSPC	1–12	0.8°	spectral imaging	400 @1.5 keV	3′	12 @6 keV

tem, BBXRT spectra were obtained from about 50 cosmic sources. Although the initial concept was to fly the Astro payload more than once, limited budgets and limited space on the shuttle make future flights difficult.

17.5 ASCA – high throughput

The Japanese spacecraft Astro-D was launched in February 1993 and promptly renamed 'Asuka', which means 'flying bird'. This has been westernised to ASCA (Advanced Satellite for Cosmology and Astrophysics). On March 28 there occurred a type II supernova in M81, a nearby galaxy shown in figure 14.2. ASCA (and ROSAT too) were quickly pointed at M81 and X-rays from the supernova were detected one week after the explosion. Nature is smiling on the Japanese space programme with the appearance of rare bright supernovae just after the launch of both Ginga and ASCA! ASCA utilises large BBXRT-type mirrors (supplied by GSFC) to achieve large collecting area and 3 arcminute spatial resolution. There are four mirror assemblies. Two focus on CCD arrays (from MIT) and two on imaging gas scintillation proportional counters (IGSPC) (from the University of Tokyo).

The function of both detector systems is imaging spectroscopy and the two systems are complementary. Spectral information is recorded for all locations within the image. The CCD arrays have superior spectral resolving power but no timing capability below 0.1 s. They are also susceptible to radiation damage from intense particle bombardment in the radiation belts.

The IGSPC, with a larger field of view, is not as vulnerable to radiation damage and is capable of measuring event times to 0.1 ms. Spectral resolution, however, is a factor of 4 worse than that of the CCD detector. Even so, the IGSPC spectral capability is a factor of 2 better than conventional proportional counters (used e.g. on ROSAT).

17.6 Spectrum-X-Gamma – high throughput

Spectrum-X-Gamma carries at least nine separate experiments, some of which have more than one detector. Fourteen countries are participating, truly an international effort, and difficult to describe briefly. The mission is primarily spectral imaging over a broad energy range. Optical and EUV telescopes are included in the payload to measure flux from the sources under observation. There are also all-sky monitors to observe transient and gamma-burst sources. The spacecraft carries 2.85 tonnes of scientific payload (table 17.7) and, like Granat, will be placed in a four-day elliptical orbit.

SODART, the largest instrument, is a joint effort of Russia, Denmark, Finland, USA, Italy, Germany, and Poland. It has two identical mirror modules. One will image on either a position sensitive proportional counter or an array of solid state detectors. The other will be devoted to Bragg crystal spectroscopy and polarimetry. Denmark is supplying the two thin-foil mirror systems.

Russia is responsible for the whole design and production of the telescope including two slides where eight focal plane detectors are situated. There are four gas imaging detectors from Denmark and two from Russia, and the USA and Italy are supplying a polarimeter. An array of silicon chips from Finland is held at low temperature by a Russian cooling system. A panel 100 cm by 60 cm in size can be placed in front of one of the mirror modules. Bragg crystals from Russia and Germany line the two sides of the panel. Three types of crystals allow high resolution spectroscopy in the vicinity of emission lines from oxygen, silicon, and iron.

JET-X is smaller but more precise. It consists of two co-aligned telescopes focused on CCD detector arrays. Each telescope consists of an array of nested electroformed mirror shells. The instrument is being built by a consortium

Table 17.7 X-ray Instruments on Spectrum X-gamma

Detector	Energy band (keV)	Field of view	Purpose	Effective area (cm²)	Spatial resolution	Spectral resolving power, $E/\Delta E$
SODART (2 modules) (8 focal-plane detectors)	0.1–20	0.7°	imaging spectra timing polarimetry	1600 @1 keV 1 module	2′	40 @6 keV 7 @ 6 kev PC
Bragg panel	3 bands @ 0.5, 2, 7	0.7°	spectroscopy			
JET-X (2 modules)	0.2–10	0.7°	spectroscopy	360 @1.5 keV	20″	50 @8 keV
Optical telescope			coaligned with JET-X	can detect mag = 22		
MART	8–50	8°	coded mask imaging	1200	5′	
EUVITA (8 modules)	0.03–0.13	1°	EUV imaging	10 each		
MOXE (8 modules)	3–12	4π	monitor	3 each	0.8°	
SPIN	10–10000	4π	gamma burst			

of institutions and the emphasis is on high sensitivity and high spectral resolution in the vicinity of the 7 keV Fe line complex.

MART is a coded-mask telescope for high energy X-rays and points in the same direction as the two large telescope arrays. Thus a single pointing of the spacecraft will measure spectra of sources in the field from 0.2 to 100 keV.

EUVITA is a set of 8 small normal-incidence telescopes. Each telescope mirror has a multilayer coating giving high sensitivity in a narrow waveband (very similar to the ALEXIS instrument described in section 17.11).

17.7 SAX − broad band spectroscopy

This is a major Italian programme with participation of the Netherlands. Several instruments of modest size will carry out observations over a broad energy band, 0.1 − 200 keV. Instruments will be coaligned and will be pointed with an accuracy of 2 arcminutes. Long observations (10⁵ s) of individual sources are planned. Observations will yield source properties through timing and spectroscopy.

The broad energy coverage is designed for accurate determination of spectral indices. The instruments are described in table 17.8. To collect photons below 10 keV, there will be four nested, gold-coated mirror/concentrator assemblies focusing on position-sensitive gas scintillation proportional counters. A high energy gas scintillation proportional counter, the HEGSPC, will extend coverage to 120 keV and a phoswich scintillation detector, the PDS, will record photons with energies as high as 200 keV. Overlapping coverage of these detectors will minimise cross-calibration problems.

For long-term monitoring of strong sources and detection of transients, two wide field cameras, the WFC, will complete the payload. Each will utilise a xenon-filled position-sensitive proportional counter behind a coded mask.

17.8 XTE − timing

The X-ray Timing Explorer (XTE) mission was designed for high resolution X-ray timing. It will monitor, with millisecond accuracy, the emission of strong and moderately strong X-ray sources which are reasonably well separated from their neighbours. It is also equipped to detect and monitor 'transient' sources which appear unexpectedly, like novae, and are bright for days

Table 17.8 Instruments on SAX

Detector	Energy band (keV)	Field of view	Purpose	Effective area (cm²)	Spatial resolution	Spectral resolving power, $E/\Delta E$
Concentrators	0.1–10	0.5°	imaging spectra	200 @7 keV	1′	12 @6 keV
HPGSPC	3–120	1.0°	spectra timing	280 @60 keV	1°	30 @60 keV
PDS	15–200	1.5°	spectra timing	680 @20 keV	1.5°	6 @60 keV
WFCS	2–30	20°	monitor	500	5′	5 @6 keV

or weeks. The physics of sources will be studied through information extracted from precisely measured light curves. This is a 'small' instrument carrying on the work so fruitfully pursued by EXOSAT and devoted to measurements that the larger imaging-oriented observatories will not or cannot undertake. The spacecraft will carry three sets of detectors: a large area proportional counter, a large area scintillation counter, and an all sky monitor.

The proportional counter array (PCA) consisting of conventional X-ray detectors, is being designed under the direction of Jean Swank and colleagues (GSFC) and Hale Bradt and associates (MIT). Because it has the largest collecting area, it is the most sensitive instrument on the payload. The array will consist of five identical methane–xenon filled proportional counters similar to those flown on HEAO-1. The area has been maximised to collect as many photons as possible in short time intervals. A 1 metre long honeycomb collimator restricts the field of view to 1 degree FWHM. Each detector has dimensions of approximately 0.3 × 1.0 × 1.3 metres and the total weight of the PCA detectors alone will be three-quarters of a tonne. This 'small' explorer-class mission is actually not very small physically! Alongside the PCA, pointed in the same direction and with the same field of view, will be the high energy X-ray timing experiment, HEXTE, designed at UCSD under the direction of Richard Rothschild. The detectors are caesium iodide/sodium iodide scintillators, 8 crystals viewed by 8 photomultiplier tubes, giving a large effective area and covering energies up to 200 keV.

These two detectors are capable of long-term monitoring of X-ray luminosity and of recording broad-band spectral emission as a function of time. The time resolution, of course, depends on the source strength. The satellite will be capable of measuring photon arrival time to an accuracy of 5 microseconds. It is expected to observe millisecond fluctuations from stronger sources and variations with durations of hours to days from the fainter ones.

The third detector is an all sky monitor (ASM) which uses instruments with the wonderful name of 'X-ray shadow cameras'. This is also being built at MIT by a group led by Bradt. One-dimensional coded masks cast shadows of the X-ray sky on small position-sensitive proportional counters. Each source produces a shadow with a distinctive pattern. A computer decodes the data, unscrambles the overlapping shadows, and derives the location of each source.

By rotating these detectors it is possible to scan the whole sky every 1.5 days. This will accomplish two things: first, about 50 strong sources in our galaxy will be monitored continuously, giving a record of daily flux intensity spanning several years for each source. Second, the sky will be surveyed for transient sources, such as X-ray novae, which appear unpredictably and have been observed serendipitously in the past.

The spacecraft is designed to be capable of pointing at any location in the

Table 17.9 Instruments on XTE

Detector	Energy band (keV)	Field of view	Purpose	Effective area (cm^2)	Spatial resolution	Spectral resolving power, $E/\Delta E$
PCA	2–60	1°	timing	6000 @10 keV	1°	6 @6 keV
HEXTE	20–200	1°	timing	1000 @100 keV	1°	6 @60 keV
ASM/SSC	2–10	6° × 90°	survey	90	0.1°	none

sky except within a small region close to the Sun. It can slew rapidly within 1 hour from one target to another. The principal instruments can thus be pointed at a transient source within hours of discovery and certainly within a day or two of the outburst. This pointing flexibility makes it possible to monitor sources at regular intervals rather than at times dictated by scheduling. An active galactic nucleus, for example, could be monitored once a week rather than at the six month intervals dictated by the pointing constraints of past HEAO missions. The instruments can also be pointed in the antisolar direction, a capability desirable for simultaneous optical and X-ray observations. Ground-based observatories are, of course, constrained to look during the night.

XTE, then, will monitor the emission of sources in a manner similar to EXOSAT and Ginga, but the instruments are more sensitive, the energy range covered is greater and the spacecraft more flexible. With this mission, it is hoped to quantify the behaviour of material in the vicinity of neutron stars and black holes, both galactic and extragalactic.

17.9 AXAF – high angular resolution and spectroscopy

AXAF, the Advanced X-ray Astrophysics Facility, is the US follow-on to Einstein. The telescope was designed to have four times the area of the Einstein mirror at low energies and to have considerable collecting area between 6 and 7 keV, the energy of the iron lines strongly emitted by most astrophysical sources. Capability of the observatory goes far beyond that of Einstein. There is an order of magnitude better angular resolution and imaging sensitivity is two orders of magnitude higher.

As originally planned, the AXAF payload carried five instruments and was to be launched from the shuttle. When it appeared that the high cost might prevent getting funding, the mission was split into two smaller spacecraft, each to be launched with an expendable booster. The imaging payload, AXAF-I, carries a somewhat smaller high-resolution mirror, two focal plane instruments, and a set of transmission diffraction gratings. It is now lighter and can be placed in an elliptical orbit which, because Earth occultation is less, will result in a high observing efficiency. The second payload, AXAF-S, is for spectroscopy and carries a thin foil low-resolution mirror focused on a liquid-helium-cooled calorimeter. So, *mirabile dictu*, two spacecraft can do the same science for less expense than one!

The AXAF-I mirror consists of a set of four nested reflecting surfaces, all arranged in the usual Wolter type 1 geometry (figure 2.7). The high energy response is achieved by use of small reflection angles and by coating the mirrors with iridium (atomic number 77). Improvements in mirror technology since Einstein include significant advances in grinding, polishing, alignment, and testing. Leon van Speybroeck (CfA), the AXAF telescope scientist, hopes to achieve mirrors with a resolution of 0.5 arcseconds. The combination of high resolution, large collecting area, and sensitivity to higher energy X-rays will make it possible for AXAF to study extremely faint sources in crowded fields.

Fig. 17.6 An artist's impression of AXAF. The large-area, high-resolution mirror is designed for the study of faint sources in crowded regions. It will focus X-rays on one of two instruments located in the focal plane, an arrangement similar to Einstein and ROSAT. (Courtesy of TRW, the spacecraft contractor.)

One focal plane instrument is a high resolution camera, the HRC, designed by Stephen Murray and associates (CfA). This instrument is similar to that designed for ROSAT. It will be used for high resolution imaging, fast timing measurements, and for observations requiring a combination of both.

The second focal plane instrument, the AXAF CCD imaging spectrometer (ACIS), is an array of charged coupled devices (CCDs), designed by a group under the direction of Gordon Garmire (Pennsylvania State University) and George Ricker (MIT). A two-dimensional array of these small detectors will do simultaneous imaging and spectroscopy. Pictures of extended objects can be obtained along with spectral information from each element of the picture.

Higher resolution spectroscopy will be possible using two transmission gratings (the TGS). These are formed by two sets of gold gratings placed just behind the mirrors and produce spectra dispersed in space at the focal plane. Either the CCD array or the HRC can be used to record the data. Development of the gratings is overseen by Claude Canizares (MIT) and by A. Brinkmann (Laboratory for Space Research, Utrecht).

The AXAF-S spectrometer, the XRS, is a 'quantum calorimeter', a brand new development never used in space, currently being designed and tested at GSFC by Stephen Holt and colleagues. The instrument will accurately measure the temperature rise due to energy deposited by a single X-ray photon in a small crystal of silicon. An energy resolution of 3 eV, some 50 times better than that of the Einstein SSS, is hoped for. Most of the weight is in a large cryostat containing the refrigerant necessary to cool the detector. The lifetime of the detector will probably be limited by the amount of refrigerant carried.

The two AXAFs will be capable of impressive observations. They will be able to detect most of the ~300 members of the Pleiades cluster. O stars and RS CVn stars in the Magellanic Clouds can be detected. High resolution spectra can be obtained from hundreds of stars. These spectra will determine the temperature, extent, and density of the coronae, and the dependence of physical parameters on stellar type. The structure of coronae can be derived from spectral observations of eclipsing binary systems and the size and extent of plasma-filled coronal loops can be calculated.

Maps can be made of emission from SNR in our galaxy, Not only will the morphology of the shell-like remnants be clearer, but spectral information will be obtained for all features. Tycho's remnant, for example, can be mapped in the light of X-rays from silicon ions, from sulphur ions, and even from iron ions. Thus, the distribution of different elements within the remnant can be measured. AXAF will be capable of not only seeing SNR in M31, but extended remnants such as the Cygnus Loop will be approximately 10 arcseconds in diameter and so details can be mapped. In M31 about 200 SNR should be accessible to AXAF.

Since the higher energy X-rays are capable of penetrating gas and dust, emission from the galactic centre will be seen clearly. Sources in spiral arms on the other side of the centre can be measured and some of their properties determined. The bright bulge sources in M31 are strong enough so that light curves can be measured and a search for the eclipses which identify binary systems will be possible. As can be seen in figure 6.8, these sources are strongly clustered about the centre of M31. AXAF-I in one pointing will be able to monitor the emission of approximately 50 bulge sources in M31.

Brighter binary sources in galaxies within the Virgo Cluster can also be resolved and detected as individuals, as obviously can sources in intermediate galaxies. Thus, the population of bright X-ray sources in hundreds of galaxies can be determined. Since high energy X-rays will be unaffected by obscuring material, luminosities of sources can be accurately measured. The hypothesis that these sources, or a subset of them, are 'standard candles' can be tested, and if sources are found with constant luminosity, distances to all these galaxies can be accurately determined. Since these distances are a crucial step in the derivation of the Hubble constant, the potential of these measurements is truly exciting.

X-rays from distant clusters of galaxies can be imaged and spectra measured as a function of position within the cluster. Since the emitting gas is quite hot (10^8 K), the high energy capability of AXAF gives a great sensitivity gain over that of Einstein. Furthermore, all clusters should emit the characteristic iron lines which can be used to measure the redshift directly. The spectral and spatial data combined will delineate the gravitational potential, thus measuring the distribution of dark matter within distant clusters.

AXAF-I is expected to detect quasars and active galaxies 20 times fainter than Einstein and can thus look to distances approximately 3 times farther.

Table 17.10 AXAF instruments

Detector	Energy band (keV)	Field of view	Purpose	Effective area (cm²)	Spatial resolution	Spectral resolving power, $E/\Delta E$
HRC	0.2–10	0.5°	high res. imaging	300 @2 keV	0.5″	1 @1 keV
ACIS	0.5–10	0.2° × 0.3°	spectral imaging	600 @1 keV	0.5″	50 @6 keV
TGS	0.2–10	0.5°	spectra	60 @1 keV	0.5″	800 @1 keV
XRS	0.3–10	0.017°	spectra	900 @1 keV	1′	1000 @1 keV

This is unknown territory, except that the integrated emission from many unresolved faint sources probably contributes most of the X-ray background. Deep AXAF-I observations will come close to imaging this background and will provide a sample of distant objects which record the state of the universe at early times.

Once launched, AXAF should enjoy a lifetime of at least 5 years and will operate as a true observatory, available through proposal, as is observing time at most ground-based facilities.

17.10 XMM – high throughput

The European Community hopes to fly a very large mirror array, XMM (X-ray multiple mirror), in the late 1990s. This high throughput X-ray spectroscopy mission is planned as a 'cornerstone' of ESA's space programme of the 1990s. The heart of the observatory is a nest of 58 grazing-incidence mirrors, fabricated by replication from a highly polished mandrel. The nickel surface is bonded to carbon-fibre reinforced plastic resulting in a thin-walled, paraboloid-hyperboloid reflector. Since angular resolution need be only approximately 10 arcseconds, manufacturing tolerances are not overly restrictive.

Because the walls are thin, the mirrors can be highly nested so that the wall of one mirror does not shadow the reflecting surface of the next. Incoming photons have a 50% chance of striking a reflecting surface. The small angle of incidence (long focal length) gives good reflectivity at high energies.

To increase sensitivity, three mirror assemblies are operated in parallel. A CCD array in the focal plane of each provides simultaneous imaging and spectroscopic capability over a 30′ field of view. As in AXAF this detector gives spatial resolution limited only by the telescope and spectral resolution equivalent to the Einstein SSS detector. Reflection-grating spectrometers on two of the mirrors provide good spectral resolution in the band 0.1–3.0 keV, a region rich in lines from thermal plasmas.

The capability of this observatory is impressive. In one day the grating spectrometer will be able to detect emission lines from sources as faint as the limiting sensitivity of the Einstein deep survey. The gratings and CCD detectors can be used simultaneously to obtain time-resolved spectra from 0.1 to 10 keV, covering the range of plasma temperatures from 10^6 to 10^8 K. Stars with stellar flares, and faint AGN are prime targets.

The spacecraft also carries a small optical telescope for simultaneous monitoring of optical and UV emission, to a limiting magnitude of 24.5. Thus, optical counterparts of X-ray bursts from galactic sources can be studied. The orbit planned is highly elliptical with a 24 hour period. Sources can be monitored for 16 hour uninterrupted intervals while the observatory is above the radiation belts.

The high throughput will allow data to be obtained from very faint sources which is equivalent to that discussed in this book for the brightest sources.

Table 17.11 XMM instruments

Detector	Energy band (keV)	Field of view	Purpose	Effective area (cm²)	Spatial resolution	Spectral resolving power, $E/\Delta E$
CCD	0.2–10	0.5°	spectral imaging	2800 @1.5 keV	1′	16 @1.5 keV
reflection grating	0.2–2.5	0.5°	spectra	440 @1.5 keV	1′	300 @1.5 keV

Observations of high quality spectra from many sources will lead to an understanding of the basic processes underlying cosmic X-ray production.

Although XMM can do much of what AXAF can do, it has unique capabilities and the two missions are complementary. AXAF will be used for sources where arcsecond resolution is essential. XMM will be used for spectroscopy of fainter sources not too close to their neighbours. How nice it will be to have both operational!

17.11 ALEXIS and DXS – to study the soft background

The ALEXIS payload (Array of Low Energy X-ray Imaging Sensors) is contained in a small satellite which was orbited by the air-launched Pegasus rocket. The project is under the direction of William Priedhorsky and colleagues at Los Alamos National Laboratory with collaborators from Sandia National Laboratory and the Center for EUV Astrophysics, University of California at Berkeley.

The heart of the system is an array of three wide-field normal-incidence telescopes. The mirror surfaces are 'multilayers', a technique used to enhance reflectivity in a narrow energy band. The surfaces are coated with alternate layers of molybdenum and silicon. The molybdenum reflects EUV radiation with high efficiency and the alternating layers produce constructive interference. Reflectivity is high in the EUV band. Furthermore, the first layers have been tailored for destructive interference at 1256 Å and 304 Å . Here there is a strong background from scattered sunlight and the reflectivity has been minimised. Finally, by means of filters, the sensitivity of each telescope is restricted to a narrow energy band. The band for one telescope is centered at 66 eV, the second at 72 eV, and the third at 93 eV. These energies are 'ultra-soft' to X-ray astronomers. Table 17.12 lists characteristics of the telescopes.

The purpose of the mission was to map the diffuse background, to survey for point sources, to monitor the brightest soft X-ray sources such as flare stars and cataclysmic variables and to search for transient events. The hot plasma which produces the soft background is expected to contain prominent lines from FeVIII, FeIX, FeX, and OVI (see figure 14.9) which fall in the two lowest energy bands. There should be no strong lines in the 93 eV band where continuum emission is monitored.

Although in orbit and operating, ALEXIS has a problem. One of the solar panels was detached during launch and the spin of the spacecraft is irregular. Although the orientation of the instruments is difficult to determine, data have been received and some mapping will be achieved.

The diffuse X-ray spectrometer (DXS), built at the University of Wisconsin under the direction of Wilton Sanders and colleagues is a shuttle-attached payload designed to measure the spectrum of the diffuse soft background in the energy range 0.15–0.28 keV. It is designed to observe diffuse emission and was flown for the first time in January 1993.

Table 17.12 Instruments on ALEXIS and DXS

Satellite	Energy band (keV)	Field of view	Purpose	Effective area (cm^2)	Spatial resolution	Spectral resolving power, $E/\Delta E$
ALEXIS	3 bands @ 0.066, 0.072, 0.093	33°	survey	20	12'	20
DXS	0.15–0.28	15°	spectra		15°	16 @0.16 keV 28 @0.28 keV

Two units, each consisting of a large array of curved Bragg crystals and position-sensitive proportional counters, were mounted to the sides of the shuttle bay. A four day mission was used to scan a $15° \times 150°$ band of the sky. The crystals consist of alternating layers of lead and stearate and the geometry is such that, at any given time, a section of the detector sees only a particular energy from a particular part of the sky. Rotation of the detector accomplishes a scan of both space and energy.

Since the soft background is thought to come from a gas with temperature near 10^6 K, emission is expected from the ions SiVIII, MgIX, SVIII, and NeIX. The purpose of the DXS observation is to search for and to study line emission from these elements.

17.12 The 'Great Observatories'

X-ray astronomy has become an integral part of general astronomy, no longer a discipline in itself. Sources are observed throughout the electromagnetic spectrum and all information is used to derive physical characteristics. This is, if you like, multi-wavelength astrophysics. In the US, NASA has planned future large astronomy missions as part of a comprehensive observing programme in many wavebands. The 'Great Observatories' are HST, the Hubble Space Telescope; GRO, the Gamma-ray Observatory; AXAF, the Advanced X-ray Astrophysics Facility; and SIRTF, the Space InfraRed Telescope Facility. These will be the major US astrophysics observatories for the next 20 years. Including the VLA, the Very Large Array, there will be available five state-of-the-art observatories operating across the spectrum, from radio to gamma-ray energies.

It is certain that with these instruments things will be seen that have not been seen before, both through detailed studies of particular objects and in exploration of unsurveyed regions of space. Because there will be surprises, the value of these observations cannot be predicted. We can, however, guess at the results in areas which will be extensions of previous observations. There are indeed 'important' problems which might be solved, simplified, or, at least, best attacked in the X-ray regime. We expect major advances in understanding the heating of stellar coronae, the physics of collapsed objects, ultra-high magnetic fields (on the surface of neutron stars), general relativity, X-ray 'standard candles' and the Hubble constant, the central engine in AGN, cosmology, and the origin of cosmic rays.

17.13 The last word

X-ray astronomy has been fun. At the beginning, we could hold in our heads all known data from every X-ray source. Each observation was an exploration. Even though observing times were short, e.g., a 5 minute rocket flight, things were seen that were completely new and not understood. This is still true. Einstein and ROSAT images and EXOSAT light curves have revealed new phenomena. One person, however, can no longer keep track of everything that is happening. There has been an explosion of growth in X-ray information and most of us now struggle to keep up with the knowledge concerning only one class of X-ray source.

Observations now are more difficult. Instruments are more expensive, more complicated, and there is a long time delay between design and use of the instrument. Observations must be planned with care and approved by a committee before observing time is granted. To look to 'see what is there' is not considered a valid justification for use of an instrument any more. In spite of this, we still hope for the thrill of unexpected discoveries, and they do happen. Only 5% of the sky was imaged by Einstein and really deep observations were made of only a few regions.

The instruments of future missions are being designed to answer specific issues – 'burning questions' raised largely by past X-ray observations. These answers are guaranteed, and although they will probably be somewhat different than anticipated, there will be delight and satisfaction in seeing them. X-ray astronomers, though not the free spirits they were in the past, look forward to future observations with great enthusiasm.

Abbreviations

AAO	Anglo-Australian Observatory (Epping, NSW, Australia)
AAT	Anglo-Australian Telescope (Coonabarabran, NSW, Australia)
ADC	Accretion disc corona
AGN	Active galactic nucleus
ANS	Astronomical Netherlands Satellite
AS&E	American Science and Engineering (Cambridge, MA, USA)
ART-P	Astronomical Roentgen Telescope-Position sensitive (a Granat instrument)
ASM	All-sky monitor
Astro	An astronomy payload for the Shuttle
AU	Astronomical Unit (the average Earth–Sun distance)
AURA	Association of Universities for Research in Astronomy
AXAF	Advanced X-ray Astrophysics Facility (planned to be one of NASA's Great Observatories)
BBXRT	Broad Band X-ray Telescope (Shuttle payload)
CCD	Charge coupled device
CfA	Harvard-Smithsonian Center for Astrophysics (Cambridge, MA, USA)
CFHT	Canada–France–Hawaii Telescope
CIT	California Institute of Technology (Pasadena, CA, USA)
CMA	Channel multiplier array (an EXOSAT X-ray detector)
CNR	Consiglio Nazionale delle Ricerche (Italy)
COS-B	The second Celestial Observation Satellite
CSIRO	Commonwealth Scientific and Industrial Research Organization (Australia)
CTIO	Cerro Tololo Inter-American Observatory (La Serena, Chile)
CV	Cataclysmic variable (chapter 10)
DAO	Dominion Astrophysical Observatory (Victoria, BC, Canada)
ESA	European Space Agency

ESTEC	European Space Technology Center (Noordwijk, The Netherlands)
EUV	Extreme ultraviolet
EXOSAT	European X-ray Observatory Satellite
FPCS	Focal plane crystal spectrometer (an Einstein instrument)
FWHM	Full width at half maximum (used to describe the spread of a distribution)
GCVS	General Catalog of Variable Stars
GRO	Gamma-Ray Observatory (one of NASA's Great Observatories, now called Compton Observatory)
GSFC	Goddard Space Flight Center (Greenbelt, MD, USA)
GSPC	Gas scintillation proportional counter (an X-ray detector)
HEAO	High Energy Astronomy Observatory (a series of 3 spacecraft)
HED	High energy detector (a HEAO-1 X-ray detector)
HRI	High resolution imager (an X-ray detector on both Einstein and ROSAT)
HST	Hubble Space Telescope
IGSPC	Imaging gas scintillation proportional counter (an X-ray detector)
IKI	Institut Kosmicheskikh Issledovaniy (Space Research Institute, Moscow)
IOA	Institute of Astronomy (Cambridge, England)
IPAC	Infrared Processing and Analysis Center (Pasadena, CA, USA)
IPC	Imaging proportional counter (an Einstein X-ray detector)
IRAS	Infrared Astronomical Satellite
ISAS	Institute of Space and Astronautical Science (Kanagawa, Japan)
ISM	Interstellar medium (the diffuse material in space between the stars)
ISO	Infrared Space Observatory (an ESA mission)
IUE	International Ultraviolet Explorer (a NASA/ESA mission)
JPL	Jet Propulsion Laboratory (Pasadena, CA, USA)
keV	kiloelectronvolt (a measure of energy = 1.6×10^{-16} J)
kpc	kiloparsec (a measure of distance = 3×10^{19} m or approx 3000 light years)
KPNO	Kitt Peak National Observatory (Tuscon, AZ, USA)
LAC	Large area counter (a Ginga X-ray detector)
LASL/LANL	Los Alamos Scientific (National) Laboratory (Los Alamos, NM, USA)
LE	Low Energy (an EXOSAT X-ray detector)
LLL/LLNL	Lawrence Livermore (National) Laboratory (Livermore, CA, USA)
LMC	Large Magellanic Cloud
LMXB	Low-mass X-ray binary (chapter 8)
MCP	Micro-channel plate (an optical, UV and X-ray detector)
ME	Medium Energy (an EXOSAT X-ray detector)

MIT	Massachusetts Institute of Technology (Cambridge, MA, USA)
MMT	Multiple Mirror Telescope (Mt Hopkins, AZ, USA)
MPC	Monitor proportional counter (an Einstein X-ray detector)
MPE	Max Planck Institut für Extraterrestrische Physik (Garching, Germany)
MRAO	Mullard Radio Astronomy Observatory (Cambridge, England)
MSFC	Marshall Space Flight Center (Huntsville, AL, USA)
MSSL	Mullard Space Science Laboratory (part of UCL)
MXRB	Massive X-ray binary (chapter 7)
NASA	National Aeronautics and Space Administration
NRAO	National Radio Astronomy Observatory (Charlottesville, VA, USA)
NRL	Naval Research Laboratory (Washington, DC, USA)
OAO	Orbiting Astronomical Observatory (a series of 3 spacecraft)
OGS	Objective grating spectrometer (an Einstein instrument)
OSO	Orbiting Solar Observatory (a series of 8 spacecraft)
OVV	Optically violent variable (a type of active galactic nucleus)
pc	parsec (a measure of distance $= 3 \times 10^{16}$ m or approx 3 light years)
POSS	Palomar Observatory Sky Survey
PSD	Position-sensitive detector (an EXOSAT instrument)
PSPC	Position-sensitive proportional counter (a ROSAT instrument)
QPO	Quasi-periodic oscillation (a characteristic of some X-ray binaries)
RAS	Royal Astronomical Society
RC	Resistor-capacitor network (electronic circuit)
RGO	Royal Greenwich Observatory
RMC	Rotating modulation collimator
ROSAT	Roentgen-Satellit
SAAO	South African Astronomical Observatory
SAS	Small Astronomy Satellite (a series of spacecraft)
S&T	*Sky and Telescope* magazine
SMC	Small Magellanic Cloud
SN	Supernova
SNR	Supernova remnant
SSS	Solid state spectrometer (an Einstein X-ray detector)
STScI	Space Telescope Science Institute (Baltimore, MD, USA)
TGS	Transmission grating spectrometer
TRW	Thompson Ramo Woolridge (Redondo Beach, CA, USA)
UCB	University of California at Berkeley
UCLA	University of California at Los Angeles
UCSC	University of California at Santa Cruz

UCSD	University of California at San Diego
UCL	University College London
UV	Ultraviolet
VLA	Very Large Array (a radio telescope, Socorro, NM, USA)
WFC	Wide field camera (a ROSAT telescope)
WHT	William Herschel Telescope (La Palma, Canary Islands, Spain)
XMM	X-ray multi-mirror (ESA Horizon 2000 mission)
XTE	X-ray Timing Explorer (planned NASA mission)

Selected references and additional reading

This section is by no means an attempt at a complete bibliography of all the topics covered in this book. Such an undertaking would be both voluminous and doomed to failure. Instead, we list (under each chapter heading) related articles and texts that will give the interested reader a wider or more detailed perspective on the material. Those references of a more advanced or technical nature are marked with a (*).

Chapter 1. Introduction

X-ray

R. F. Hirsch, *Glimpsing an Invisible Universe*, Cambridge University Press, Cambridge, England (1983)

W. H. Tucker and R. Giacconi, *The X-ray Universe*, Harvard University Press, Cambridge, MA (1985)

Radio

G. Haslam and R. Wielebinski, Radio Maps of the Sky, *Sky and Telescope* **63**, 230 (1982)

J. S. Hey, *The Evolution of Radio Astronomy*, Science History Publications, New York (1973)

W. T. Sullivan III, Radio Astronomy's Golden Anniversary, *Sky and Telescope* **64**, 544 (1982)

Ultraviolet

Y. Kondo (editor in chief), *Exploring the Universe with the IUE Satellite*, D. Reidel, Dordrecht (1987)

J. Pasachoff, J. Linsky, B. Haisch, and A. Boggess, IUE and the Search for a Lukewarm Corona, *Sky and Telescope* **57**, 438 (1979)

Infrared

H. J. Habing and G. Neugebauer, The Infrared Sky, *Scientific American* Nov, p. 48 (1984)

L. J. Robinson, The Frigid World of IRAS I, *Sky and Telescope* **67**, 4 (1984)

R. A. Shorn, The Frigid World of IRAS II, *Sky and Telescope* **67**, 119 (1984)

Gamma ray

G. F. Bignami, Gamma-ray Astronomy Comes of Age, *Sky and Telescope* **70**, 301 (1985)

Some early X-ray papers (in chronological order) (all (*))

R. Giacconi, H. Gursky, F. Paolini, and B. Rossi, Evidence for X-rays from Sources Outside the Solar System, *Phys. Rev. Letters* **9**, 439 (1962)

H. Gursky, R. Giacconi, F. Paolini, and B. Rossi, Further Evidence for the Existence of Galactic X-rays, *Phys. Rev. Letters* **11**, 530 (1963)

G. W. Clark, Balloon Observation of the X-ray Spectrum of the Crab Nebula above 15 keV, *Phys. Rev. Letters* **14**, 91 (1965)

L. E. Peterson, A. S. Jacobson, and R. M. Pelling, Spectrum of Crab Nebula X-rays to 120 keV, *Phys. Rev. Letters* **16**, 142 (1966)

H. Friedman, E. T. Byram, and T. A. Chubb, Distribution and Variability of Cosmic X-ray Sources, *Science* **156**, 374 (1967)

G. Chodil, *et al.*, Spectral and Location Measurements of Several Cosmic X-ray Sources Including a Variable Source in Centaurus, *Phys. Rev. Letters* **19**, 681 (1967)

P. C. Fisher, *et al.*, Observations of Galactic X-ray Sources, *Astrophysical Journal* **151**,1 (1968)

H. Bradt, S. Naranan, S. Rappaport, and G. Spada, Celestial Positions of X-ray Sources in Sagittarius, *Astrophysical Journal* **152**, 1005 (1968)

R. J. Grader, R. W. Hill, and J. P. Stoering, Soft X-rays from the Cygnus Loop, *Astrophysical Journal* **161**, 145 (1970)

T. M. Palmieri, *et al.*, Soft X-rays from Two Supernova Remnants, *Astrophysical Journal* **164**, 61 (1971)

B. A. Cooke and K. A. Pounds, Further High Sensitivity X-ray Sky Survey from the Southern Hemisphere, *Nature Phys. Sci.* **229**, 144 (1971)

R. Giacconi, *et al.*, Discovery of Periodic X-ray Pulsations in Centaurus X-3, *Astrophysical Journal* **167**, L67 (1971)

E. Schreier, *et al.*, Evidence for the Binary Nature of Centaurus X-3 from Uhuru X-ray Observations, *Astrophysical Journal* **172**, L79 (1972)

F. D. Seward, *et. al.*, Distances and Absolute Luminosities of Galactic X-ray Sources, *Astrophysical Journal* **178**, 131 (1972)

S. Rappaport, *et al.*, Possible Detection of Very Soft X-rays from SS Cygni, *Astrophysical Journal* **187**, L5 (1974)

G. W. Clark, T. H. Markert, and F. K. Li, Observation of Variable Sources in Globular clusters, *Astrophysical Journal* **199**, L93 (1975)

R. C. Catura, L. W. Acton, and H. M. Johnson, X-ray Emission From Capella, *Astrophysical Journal* **196**, L47 (1975)

J. Heise, et al, Evidence for X-ray Emission from Flare Stars Observed by ANS, *Astrophysical Journal* **202**, L73 (1975)

Chapter 2. The tools of X-ray astronomy

J. K. Beatty, ROSAT and the X-ray Universe, *Sky and Telescope* **80**, 128 (1990)

H. Bradt and B. Margon, In Search of X-ray Quasars, *Sky and Telescope* **56**, 499 (1978)

H. Bradt, T. Ohashi, and K. Pounds, X-ray Astronomy Satellites, *Annual Review of Astronomy and Astrophysics* **30**, 391 (1992) (*)

F. A. Cordova and K. O. Mason, EXOSAT, Europe's New X-ray Satellite, *Sky and Telescope* **67**, 397 (1984)

G. W. Fraser, *X-ray Detectors in Astronomy*, Cambridge University Press, Cambridge, England (1989) (*)

R. Giacconi, The Einstein X-ray Observatory, *Scientific American* Feb, p. 80 (1980)

M. Lampton, The Microchannel Image Intensifier, *Scientific American* Nov, p. 62 (1981)

G. K. Skinner, X-ray Imaging with Coded Masks, *Scientific American* Aug, p. 84 (1988)

E. Spiller and R. Feder, The Optics of Long Wavelength X-rays, *Scientific American* Nov, p. 70 (1978)

W. H. Tucker, *The Star Splitters: The High Energy Astronomy Observatories*, NASA SP-466 (1984)

Chapter 3. Supernova remnants

Supernovae

H. A. Bethe and G. Brown, How a Supernova Explodes, *Scientific American* May, p. 60 (1985)

A. Burrows, The Birth of Neutron Stars and Black Holes, *Physics Today* **40**, 28 (1987)

D. H. Clark and F. R. Stephenson, *The Historical Supernovae*, Pergamon Press, Oxford (1977)

D. H. Clark, *Superstars*, McGraw-Hill, New York (1984)

D. Helfand, Bang: The Supernova of 1987, *Physics Today* **40**, 24 (1987)

F. Reddy, Supernovae: Still a Challenge, *Sky and Telescope* **66**, 485 (1983)

J. C. Wheeler and R. P. Harkness, Helium-rich Supernovas *Scientific American* Nov, p. 50 (1987)

Supernova remnants

J. Danziger and P. Gorenstein (editors), *X-ray Emission from Supernova Remnants*, D. Reidel, Dordrecht (1983) (*)

G. Greenstein, *Frozen Star*, Freundlich Books, New York (1984)

F. D. Seward, A Trip to the Crab Nebula, *J. British Interplanetary Soc.* **31**, 83 (1978)

F. D. Seward, P. Gorenstein, and W. H. Tucker, Young Supernova Remnants, *Scientific American* Aug, p. 88 (1985)

F. D. Seward, Neutron Stars in Supernova Remnants, *Sky and Telescope* **71**, 6 (1986)

F. D. Seward, Einstein Observations of Galactic Supernova Remnants, *Astrophysical Journal Supplement Series* **73**, 781 (1990) (*)

K. W. Weiler, A New Look at Supernova Remnants, *Sky and Telescope* **58**, 415 (1979)

K. W. Weiler and R. A. Sramek, Supernovae and Supernova Remnants, *Annual Review of Astronomy and Astrophysics* **26**, 295 (1988) (*)

Chapter 4. Active stellar coronae

R. W. Chapman, NASA's Search for the Solar Connection - I, *Sky and Telescope* **58**, 118 (1979)   ... - II, *ibid* **58**, 223 (1979)

J. B. Kahler, Origins of the Spectral Sequence, *Sky and Telescope* **71**, 129 (1986)

B. O'Leary, The Stormy Sun, *Sky and Telescope* **59**, 199 (1980)

R. Pallavicini, Stellar Coronae, the EXOSAT Picture, *X-ray Astronomy with EXOSAT*, Memorie della Societa Astronomica Italiana, **59**, 71 (1988) (*)

R. A. Stern, Stellar Coronas, X-rays, and Einstein, *Sky and Telescope* **68**, 24 (1984)

M. Zeilik, P. A. Feldman and F. Walter, The Strange RS Canum Venaticorum Binary Stars, *Sky and Telescope* **57**, 132 (1979)

Chapter 5. Early-type stars and superbubbles

W. Cash and P. Charles, Stalking the Cygnus Superbubble, *Sky and Telescope* **59**, 188 (1980)

J. B. Kaler, The Spectacular O Stars, *Sky and Telescope* **74**, 465 (1987)

P. Maffei, *Monsters in the Sky*, Chapter 4, translated by M. and R. Giacconi, MIT Press, Cambridge, MA (1980)

D. Malin, The Splendor of Eta Carinae, *Sky and Telescope* **73**, 14 (1987)

M. H. Schneps and N. Wright, A Bubble in Space: The Shell of NGC 2359, *Sky and Telescope* **59**, 196 (1980)

N. R. Walborn, T. R. Gull, and K. Davidson, Eta Carinae's Numbered Days, *Sky and Telescope* **64**, 16 (1982)

Chapter 6. Normal galaxies

S. Bowyer, B. Margon, M. Lampton, and R. Cruddace, Observation of X-rays from M31, *Astrophysical Journal* **190**, 285 (1974) (*)

G. Fabbiano, X-rays from Normal Galaxies, *Annual Review of Astronomy and Astrophysics* **27**, 87 (1989) (*)

K. S. Long, D. A. Grabelsky, and D. J. Helfand, A Soft X-ray Study of the Large Magellanic Cloud, *Astrophysical Journal* **248**, 925 (1981) (*)

K. S. Long and L. P. Van Speybroeck, X-ray Emission from Normal Galaxies, *Accretion Driven Stellar X-ray Sources*, W. Lewin and E. van den Heuvel, eds., Cambridge University Press, Cambridge, England (1983) (*)

Hans Mark, *et. al.*, Detection of X-rays from the Large Magellanic Cloud, *Astrophysical Journal* **155**, L143 (1969) (*)

D. S. Mathewson, *et al.*, Supernova Remnants in the Magellanic Clouds, *Astrophysical Journal Supplement Series* **51**, 345 (1983) (*)

D. Mathewson, The Clouds of Magellan, *Scientific American* Apr, p. 106 (1985)

G. K. Skinner, *et al.*, Hard X-ray Images of the Galactic Center, *Nature* **330**, 544 (1987) (*)

M. Watson, R. Willingale, J. Grindlay, and P. Hertz, An X-ray Study of the Galactic Center, *Astrophysical Journal* **250**, 142 (1982) (*)

Chapter 7. Massive X-ray binaries

P. Ghosh and F. K. Lamb, Plasma Physics of Accreting Neutron Stars, in *Neutron Stars: Theory and Observation*, 363, Kluwer Academic Publishers, The Netherlands (1991) (*)

P. K. MacKeown and T. C. Weekes, Cosmic Rays from Cygnus X-3 *Scientific American* Nov, p. 60 (1985)

S. A. Rappaport and P. C. Joss, X-ray Pulsars in Massive Binary Systems, *Accretion Driven Stellar X-ray Sources*, W. Lewin and E. van den Heuvel, eds., Cambridge University Press, Cambridge, England (1983) (*)

D. H. Smith, Cygnus X-3: Cosmic-Ray Powerhouse, *Sky and Telescope* **69**, 497 (1985)

V. Trimble, A Field Guide to Close Binary Stars, *Sky and Telescope* **68**, 306 (1984)

Chapter 8. Low-mass X-ray binaries

P. A. Charles, The Optical and X-ray Outbursts of Aquila X-1, *Sky and Telescope* **59**, 188 (1980)

J. Frank, A. King and D. Raine, *Accretion Power in Astrophysics*, 2nd edition, Cambridge University Press, Cambridge, England (1992) (*)

F. K. Lamb, Unified Model of X-ray Spectra and QPOs in Low Mass Neutron Star Binaries, in *Neutron Stars: Theory and Observation*, 445, Kluwer Academic Publishers, The Netherlands (1991) (*)

W. H. G. Lewin, The Sources of Celestial X-Ray Bursts, *Scientific American* May, p. 72 (1981)

W. H. G. Lewin and J. van Paradijs, What are X-ray Bursters?, *Sky and Telescope* **57**, 446 (1979)

B. E. Schaefer, Gamma-Ray Bursters, *Scientific American* Feb, p. 52 (1985)

M. van der Klis, Quasi-Periodic Oscillations, *Scientific American* Nov, p. 50 (1988)

J. van Paradijs, Optical Observations of Compact Galactic X-ray Sources, *Accretion Driven Stellar X-ray Sources*, W. Lewin and E. van den Heuvel, eds., Cambridge University Press, Cambridge, England (1983) (*)

Chapter 9. X-ray binaries in globular clusters

P. A. Charles, Globular Cluster X-ray Binaries, *ESA SP*-296, 129 (1989) (*)

G. W. Clark, X-Ray Stars in Globular Clusters, *Scientific American* Oct, p. 42 (1977)

F. Verbunt, X-ray Binaries and Millisecond Pulsars in Globular Clusters, *The Physics of Compact Objects*, N. E. White and L. G. Filipov, eds., Advances in Space Research, **8**, 293 (1988) (*)

Chapter 10. Cataclysmic variable stars

K. Beuermann, X-ray Properties of AM Herculis Binaries, *The Physics of Compact Objects*, N. E. White and L. G. Filipov, eds., Advances in Space Research, **8**, 283 (1988) (*)

P. A. Charles, The Mysterious SU UMa Stars, *Sky and Telescope* **79**, 607 (1990)

F. A. Cordova and K. O. Mason, Accreting Degenerate Dwarfs in Close Binary Systems, *Accretion Driven Stellar X-ray Sources*, W. Lewin and E. van den Heuvel, eds., Cambridge University Press, Cambridge, England (1983) (*)

M. Cropper, Optical Properties of AM Her Stars, *The Physics of Compact Objects*, N. E. White and L. G. Filipov, eds., Advances in Space Research, **8**, 273 (1988) (*)

K. O. Mason and F. A. Cordova, Satellite Observations of Cataclysmic Variables, *Sky and Telescope* **57**, 446 (1982)

K. O. Mason, S. R. Rosen and C. Hellier, The Accretion Geometry of Intermediate Polars (DQ Her Stars), *The Physics of Compact Objects*, N. E. White and L. G. Filipov, eds., Advances in Space Research, **8**, 293 (1988) (*)

J. P. Osborne, Magnetic Cataclysmic Variables Seen By EXOSAT, *X-ray Astronomy with EXOSAT*, Memorie della Societa Astronomica Italiana, **59**, 117 (1988) (*)

J. E. Pringle and R. A. Wade (editors), *Interacting Binary Stars*, Cambridge University Press, Cambridge, England (1985)

Chapter 11. Are there black holes in our galaxy?

J. McClintock, Do Black Holes Exist? *Sky and Telescope* **75**, 28 (1988)

J. E. McClintock, Black Holes in the Galaxy, *Texas-ESO-CERN Symposium on Relativistic Astrophysics, Proc. N.Y. Academy of Sciences* (1991) (*)

S. L. Shapiro and S. A. Teukolsky, *Black Holes, White Dwarfs and Neutron Stars, The Physics of Compact Objects*, Wiley-Interscience, New York (1983) (*)

Y. Tanaka, Black Hole X-ray Binaries, in *The Ginga Memorial Symposium*, ISAS, Japan (1992) (*)

Chapter 12. SS433–the link with AGN

D. H. Clark, *The Quest for SS433*, Viking (1985)

B. Margon, The Bizarre Spectrum of SS433, *Scientific American* Oct, p. 54 (1980)

B. Margon, Observations of SS433, *Annual Reviews of Astronomy and Astrophysics* **22**, 507 (1984) (*)

Chapter 13. Active galactic nuclei

R. D. Blandford, M. C. Begelman, and M. J. Rees, Cosmic Jets, *Scientific American* May, p. 124 (1982)

E. D. Feigelson and E. J. Schreier, The X-ray Jets of Centaurus A and M87, *Sky and Telescope* **65**, 6 (1983)

J. Frank, A. King and D. Raine, *Accretion Power in Astrophysics*, 2nd edition, Cambridge University Press, Cambridge, England (1992) (*)

I. McHardy, EXOSAT Observations of Variability in Active Galactic Nuclei, *X-ray Astronomy with EXOSAT*, Memorie della Societa Astronomica Italiana, **59**, 239 (1988) (*)

K. A. Pounds and T. J. Turner, X-ray Spectra of Seyfert-type AGN Observed with EXOSAT, *X-ray Astronomy with EXOSAT*, Memorie della Societa Astronomica Italiana, **59**, 261 (1988) (*)

Chapter 14. Clusters of galaxies

Neta A. Bahcall, Clusters of Galaxies, *Annual Review of Astronomy and Astrophysics* **15**, 505 (1977) (*)

J. O. Burns, Dark Matter in the Universe, *Sky and Telescope* **68**, 396 (1984)

E. T. Byram, T. A. Chubb, and H. Friedman, Cosmic X-ray Sources, Galactic and Extragalactic, *Science* **152**, 66 (1966)

A. C. Fabian, P. E. J. Nulsen, and C. R. Canizares, Cooling Flows in Clusters of Galaxies, *Nature* **310**, 733 (1984) (*)

D. Fabricant and P. Gorenstein, Further Evidence for M87's Massive, Dark Halo, *Astrophysical Journal* **267**, 535 (1983) (*)

D. Fabricant, G. Rybicki, and P. Gorenstein, X-ray Measurements of the

Nonspherical Mass Distribution in the Cluster of Galaxies A2256, *Astrophysical Journal* **286**, 186 (1984) (*)

W. Forman and C. Jones, X-ray Imaging Observations of Clusters of Galaxies, *Annual Review of Astronomy and Astrophysics* **20**, 547 (1982) (*)

H. Gursky *et al.*, A Strong X-ray Source in the Coma Cluster Observed by Uhuru, *Astrophysical Journal* **167**, L81 (1971) (*)

L. M. Krauss, Dark Matter in the Universe, *Scientific American* Dec, p. 58 (1986)

Chapter 15. X-rays from planets

I. Adler et al, Apollo 15 Geochemical X-ray Fluorescence Experiment: Preliminary Report, *Science* **175**, 436 (1972) (*)

J. K. Beatty, The Far-Out Worlds of Voyager 1–II, *Sky and Telescope* **57**, 516 (1979)

H. Fink, J. Schmitt, and F. R. Harnden Jr., The Scattered X-ray Background in Low-Earth Orbit, *Astronomy and Astrophysics* **193**, 345 (1988) (*)

R. J. Grader, R. A. Hill, and F. D. Seward, X-ray Airglow in the Daytime Sky, *Journal of Geophysical Research* **73**, 7149 (1968) (*)

J. R. Harries and R. J. Francey, Observation of Cen XR-2, Sco XR-1, and Terrestrial X-rays, *Australian Journal of Physics* **21**, 715 (1968) (*)

A. E. Metzger *et al.*, The Detection of X-rays from Jupiter, *Journal of Geophysical Research* **88**, 7731 (1983) (*)

J. H. M. M. Schmitt *et al.*, A Soft X-ray Image of the Moon, *Nature* **349**, 583 (1991) (*)

F. D. Seward, B. Horton, G. Pollard, and P. W. Sanford, Argon Flourescent X-rays in the Earth's Atmosphere During Solar Flares, *Nature* **264**, 421 (1976) (*)

Chapter 16. The Diffuse X-ray background

S. Anderson and B. Margon, The X-ray Properties of High Redshift Quasi-stellar Objects, *Astrophysical Journal* **314**, 131 (1987) (*)

E. A. Boldt, The Cosmic X-ray Background, *Physics Reports* **146**, 215 (1987) (*)

D. N. Burrows and J. A. Mendenhall, Soft X-ray Shadowing by the Draco Cloud *Nature* **351**, 629 (1991) (*)

I. Gioia, *et al.*, The Einstein Observatory Extended Medium Sensitivity Survey I: X-ray Data and Analysis, *Astrophysical Journal Supplement Series* **72**, 567 (1990) (*)

R. E. Griffiths, et al, The Optical Identification Content of the Einstein Observatory Deep X-ray Survey of a Region in Pavo, *Astrophysical Journal* **269**, 375 (1983) (*)

D. Iwan, et al, A Large Scale Height Galactic Component of the Diffuse 2–60 keV Background, *Astrophysical Journal* **260**, 111 (1982) (*)

D. McCammon and W. Sanders, The Soft X-ray Background and its Origins, *Annual Review of Astronomy and Astrophysics* **28**, 657 (1990) (*)

B. Margon, The Origin of the Cosmic X-ray Background, *Scientific American*, Jan, p. 104 (1983)

F. E. Marshall, *et al.*, The Diffuse X-ray Background Spectrum from 3 to 50 keV, *Astrophysical Journal* **235**, 4 (1980) (*)

F. A. Primini, *et al.*, The Einstein Observatory Extended Deep X-ray Survey, *Astrophysical Journal* **374**, 440 (1991) (*)

J. Stocke, *et al.*, The Einstein Observatory Extended Medium Sensitivity Survey II: The Optical Identifications, *Astrophysical Journal Supplement Series* **76**, 813 (1991) (*)

Chapter 17. Some major observatories of the 1990s

J. K. Beatty, The First 100 Days of MIR, *Sky and Telescope* **72**, 134 (1986)

J. K. Beatty, ROSAT and the X-ray Universe, *Sky and Telescope* **80**, 128 (1990)

W. P. Blair and T. R. Gull, Astro: Observatory in a Shuttle, *Sky and Telescope* **79**, 591 (1990)

H. Bradt, T. Ohashi, and K. Pounds, X-ray Astronomy Satellites, *Annual Review of Astronomy and Astrophysics* **30**, 391 (1992) (*)

B. Margon, Exploring the High Energy Universe, *Sky and Telescope* **82**, 607 (1991)

Index

Exploring The X-ray Universe describes the view of the stars and galaxies that is obtained through X-ray telescopes. X-rays, which are invisible to human sight, are created in the cores of active galaxies, in cataclysmic stellar accretion onto white dwarfs, neutron stars and black holes, and in streams of gas expelled by the Sun and stars. The window on the heavens used by X-ray astronomers shows the great drama of cosmic violence on the grandest scale.

This account of X-ray astronomy incorporates the latest findings from observatories that operate in space. These include the Einstein Observatory operated by NASA, the EXOSAT satellite of the European Space Agency, ROSAT and Japan's Ginga Satellite. The book covers the entire field, with chapters on stars, supernova remnants, X-ray binaries, normal and active galaxies, clusters of galaxies, the diffuse X-ray background, and more. The authors review basic principles, include the necessary historical background, and explain exactly what we know from X-ray observations of the universe.

Copiously illustrated, the book is a complete account of X-ray astronomy. Although written for the nontechnical reader, the technical content is high. There are over 200 illustrations, many in full colour, from X-ray, optical and infrared telescopes. Many of the X-ray images are previously unpublished and appear for the first time in this book. Clear diagrams support the text, and more specialised topics are covered in separate boxed features.

The book is accessible to anyone with a basic knowledge of science and some understanding of general astronomy. It is suitable for amateur astronomers, students interested in the high-energy universe, and professionals needing an overview of the field.

Philip Charles lectures in astronomy at the University of Oxford, where he is a Fellow of St Hugh's College and Head of Astrophysics. Recently he spent five years with the Royal Greenwich Observatory, and he has used telescopes in Australia, Hawaii, California and the Canary Islands in the course of his career. Frederick Seward is an astrophysicist at the Smithsonian Astrophysical Observatory in Cambridge, Massachusetts, where he was Director of the Einstein Observatory Guest Observer Programme. He conducted pioneering rocket observations of cosmic X-rays when he was on the staff of the Lawrence Livermore Laboratory, California.